Parametric Programming for Computer Numerical Control Machine Tools and Touch Probes

Parametric Programming for Computer Numerical Control Machine Tools and Touch Probes

CNC's best-kept secret

By Mike Lynch

Society of Manufacturing Engineers
One SME Drive, P.O. Box 930
Dearborn, Michigan 48121

98765432

Library of Congress Catalog Card Number: 96-070115
International Standard Book Number: 0-87263-481-7

Additional copies may be obtained by contacting:

Society of Manufacturing Engineers
Customer Service
One SME Drive
Dearborn, Michigan 48121
1-800-733-4763

SME staff who participated in producing this book:

Donald A. Peterson, Senior Editor
Dorothy M. Wylo, Production Assistant
Rosemary K. Csizmadia, Operations Administrator
Karen M. Wilhelm, Manager, Book Publishing
Cover design by: Judy D. Munro, Manager, Graphic Services

Printed in the United States of America

For Cheryl

Acknowledgments

The author would like to thank the following people for their help during the preparation and validation of this manuscript.

Kathleen Holst, *Fadal Engineering*
Don Nogy, *Ellison Machinery Company*
Jeff Hickernell, *Ellison Machinery Company*
Michael Sterioff, *Marposs Corporation*
Dana Scott, *Okuma America*

Table of Contents

Preface

Over the years, the ease-of-use, flexibility, and capabilities of CNC machine tools and controls have improved tremendously. The first numerically controlled (NC) machines (before computers were incorporated in the machine control) were cumbersome, inflexible, and weak. The typical NC program had to be written in a very strict format and coded onto 1-in. (25.4-mm)-wide tape or punched cards to get the machine tool to perform even the most basic tasks. If a programming mistake was made, the tape or cards had to be changed. Many features we now take for granted, like simple program editing capabilities, were on the horizon, yet to be developed.

Control manufacturers have been constantly striving to make their controls as easy to work with as possible. Most of the limitations that made early NC machines difficult and cumbersome to work with have been overcome. For example, because we now use tool-length and cutter-radius compensation, the machining center programmer no longer has to deal with a cutter's length and radius when writing a program. In like manner, fixture offsets keep programmers from having to deal with the program zero assignment until the setup is made. Motion interpolation types like linear, circular, helical, polar coordinate, and spline allow the programmer to define even the most complex motions with relatively simple commands. And many of the canned cycles available with today's CNC controls rival what can be done with the best computer-aided manufacturing (CAM) systems.

THE BEST-KEPT SECRET OF CNC

Some CNC technology developments are highly publicized and their application is so obvious that almost everyone immediately recognizes their value. Decimal point programming is such a feature. Its value lies in the elimination of the need for programming decimal values with an inflexible format. The command *X1.25*, for example, is obviously easier to read and understand than *X12500*. When introduced, decimal point programming gained immediate and universal approval among CNC users; nearly everyone wanted this feature to be included with all future CNC machine purchases.

Other CNC control technology developments haven't drawn as much attention or their proper application hasn't been so easily understood. This has led to confusion and controversy among CNC users. *Conversational controls* (also called shop-floor programmed controls), for example, are great for those CNC users who for various reasons (small lot sizes, no repeat business, limited CNC staff, etc.) must prepare programs during setup. However, these same controls are not as well accepted in companies that run medium and large lot sizes, or those that enjoy a great deal of repeat business, or in companies that can afford to adequately staff their CNC environment.

Even more discouraging are certain CNC control technology developments that are largely overlooked, regardless of their potential benefit to CNC users. Such oversight can lead to dramatic under-utilization of CNC equipment for companies that have need of the feature, since no user can begin to utilize a CNC feature if they are not aware it exists. *Parametric programming* is one such feature. Though its capabilities have been available from CNC control manufacturers for well over 15 years, and though almost all CNC users can greatly benefit in one way or another from it, few people in this industry even know what parametric programming is, let alone how to reap its benefits. The first objective of this text is to correct this oversight. Early in the text, you will be exposed to countless applications for parametric programming, and surely you will recognize some that can help in your own CNC environment.

Once you see all that is possible with this very helpful programming tool, you may wonder why you were not exposed to it long ago. In fact, given the broad range of applications for parametric programming, it is likely that you will discover countless applications of this best-kept secret of CNC that will streamline your CNC environment in many ways.

WHY PARAMETRIC PROGRAMMING IS SUCH A SECRET

We do not wish to imply that keeping parametric programming a secret was or is intentional. In fact, the reason why most machine tool builders and control manufacturers do not heavily publicize this feature is that *parametric programming was not originally developed for use by CNC users*; it was developed by CNC control manufacturers to allow machine tool builders and accessory device manufacturers to integrate higher level programming capabilities at G-code level.

Many probing systems, for example, must be programmed at G-code level. However, there are a great many things a probe must do that cannot be handled with normal G-code level programming commands. For example, once a probe has contacted a surface, the surface location must be attained and used within the CNC program. Can you think of any G-code level programming command that allows you to determine the machine's position and use it within the program? There isn't any. Without parametric programming, this function is simply not available within the G-code level programming functions of most controls.

While probing systems are generally accepted as the machine accessory that originally prompted CNC control manufacturers to develop parametric programming functions, certain other machine tool accessories require a higher level of programming functions than can be found in the G-code level commands of most CNC controls. Such accessories as post-process gaging systems, digitizing systems, adaptive control systems, tool breakage detection systems, and tool- life management systems outpace G-code capability.

That parametric programming was not originally developed with the typical CNC user in mind also explains why some versions of parametric programming tend to be rather cumbersome to work with. When compared to today's higher level computer programming languages, like BASIC, C language, and PASCAL, many versions of parametric programming are somewhat crude.

Because parametric programming was not originally developed for CNC users (and hence emerged as somewhat cumbersome to work with) many machine tool builders today are reluctant to teach their end users how to use it. Admittedly, beginners with no computer programming experience may find it difficult to get comfortable with parametric programming. This is why we stress so heavily the *applications* of parametric programming in the early stages of this text. Our purpose is to whet your appetite for what can be done with this very powerful programming tool; we thereby hope to instill in you the motivation to stick with it until you have mastered all that is available. When you learn what can be done, you will want to learn how to do it.

OPTION ON SOME CONTROLS

Some control manufacturers contend that parametric programming is only required when a special accessory device (like a probing system) is attached to the machine tool, and they make parametric programming an optional feature. They contend that they are actually saving their customers money by not forcing them to purchase features they do not need. But as you will see, many applications for parametric programming will save time, reduce effort, and minimize the potential for costly mistakes. These benefits, of course, result in a direct cost savings. In many cases, the cost to have parametric programming added to your control(s) (if it is not a standard feature from your control manufacturer) can be easily justified in light of the benefits it can provide.

SCOPE

Parametric programming encompasses a broad field with almost countless possibilities. To attempt to adequately address them all in a single book would be futile, therefore we need to qualify and limit the scope of this text.

VERSIONS OF PARAMETRIC PROGRAMMING

Just as there are many versions of computer programming languages (BASIC, FORTRAN, C language, PASCAL, etc.), there are numerous versions of parametric programming. In fact, with the exception of Fanuc-compatible controls, almost every control manufacturer has its own special version of parametric programming. And just as the different computer programming languages vary with regard to complexity and capability, so do the different versions of parametric programming. Some CNC control manufacturers even have multiple versions of parametric programming within their own product lines.

It would be impossible for this text to show every technique available from every version of parametric programming. Instead, our presentations will be heavily *application* based. We'll begin by ensuring that you understand *what can be done*. After we discuss an application or parametric programming feature in general terms, we will show specific examples in three popular versions of parametric programming. If you can understand the general presentations, and if you can see how features of parametric programming are applied in three popular versions, you should be able to adapt what you know to virtually any version of parametric programming.

Custom macro B

Much thought went into determining *which* specific versions of parametric programming would be addressed in this text. Since custom macro B is the most popular version of parametric programming (several CNC control manufacturers supply custom macro B), and since several other versions of parametric programming strongly resemble custom macro B, it was the obvious choice.

Unfortunately, custom macro B was developed more than 15 years ago and has not changed substantially in that time. As anyone familiar with computer technology knows, a lot has happened in the last 15 years to help computer programmers. If you have been working with the latest version of BASIC or C language, be ready to step back a few years if you are going to be working with custom macro B. While custom macro B is still a powerful and formidable version of parametric programming, programming with it is crude by today's programming standards.

Okuma's user task 2

In similar fashion, Okuma's user task 2 was developed more than 10 years ago and is representative of CNC control manufacturer-based parametric programming versions. You will find that many of the functions of user task 2 closely resemble custom macro B.

Fadal's macro

The current trend in CNC technology is toward personal computer-based CNC controls. Many of the PC-based CNC control manufacturers incorporate the best features of *current* computer programming languages in their versions of parametric programming. Fadal's version, called *macro*, resembles more current versions of BASIC. When it comes to the computer-related features, it will be quite easy to adapt to Fadal's macro if you have experience with current versions of BASIC. Several other control manufacturers utilize current versions of BASIC in their version of parametric programming, though they may not be completely compatible with Fadal's macro.

Since we offer comprehensive descriptions of these three versions of parametric programming in Part One of this text, you must first determine which discussions you will follow. If, of course, you have one of these versions of parametric programming in one or more of your controls, your choice is obvious. However, if you have machines with another version, you must choose the one that most closely resembles the version of parametric programming you will be using. It is our sincere hope that if you understand the general application (you know what you *want* to do), and if you see how it is specifically done in three versions of parametric programming we feature in this text, you will be able to adapt what you know to your own specific version of parametric programming.

MACHINE TYPES

Excellent applications for parametric programming exist for just about every kind of CNC machine tool in existence today. However, in our discussions we limit our presentations to include the two most popular machine types, the CNC machining center and CNC turning center. Again, with an understanding of how parametric programming can be applied to these popular machines, you should be able to adapt what you know to other forms of CNC equipment.

PREREQUISITES

While we assume nothing of your current knowledge of parametric programming, we do assume you possess a firm understanding of *manual* programming for the CNC equipment you will be working with. In almost all applications, you will be merging parametric programming techniques with manual programming. We will explain only the parametric programming functions, assuming that you understand the G-code level manual programming commands being used.

An understanding of computer programming also helps with parametric programming. We do not assume you have this background and will explain the basics of each computer-programming-like feature we address. For readers with absolutely no prior computer programming experience, we recommend you purchase one of the excellent beginner's books on BASIC to reinforce our presentations. These can be found in your local bookstore or library.

PRESENTATION METHOD

You may have picked up your CNC control manufacturer's programming manual and tried to read a section on parametric programming. If you have, you likely have become somewhat frustrated, since most control manufacturers' manuals tend to be written in *reference* format. Like a disk operating system (DOS) reference manual for personal computers, the parametric programming section of most control manufacturers' manuals tend to assume you already know what you want to do and therefore give only the correct *syntax* for each word or feature. Few explain the implications of *when* a given word or feature could be helpful. Fewer still show comprehensive examples for all features.

By comparison, the text herein will be *tutorial* in nature. We will, of course, show the correct syntax for each word or command we show, but more importantly, we will stress the implications of why each feature can be helpful. We also show many example programs. In addition, when you finish with this text, you should be able to easily understand your control manufacturer's descriptions relative to parametric programming.

ORGANIZATION OF THE TEXT

This book is structured into two parts. Part One, *Parametric Programming for Machine Tools*, introduces you to the many features of parametric programming and describes how to apply them to improve the general utilization of your CNC machine tools.

In Chapter 1 we introduce the various applications of parametric programming. Chapter 2 describes variable techniques. Chapter 3 discusses arithmetic capabilities, and Chapter 4 presents program flow and logic commands. In Chapter 5 we discuss the CNC-related features of parametric programming. Chapter 6 shows how to approach and verify parametric programs in each of the five application categories.

Part Two, *Parametric Programming for Touch Probes*, is a text that shows you how to write parametric programs to drive probing systems. Chapter 7 introduces you to probing, showing the various probe types and their applications, and how probes work. In Chapter 8 the new programming commands for probing systems are defined. Chapter 10 focuses exclusively on how to write programs for spindle probes, since they require the most from probe users in the way of custom programming.

PART ONE

PARAMETRIC PROGRAMMING FOR MACHINE TOOLS

Part One of this text will be devoted to showing you how parametric programming can dramatically streamline the utilization of your CNC machine tools. Since parametric programming is one of the most advanced features of CNC, and since it has been overlooked by so many CNC users for so long, we begin in Chapter 1 by presenting a comprehensive view of what parametric programming is and just what can be done with this powerful programming tool.

From that point, we present computer-programming-related features of parametric programming, including variables (Chapter 2), arithmetic (Chapter 3), and program flow control (Chapter 4). In Chapter 5 we go into the CNC-related features of parametric programming, and in Chapter 6, we show how to approach and verify parametric programs.

For each step through the parametric programming journey there will be examples to stress the points being made. When finished with Part One, you should be ready (and anxious) to tackle applications in at least four of the five application categories.

Chapter 1

Introduction to Parametric Programming

Parametric programming is the best-kept secret of CNC. Few in this industry even know what parametric programming is; fewer still know how to apply it. What is mystifying is that almost all CNC users have excellent applications for parametric programming, and as a consequence, are underutilizing their CNC equipment. It is quite likely that your own CNC machine tool utilization can be dramatically improved by incorporating parametric programming techniques.

WHAT IS PARAMETRIC PROGRAMMING?

Parametric programming goes by many names. Fanuc (or any control manufacturer that claims to be 100% Fanuc-compatible) calls it *custom macro*. Fadal calls it *macro*. Okuma calls it *user task*. Sodick calls it *Q routine*. Kearney & Trecker calls it *advanced programming language* (APL) for their Gemini controls. Sharnoa calls it *arithmetics*. Some control manufacturers have parametric programming capabilities but have not given it any special name. Bridgeport's Boss Series controls, for example, have excellent parametric programming functions, yet they are simply included with other G-code level programming functions.

Even within a given control manufacturer's product line there may be variations in parametric programming functions. Fanuc, for example, has custom macro *version A and version B*, with version B being more powerful and easier to use. Which version of custom macro you have is based on which control model you purchase within Fanuc's product line. In similar fashion, Okuma offers user task 1 and user task 2, with user task 2 being more powerful and easier to use.

While the variations from one version of parametric programming to another lead to differences in specific usage techniques, the broader applications and uses for parametric programming remain remarkably similar. This is evidenced by the fact that the majority of specific applications described in this text can be adapted to almost every version of parametric programming just mentioned.

Just as a given software application can be handled by a variety of computer programming languages (BASIC, C language, PASCAL, etc.), so can a given CNC application be handled by a variety of parametric programming versions.

COMPARISON TO SUBPROGRAMMING

The best way to get comfortable with any complex subject is to compare it to simpler topics with which one may already be familiar. Parametric programming is no exception. If, for example, you have worked with the subprogramming functions of your control, you have scratched the surface of what can be done with parametric programming.

All CNC controls have subprogramming functions to allow commands within the CNC program to be repeated. This minimizes the number of commands that must be given in the CNC program. If, for instance, five identical pockets are to be milled in five workpieces on a machining center during the same cutting cycle, it would be cumbersome to program each pocket independently. Instead, only the programming commands necessary to machine *one* of the pockets need be programmed. These redundant commands can be executed five times to machine the five pockets, eliminating many cumbersome, lengthy, and error-prone commands.

While the specific commands related to subprogramming vary from one control manufacturer to another, Fanuc uses an M98 to call a subprogram. A P word within the M98 specifies the subprogram number and an L word specifies the number of executions of the subprogram. With this control, the command

```
N050 M98 P1000 L5
```

tells this control to execute subprogram O1000 five times. As long as program O1000 contains the commands needed to correctly machine one of the pockets, the program's length can be shortened. In addition, the potential for mistakes is greatly reduced.

Subprogramming techniques are obviously very helpful. However, if anything changes from one pocket to the next (width, height, depth, e.g.), simple subprogramming techniques *cannot* be used. Without parametric programming, each pocket must be programmed independently. In addition to giving the programmer the ability to repeat redundant commands, parametric programming allows *anything* to change from one execution of the parametric program to the next. In the pocket example, if any pocket attributes (parameters) change from one pocket to the next (width, height, depth, etc.) these variations can be easily handled within the parametric program.

In this sense, parametric programming gives the programmer the ability to write a general-purpose subprogram. *If you have ever found yourself wishing you had the ability to write general-purpose subprograms, you have an application for parametric programming.*

As we discuss in the next chapter, the things that change from one pocket to the next are called *arguments*. Some versions of parametric programming let you name the arguments being passed to the parametric program in a very logical manner. Fanuc's custom macro B, for example, uses a G65 command to call the parametric program. Letters of the alphabet can be included in this command to specify argument values. In the command

```
N050 G65 P1000 X2.0 Y1.5 W4.0 H2.0 D.25
```

G65 tells the control this is a custom macro *call statement*. The P word still specifies the program number of the parametric program that does the pocket milling. The letters *X* and *Y* are used to specify the lower left-hand corner position of this pocket along the *X* and *Y* axes; W is used to specify the pocket width, H the height of the pocket, and D the pocket depth.

Notice how logical this version of parametric programming makes the entry of values that change from one pocket to another. Anyone can easily recognize the meanings of *X*, *Y*, W, H, and D. If another pocket of a different size must be machined, another G65 command can be easily specified that contains different argument values.

While other versions of parametric programming may not make it quite this logical, they still allow anything to change from one execution of the parametric program to the next. With Fadal's macro, for example, V-series variables (from V1 to V100) can be used to specify the values of arguments. The pocket milling example shown earlier would look like this:

```
N005 #V1 = 2.0 'X position
N010 #V2 = 1.5 'Y position
N015 #V3 = 4.0 'Width
N020 #V4 = 2.0 'Height
N025 #V5 = 0.25 'Depth
N035 M98 P100 (Call parametric program O100)
```

Notice that the meaning of each argument would not be easily understood without the message following the apostrophe or in parentheses in each command. For this version of parametric programming, the programmer must document in this manner, especially if the operator or setup person is expected to modify these values at any time during the setup or production run. A detailed discussion on arguments and their documentation appears in Chapter 2.

COMPARISON TO CANNED CYCLES

All CNC control manufacturers do their best to create a series of special programming features to minimize the work a programmer must do. Almost all CNC machining center controls, for instance, come with a standard set of hole machining "canned cycles" (typically specified by G81-G89). Some machining center controls also have certain milling functions such as circle pocket milling, slot milling, thread milling, and face milling. Turning centers commonly come with a set of canned cycles for rough and finish turning and boring, grooving, hole machining, and threading.

The commands incorporated in one popular machining center control to drill a series of holes follow.

.
.
.

N065 G54 G90 S400 M03 (Select coordinate system, absolute mode, and start spindle)
N075 G00 X1.0 Y2.0 (Rapid to first hole location)
N080 G43 H01 Z0.1 (Instate tool-length compensation, move to *Z* approach position)
N085 G81 R0.1 Z – 0.75 F4.5 (Drill first hole)
N090 X3.0 (Drill second hole)
N095 X5.0 (Drill third hole)
N100 X7.0 (Drill fourth hole)
N105 G80 (Cancel cycle)
N110 G91 G28 Z0 M19 (Return to *Z* axis reference position)

.
.
.

In line N085, the first hole is completely machined based on the control's G81 function, and the words are included in the command (R, Z, F, etc.). The control will perform a series of strictly planned motions based on the canned cycle's design. In the case of G81, the control will first rapid the tool to the *XY* position. Next it will rapid the tool to the R plane, plunge the tool to the hole bottom, and retract the tool from the hole. With G81, four movements are generated with one command. With other canned cycles (like peck drilling) many movements can be caused by one command.

Notice how similar the G81 command format is to that of the pocket milling example calling command shown earlier (for version B custom macro). The R, Z, and F words in the G81 (or any canned cycle) are like the arguments being passed to the parametric program. You can think of all canned cycles as being like parametric programs written and maintained by the CNC control manufacturer.

If your control lacks a canned cycle that you need, or if you do not agree with how a given canned cycle functions, you can develop a parametric program to handle the application. In essence, you can create your own canned cycles. *If you have ever wanted the ability to create your own canned cycles, you have an application for parametric programming.*

COMPARISON TO COMPUTER PROGRAMMING

If you have had experience with any computer programming language, you already know much of what is available with parametric programming. Many of the computer-related features of parametric programming closely resemble the computer programming language BASIC (or just about any other computer programming language). These features—variables, arithmetic, logic, and looping—are explained in detail later in Part One of this book. For now, it is enough to know that what can be done in BASIC programs can be done in parametric programs. Again, if you have had no previous experience programming in BASIC, we recommend that you get one of the many excellent beginner's books on BASIC that are available. A working comprehension of BASIC will reinforce the presentations we make for the computer-related features of parametric programming. *If you have ever found yourself wishing you could include computer-programming-like commands in your CNC program, you have an application for parametric programming.*

APPLICATION CATEGORIES

As stated in the Preface, there are countless applications for parametric programming, and almost every CNC user has use for at least some of them. In this discussion, we organize all applications for parametric programming into five basic categories.

Given the vast potential benefits afforded by parametric programming, *all CNC people should be able to recognize parametric programming applications*. However, in reality, most do not. The vast majority of CNC people have never even heard of parametric programming. Of those who have, few can recognize good applications (or have misconceptions of what really can be done). And fewer still can actually develop a parametric program once an application is recognized.

The first step to doing anything is knowing it is possible. Before you can utilize the automatic-frequency-searching function of your stereo's FM tuner, you must first know it exists. In like manner, before you can solve any problem with parametric programming, you must first know that there is such a programming technique and *which* problems can be solved. In this discussion, you will see several applications that cannot be handled with normal G-code level programming techniques. In fact, many of the applications of parametric programming are so remarkable that you may have trouble believing some of the claims we make. While we do not actually show *how* to handle these applications in this introductory discussion, we do address them in greater detail later in the book.

FAMILIES OF PARTS

Many CNC users machine groups of very similar workpieces. These groups are commonly called *part families*. Generally speaking, all workpieces in a part family closely resemble one another and require similar (if not identical) machining operations.

In perfect part families, only the size of each workpiece changes. Bolts, screws, nuts, washers, and pins, for example, are made in a variety of sizes to suit the needs of industry. The hex-shaped sockets a hand tool manufacturer makes are made in various sizes to accept changing bolt and nut sizes. The rings a piston ring manufacturer makes are made in various sizes and used with a variety of piston sizes. The list of common part families is virtually endless.

The similarity among workpieces in a family usually corresponds to the similarity of the *machining process* used to make each workpiece. For perfect part families, the same machining process and tooling can be used to machine all workpieces in a family. With conventional CNC programming techniques, this can result in a very large number of similar CNC programs. *If you find yourself slightly modifying one CNC program to create another, you have a family-of-parts application for parametric programming.*

If you do any repeat business, each hard-and-fixed CNC program must be *maintained.* The larger the number of workpieces in the family, the more CNC programs are involved. If an engineering change is made to a part family containing 30 workpieces, 30 hard-and-fixed CNC programs must be changed. Additionally, the same verification procedures required for new CNC programs must be repeated 30 times as the modified programs are run for the first time. Many CNC programmers, setup people, and operators dread engineering changes for these reasons.

In similar fashion, if you wish to make a CNC program change to improve quality, increase tool life, reduce machining time, or to achieve any improvement in the manufacturing process, the programs for all workpieces in the part family must be changed. This dramatically limits flexibility, time-to-market, and your ability to try new things, since even minor improvements will involve a great deal of time and work. All too often, process-improving changes are not incorporated owing to the work involved in modifying all of the programs affected by the changes.

With many family-of-parts parametric programming applications, only one base program is required to machine all workpieces in the entire part family. When machining a specific workpiece in the family, arguments (variables) will specify the value of each changing element within the part family. In essence, the parametric program within the CNC control is told *which* workpiece is currently being machined.

How these changing arguments are specified varies with application. In some companies, the CNC operator or setup person manually changes these arguments during setup. With Fadal's macro language, it is even possible to prompt for arguments as the program is executed. This function is programmed much like the INPUT statement of BASIC.

In other companies, a CNC programmer programs the changes. Indeed, for perfect family-of-parts applications, a CNC program containing the arguments even can be automatically created by the production control department as a production order is issued. There is more on how arguments can be passed to the parametric program in later chapters.

As you can readily comprehend, this dramatically simplifies the long-term programming of individual workpieces (once the parametric program is developed) and allows for unlimited modifications right at the CNC control, providing a dramatic increase in flexibility. Indeed, going from one size workpiece to another is so simple that many companies allow their CNC operators to make the needed changes. Additionally, modifications that affect the entire part family (engineering and optimizing changes) are much easier to make. Only one program need be changed.

How are your prints dimensioned?

Some companies utilize *variable dimensioning techniques* for dimensioning a family of parts. The design engineer will dimension values that change with a letter of the alphabet. Any person viewing the drawing will determine the value of a given dimension by referencing a chart included on the drawing. Based on knowing the workpiece's part number, anyone can find the values of each variable dimension. Figure 1-1 shows an example of this kind of family-of-parts dimensioning. *If your company uses variable dimensioning techniques, you have a family-of-parts application for parametric programming.*

Notice in Figure 1-1 that dimensions are specified with letters (A, B, C, etc.). Some versions of parametric programming even allow you to designate the value of each argument to be passed to the parametric program with the same letters used on the print. To specify that part number SC-0875 in Figure 1-1 is being made, for example, one way of doing so using Fanuc's custom macro B is:

```
N060 G65 P1000 A1.375 B0.875 C0.437 D0.2187
```

In this example, notice how arguments A, B, C, and D directly correspond to print dimensions (D specifies the top hole diameter, which in turn, determines how the rest of the hole must be machined). Though there may be other changing attributes to be accommodated by the parametric programmer (speed and speed variations, tool station numbers, etc.), this example command demonstrates precisely how easy it is to specify which workpiece is to be machined.

The more workpieces in a part family, the easier it is to justify parametric programming techniques. Keep in mind, however, that parametric programs do take longer to write than conventional CNC programs. From a strictly program-

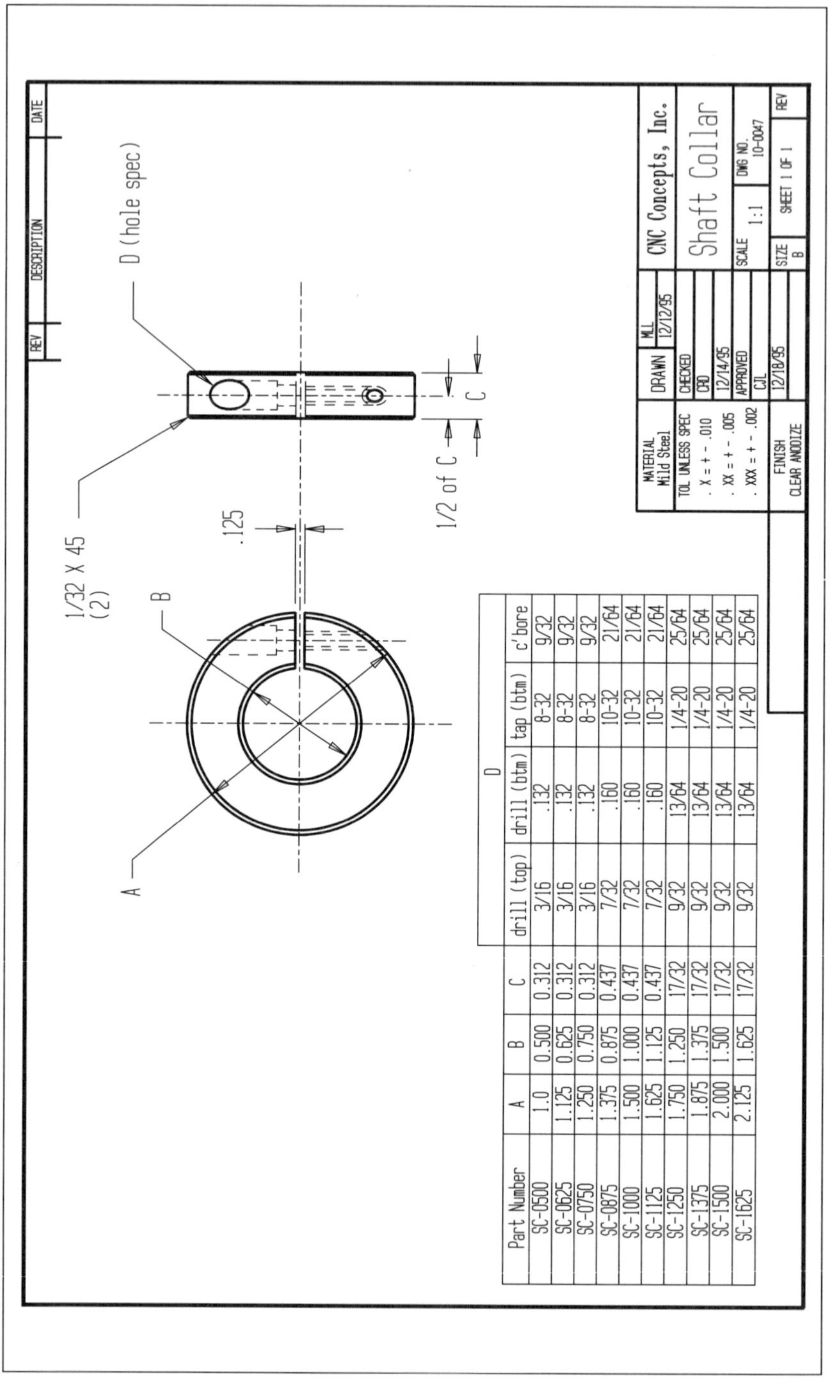

Part Number	A	B	C	D			
				drill (top)	drill (btm)	tap (btm)	c'bore
SC-0500	1.0	0.500	0.312	3/16	.132	8-32	9/32
SC-0625	1.125	0.625	0.312	3/16	.132	8-32	9/32
SC-0750	1.250	0.750	0.312	3/16	.132	8-32	9/32
SC-0875	1.375	0.875	0.437	7/32	.160	10-32	21/64
SC-1000	1.500	1.000	0.437	7/32	.160	10-32	21/64
SC-1125	1.625	1.125	0.437	7/32	.160	10-32	21/64
SC-1250	1.750	1.250	17/32	9/32	13/64	1/4-20	25/64
SC-1375	1.875	1.375	17/32	9/32	13/64	1/4-20	25/64
SC-1500	2.000	1.500	17/32	9/32	13/64	1/4-20	25/64
SC-1625	2.125	1.625	17/32	9/32	13/64	1/4-20	25/64

Figure 1-1. Typical CAD drawing showing variable dimensioning techniques commonly used by companies that machine part families.

ming-time-based standpoint, it may be hard to justify writing a parametric program for part families having only a few workpieces. Even for simple workpieces, it will take from three to five times as long to write the parametric program as it will to write a hard-and-fixed CNC program for one workpiece in the part family.

Family-of-parts program structure

At the risk of getting deeper into specific issues than is customary in an introduction of application categories, we introduce an important point relative to family-of-parts applications. Most family-of-parts applications for parametric programming require just one program, commonly called the *main program.* While other programs could be involved, *the main program is the parametric program* for part family applications. The arguments that specify the values of each changing element of the part family are commonly listed at the very beginning of this program and are referenced later in the program when they are needed. Different methods of argument assignment are covered in later chapters.

USER-CREATED CANNED CYCLES

Even if your company's products contain no part families, it is quite likely that you have at least some similar machining operations that occur on several workpieces. Parametric programming can dramatically streamline the programming of these repeated machining operations. Many machining operations are similar in nature and can be easily handled with parametric programming techniques. Here are some examples you should easily recognize:

- In machining centers
  - Thread milling,
  - Round pocket milling,
  - Rectangular pocket milling,
  - Circle milling,
  - Face milling,
  - Keyway milling,
  - Slot milling,
  - Hole machining (drilling, tapping, reaming, boring, etc.),
  - Hole pattern machining (bolt-hole circle, grid pattern, window pattern, etc.).
- In turning centers
  - Groove necking,
  - Tapping (many turning centers do not have this cycle),
  - Deep-hole peck drilling (many turning centers do not have this cycle),
  - Knurling.

These machining operations are quite common and many control manufacturers have developed canned cycles to handle at least some of them. While control-manufacturer-created canned cycles are extremely helpful, they are, of course, fixed in their function. If you do not like the way they work, there is little you can do about it. A good example is the machining center G83 deep-hole drilling cycle which is quite limited on the majority of controls. Most controls do not allow a change in the depth of each successive peck as the hole gets deeper. Most also do not allow you to combine the chip-breaking cycle (G73 on many controls) with the deep-hole drilling cycle. And the method by which the number of passes and the depth of each pass are calculated may not be in line with what you are comfortable with. *If you have ever found yourself wishing that your control's canned cycles worked differently, you have a user-created canned-cycle application for parametric programming.* Parametric programming actually gives you a way to improve the canned cycles that come with your machine tool.

In addition to modifying the method by which your *current* canned cycles work, with parametric programming you have the additional capability of creating your *own* canned cycles. Most machining center controls, for example, do not have a canned cycle for thread milling. If you perform thread milling on a regular basis and if your control does not have a thread milling cycle, you are likely writing many tedious, redundant, and error-prone commands. With parametric programming, you can create your own thread-milling canned cycle.

In similar fashion, most turning center controls do not include an adequate canned cycle to machine grooves. If you must machine neck grooves in many workpieces and if your control does not have a canned cycle for grooving, again, you must write many tedious, redundant, and error-prone commands. With parametric programming, you can create your own grooving cycle.

Many companies perform rather unusual operations that are specific to only their own products and manufacturing processes. Understandably, control manufacturers won't consider creating canned cycles for machining operations that are not helpful to the majority of their users. Relatively few machining center users, for example, machine dovetails. This operation normally requires a number of successive milling passes with a dovetail cutter. As when chasing a thread on a turning center, the number of passes and the depth per pass changes based on dovetail size, cutter material and rigidity, and workpiece material. If you machine dovetails, you have likely found there is no standard canned cycle to help with these tedious and error-prone commands. Wouldn't it be nice if you could complete this operation with one simple command? With parametric programming, you can.

As you consider your own applications for user-created canned-cycle parametric programs, be sure to take into account *all* of your company's special machining operations. Remember, almost anything is possible—even your most obscure machining operations are fair game. If you are writing more than a few hard-and-fixed CNC commands to perform a given machining operation among many workpieces, consider parametric programming techniques; they can streamline your programming process.

Parametric programs in this category are typically handled like subprograms. One parametric program per cycle must be stored in the control's memory. The *main* program (the program that does the bulk of the machining operations) will include a *call statement* that invokes the parametric program. With Fanuc's custom macro, for example, a G65 command invokes the parametric program. The P word within this command specifies the program number to be called. A set of letter address arguments passes data to the parametric program. To illustrate, a sample program is included here commanding that four threads be milled on a machining center. It assumes the thread-milling parametric program (program number O1000) has already been written and verified (we will show this parametric program later in Part One).

O0001 (Program number)
N005 G54 G90 S500 M03 (Select coordinate system, absolute mode, start spindle)
N010 G00 X3.0 Y3.0 (Rapid to first hole location)
N015 G43 H01 Z0.1 (Instate tool length compensation, rapid to approach position)
N020 G65 P1000 X3.0 Y3.0 D0.75 Q0.125 T1.0 M3.0 R1.0 F5.0 (Mill first thread)
N025 G65 P1000 X7.0 Y3.0 D0.75 Q0.125 T1.0 M3.0 R1.0 F5.0 (Mill second thread)
N030 G65 P1000 X7.0 Y7.0 D0.75 Q0.125 T1.0 M3.0 R1.0 F5.0 (Mill third thread)
N035 G65 P1000 X3.0 Y7.0 D0.75 Q0.125 T1.0 M3.0 R1.0 F5.0 (Mill fourth thread)
N040 G91 G28 Z0 (Return to Z reference position)
N045 M30 (End of program)

Note again how similar these commands are to the commands that call any canned cycle. Prior to invoking the parametric program for the first time, the spindle is started and the tool is positioned up close to the first hole to be machined. Tool-length compensation is also instated. These same commands would be given prior to commanding a hole-machining canned cycle. *Take special note that, like any canned cycle, user-created canned-cycle parametric programs will only machine a small portion of the workpiece. The rest of the machining is still done by the main program.*

Lines N020 through N035 machine the four threads. The G65 and P1000 specify that parametric program O1000 is being invoked. The letter address arguments are chosen by the parametric programmer in as logical a manner as possible. In this example, *X* and *Y* specify the center position of the hole to be thread milled. The *Z* position is the top surface into which the thread is being machined. The letter D identifies the thickness of the workpiece (depth), Q is the thread pitch, T is the thread-milling cutter (tool) diameter, M is the major diameter of the thread, R is the approach radius size, and F identifies the machining feedrate.

Note that the only things that change in lines N025 through N035 are the *X* and *Y* positions of the other three holes being machined (four identical threads are being milled). Since the custom macro-invoking G65 command is a one-shot G-code, all other arguments must be included in each command to specify their current values. While you may still be wondering what the parametric program for thread milling will look like, the key point at this time is the second parametric programming application category. You should now be able to easily recognize parametric programming applications as they relate to *user-created canned cycles.*

UTILITIES

The application category that tends to be the most overlooked is utilities. Even people who consider themselves to be well versed in parametric programming may not be aware of some of the countless utility applications for the technique.

There is a good reason for this. Most of these applications incorporate little-known CNC features of parametric programming. And before you can incorporate any new technique, you must of course know it is possible to do so. While we discuss CNC features of parametric programming in great detail later in Part One, here we introduce you to some of the remarkable possibilities.

Utility applications can reduce setup and program verification time, catch mistakes in data entry, make the machine easier and safer to run, emulate costly options, and in general, facilitate CNC machine tool usage. In fact, *any specific problem you are having with your CNC machine tool can probably be resolved in some way through the use of a utility parametric program.* That's a pretty bold statement. Let's look at a few specific examples of utility applications to support this claim.

An automatic tool changer test

ABC company is having problems in the form of automatic tool changer crashes on their vertical machining center. They have a very large fixture on the machine located in the middle of the table. The tool changer will crash the tool into the fixture if the fixture is under the spindle when the tool change command (M06) is given. For reasons unknown, the programmer/operator/setup person continually forgets to position the table properly prior to the tool change, resulting in several tool changer crashes.

Two CNC features of parametric programming help solve this problem. First, most versions of parametric programming allow *access to the machine's current position* (relative to the machine's reference position). This means the utility parametric program can test to see if the fixture is under the spindle when the tool change is commanded. Some versions of parametric programming even allow M codes to invoke parametric programs, meaning this test can be done by a parametric program invoked by the M06 tool-change command itself.

Second, most versions of parametric programming allow alarms to be generated. Like program storage alarms, alarms generated from within parametric programs put the machine into a true alarm state and halt the program. Everything necessary to clear any other alarm must be done to clear a parametric program-generated alarm. With these abilities, the utility parametric program can generate an alarm to stop the tool change from occurring if it finds that the fixture is under the spindle when any tool change is commanded, thus eliminating the possibility for future automatic tool changer crashes.

A part counter

DEF Company has a bar feeder on its turning center. The company's workpieces are relatively small and more than 200 of them can be machined per 12-ft. (3.66-m) bar. It is quite common that a production run is completed right in the middle of the bar, but since the machine runs unattended, the balance of the bar is run and many *extra* workpieces are machined. While the company does have a control-based part counter on its turning center, all it does is count parts; the operator still must monitor the number of workpieces that have been machined.

With the alarm generation capability mentioned, a utility parametric program can be created to replace the part counter included with the control. Not only will this utility parametric program count parts, but it will allow its user to specify how many parts must be machined in the production run. When the production run is completed, the parametric program will put the machine in alarm state, halting the production run at the specified number of workpieces.

A jaw-boring program

GHI company has a turning center with a three-jaw chuck. Almost all of the company's jobs require soft jaws, meaning they do a great deal of jaw boring. Until recently, the setup people bored jaws manually using the machine's handwheel. It was tedious, time consuming, and error prone. If a mistake was made and the jaws were bored oversize, the process had to be repeated. It was not unusual for the jaw-boring process alone to take well over 30 minutes. A utility jaw-boring parametric program now allows the setup person to set a few arguments (variables) to specify the size of the jaws currently being bored. The program then runs unattended, completely boring the soft jaws in less than 10 minutes, freeing the setup person to do other setup tasks.

A manual corner pickup utility

JKL Company makes numerous vise setups on its CNC machining center. The size of the workpiece being machined changes from setup to setup, and the setup person must measure the new location of program zero in every setup. Since this machine does not have a spindle probe, the setup person uses an edge-finder to locate the program zero position. Once found, three values (*X*, *Y*, and *Z*) must be entered into the corresponding fixture offset. Even with this relatively experienced setup person, this error-prone task takes about 10 to 15 minutes per setup.

Another CNC feature of most versions of parametric programming is access to offsets. Offset values of all kinds can be accessed (read and write capabilities) from within the parametric program. This, combined with the ability to determine machine position relative to the machine's reference position, means a utility parametric program can determine and store each fixture offset value.

To do this, the operator will first (manually) touch the edge finder to a program zero surface. The utility program will then determine the machine's current position. By knowing the edge-finder radius, the program can then calculate the distance from the machine's reference position to the surface (the fixture offset value). Finally, it will automatically store this value into the fixture offset. This process is repeated for the other axes (the *Z* axis measurement will require knowing the *length* of the edge finder). With this corner-finding utility parametric program, program zero measurement and fixture offset entry are much less tedious and error prone, and combined, now take less than 5 minutes.

These examples represent only a few of the countless utility applications for parametric programming; in fact, they barely scratch the surface of what can be done. Throughout this text we show many more utility applications. You will find that your only limitations to what can be done are your ingenuity, determination, and imagination. Truly, if you question yourself as to whether something is possible, immediately answer *"yes it can."* Then figure out how to do it.

GEOMETRIC SHAPES

All versions of parametric programming allow arithmetic calculations. While the level of arithmetic capabilities varies, almost all allow basic functions like addition, subtraction, multiplication, and division, as well as certain higher level functions such as sine, cosine, tangent, square root, and rounding. This, combined with the ability to perform logic and looping, gives the programmer the power to create parametric programs capable of machining highly complex geometric shapes. Indeed, any shape that can be defined with an arithmetic calculation can be machined with a parametric program.

While parametric programs for standard shapes like spheres, pyramids, ellipses, and splines can be created with relative ease, the difficulty of writing parametric programs for applications in this category is directly related to the complexity of the workpiece to be machined. When it comes right down to it, it is usually easier to utilize a computer-aided manufacturing (CAM) system for highly complex work. However, you should know that parametric programming for complex shapes has two advantages over even the most sophisticated CAM systems.

First, CNC programs for complex shapes generated by CAM systems tend to be very long. It is not unusual for a program that simply machines a spherical shape to include hundreds (if not thousands) of commands. The comparable sphere milling parametric program will consist of about 50 commands.

Second, parametric programming gives users the ability to create programs specific to their own applications. The parametric program for the circular screw machine cam shown in Figure 1-2, for example, is very easy to use. While there are CAM systems capable of creating CNC programs to machine a circular cam, there are also many that do not, and none make it as simple as a parametric program developed specifically for that purpose.

No computer-aided manufacturing system we are aware of makes it as simple for the user to input the values needed to machine the circular cam as the following example command using custom macro's argument assignment No. 2. (The actual parametric program for this macro application will be shown later in this text.)

```
N010 G65 P1002 A1.75 B-0.5 C0.1 I12.0 J1.75 I35.0 J1.625 I175.0 J1.1 I25.0
J1.75 I114.0 J1.75
```

With this method of argument assignment, A, B, and C allow the programmer to specify three unrelated values. In this case (and as Figure 1-2 shows), A is the starting radius, B represents the *Z* surface to mill, and C represents the incremental calculation angle. With C set to 0.1, a resulting *XY* position will be calculated every 0.1 degree, meaning as many as 3,600 positions will be calculated.

Then, up to 10 sets of I, J, and K can be programmed, one per rise or fall. In this case, I represents the incremental angular length of the current rise or fall. Value J represents the ending radius for the current rise or fall; K is not needed in this application.

Though this description is still introductory and you are likely not totally at ease with exactly how this works, you should at least be able to see how easy it is to specify the various rises and falls of circular cams with this parametric program. One command does the machining of the entire cam. Again, no CAM system that we know of can match this ease of use. *If you encounter a limitation in your CAM system, or if you need to improve the method of data input, you can develop a parametric program to machine the shape.* In essence, parametric programming gives you the ability to generate your own CNC control-based CAM system.

Need more motion types?

CNC control manufacturers strive to provide their users with all the motion types they need. Machining centers commonly come with linear and circular interpolation. If thread milling is to be done, helical interpolation also can be provided. In similar fashion, turning centers also come with linear and circular interpolation. If *C* axis with live tooling is available on the turning center (three-axis turning

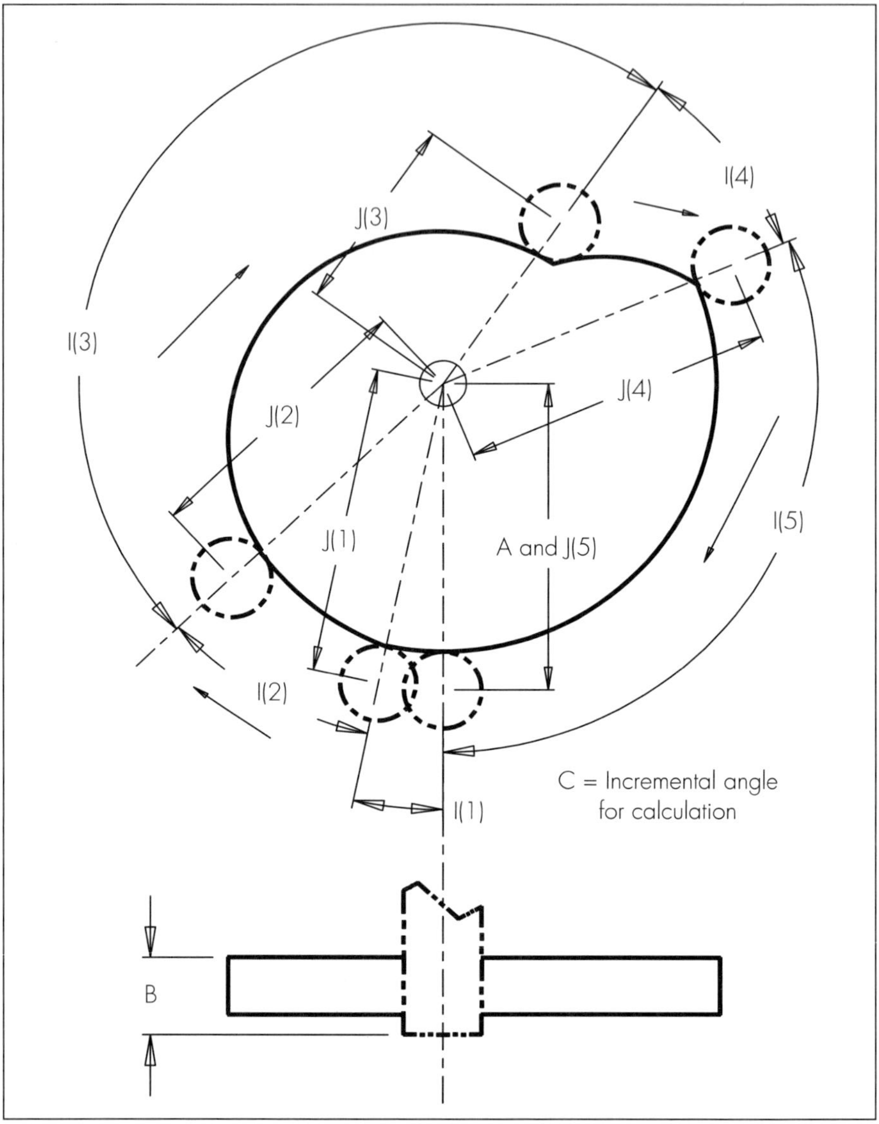

Figure 1-2. Parametric programming allows very specific control over how data is entered and machining is done. The rises and falls of a circular screw machine cam would present special challenges to even the most sophisticated CAM system.

center), most turning center control manufacturers also can provide polar coordinate interpolation to allow flats and contours to be machined around the face of the workpiece.

While the interpolation types provided by control manufacturers are usually adequate for most CNC applications, there are times when you need more. If, for example, you need to mill a tapered thread on a machining center, and if the thread quality is quite critical, normal helical interpolation will not suffice. Since standard helical interpolation causes a *circular XY* motion, and since the thread milling cutter is tapered, helical interpolation will cause the diameter being machined to change as the milling cutter moves into the hole. The thread will be milled in the form of a *spiral.*

To counteract this problem, the motion during tapered thread milling must not be in the form of a circular *XY* motion. Instead, the radius of the circular motion must change as the thread is machined. If machining occurs from top to bottom on a thread in a hole, this radius must get progressively smaller. The motion required is spiral, not circular. Few controls offer spiral interpolation. When faced with this limitation, many CNC users simply live with the nasty witness mark left in the hole by the thread-milling cutter. Others use their CAM systems to generate the lengthy series of point-to-point motions needed to cause the milling cutter to correctly machine the thread.

The best alternative, however, is to utilize parametric programming and create *spiral interpolation.* Armed with the tools of parametric programming, a programmer can easily cause the control to calculate the amount of radius change per degree (or portion of a degree) of circular motion based on the taper angle. A series of tiny straight-line movements can be generated in this manner, resulting in a spiral motion.

The spiral motion is just one example of a motion you may need to generate that is not included as an interpolation type in your CNC control. Others not commonly available from CNC control manufacturers that can be handled with parametric programming include polar coordinate interpolation, ellipse interpolation, parabolic interpolation, and spline interpolation. Again, *if the motion can be defined with an arithmetic expression, it can be commanded by a parametric program.*

DRIVING ACCESSORY DEVICES

Certain devices associated with CNC machine tools require a higher level of programming than can be found with standard G-code level manual programming. Most probing systems, for example, are programmed at G-code level, yet require sophisticated programming functions. When a probe stylus contacts a surface, the programmer must have a way to locate the surface contacted. When located, it is likely that some calculation will be necessary to determine if the surface is where it should be. If it is not where it should be, possibly an alarm should be sounded. These are all functions that are possible only through the use of parametric programming. (In Part Two of this text we focus exclusively on programming probing systems; hence we minimize our discussions of them here. It is important to note, however, that *you must understand parametric programming functions in order to write probing programs.*)

Other accessory devices also require parametric programming capability. Post-process gaging systems, for example, must have a way to feed back data to the control as to which dimensions must be modified through offset changes. The parametric program will receive this data and make the necessary changes. Some tool-life management systems use parametric programming as well to specify tool group criteria. And some adaptive control systems use parametric programming to specify machining criteria.

Even simple M-code functions, like hydraulic clamping devices, indexing devices, and automatic doors can be interfaced using some versions of parametric programming. Fanuc's custom macro, for example, allows the user to actually create new M-code functions and interface them with outside devices. This powerful version of parametric programming even allows an M-code confirmation signal to confirm that the action commanded by the M-code is finished before the program continues.

LIMITATIONS

To this point we have focused on what you can do with parametric programming, intentionally promoting this powerful programming tool to motivate you to learn more about it. And after reading about the application categories and the benefits you can expect, we hope you are anxious to know more. However, to be perfectly fair we must reveal the limitations of parametric programming as well.

As with any expenditure of time and money, you will have to justify the time you take to develop parametric programs for your applications. You may even have to justify the purchase of the parametric programming feature itself if it is not currently available on your control(s). The applications and benefits are many and you can use these in your justification. But to justify properly, you also must know the limitations.

COMMONLY AN OPTION THAT MUST BE PURCHASED

As stated in the Preface, certain CNC control manufacturers consider parametric programming to be an option that must be purchased at additional charge. While the price is usually relatively low and easy to justify in light of the benefits you can expect, the charges can add up fast if you have several machines to which you need to add the feature.

Keep in mind that some machine tool builders include parametric programming with their machines even if it is normally considered an option, especially if the machine has a special accessory that requires parametric programming. If, for example, your machine has a probing system, it is quite likely that parametric programming is being used to drive it. You may have parametric programming capabilities in some of your controls and not even know it.

COMPATIBILITY

While you may have several excellent applications for parametric programming, you must be concerned with how many different controls are involved. As stated, there are substantial differences among the various versions of parametric programming. If you have several similar CNC machine tools that have controls from several different control manufacturers, it can be difficult to develop, verify, and maintain the related parametric programs for a given application among all the machine tools you own.

PARAMETRIC PROGRAMS MUST RESIDE IN THE CONTROL

As stated earlier in this chapter, parametric programs for user-created canned-cycle applications are treated like subprograms. They must reside in the control's memory at the time they are invoked. If, for example, you have an often-used parametric program for thread milling, the program must be kept in the control on a permanent basis.

While the current trend in CNC control technology is toward PC-based controls that virtually eliminate program storage problems, many CNC controls are in use by manufacturing companies today that have very little program storage capacity. And adding more memory can be costly if memory is purchased directly from the control manufacturer.

EXECUTION TIME FOR CERTAIN COMMANDS

In the CNC world, time is the foe of productivity, and "Any time you make the control think, it takes time." Of course, CNC controls cannot actually think, but many of the calculations and much of the logic they perform resemble thinking. Some parametric programs are filled with commands that make the control perform calculations and logic. Any complex arithmetic expression that must be repeated many times, any logic statement that causes the control to make a decision, or any looping command that causes the control to repeat a series of commands will add to program execution time. While newer 32-bit controls can execute these number-crunching commands at lightning speed, and while there are techniques a parametric programmer can use to minimize "thinking time," the effects of parametric programming techniques on program execution time simply cannot be ignored.

Use your production quantities to help you judge whether it is wise to incorporate parametric programming. For relatively low production quantities (under 500 workpieces) the convenience-related benefits of using parametric programming will easily outweigh the minor increase in production time. As quantities grow, however, you must be on the lookout for those times when parametric programming techniques add to cycle time. For ultra-high production lot sizes (more than 5,000 workpieces), the effects of parametric programming on cycle time may be prohibitive.

PROGRAM PREPARATION TIME

As we mentioned, parametric programs tend to be more difficult to create than standard G-code level CNC programs. For simple family-of-parts applications, it commonly takes three to five times longer to write a parametric program (for an experienced parametric programmer) than it does to write one hard-and-fixed CNC program for one part in the family. From a strictly time-based justification standpoint, it may be hard to justify developing a parametric program for families of parts containing fewer than five workpieces.

For applications in other categories, however, the time it takes to prepare and verify the parametric program may be acceptable regardless of how long it takes. Consider the utility parametric program mentioned earlier that keeps the machining center automatic tool changer from crashing into a large fixture mounted on the table. One crash could cost hundreds (if not thousands) of dollars to repair, not to mention lost production time for repair and the potential for injury to the operator. In this case, the time and cost of preparing the parametric program will be very easy to justify, regardless of how long it takes.

SKILL REQUIRED

While the ultimate goal with parametric programming is to make the *long-term* utilization of your CNC machine tools easier, parametric programs take more skill to develop than standard G-code level CNC programs. It is not unusual for only one person in a company to really dig in and learn how to use parametric programming techniques. Of course, this person is the only one who can maintain those programs, and everyone else tends to be somewhat reluctant to work with them. If this individual is not available, or worse, leaves the company, it leads to many problems. When you decide to incorporate parametric programming techniques, you must be willing to train several of your CNC people so everyone understands how your parametric programs function.

INTRODUCTION TO FEATURE TYPES

To this point, we have been heavily discussing *what* can be done with parametric programming. We have not been explaining *how* to handle parametric programming applications. And if this chapter is your first exposure to parametric programming, you are likely anxious to learn more about how to handle the applications we have introduced. This involves knowing all of the features you have available to you with parametric programming.

We begin specific discussions on each feature in Chapter 2; here, however, we introduce the two feature types. We do little more than define each feature type, but this should prepare you for the more detailed discussions in Chapter 2. While different versions of parametric programs vary with regard to how these features are used and even as to which features are available, we introduce the most common.

COMPUTER-RELATED FEATURES

Computer-related features of parametric programming resemble features of any computer programming language. They give the CNC control an ability to perform number-crunching functions.

Variables

Most applications for parametric programs involve variables. Variables *are temporary storage locations for numbers,* and give parametric programming much of its flexibility. Like the memory in an electronic calculator, a value can be stored in a variable and used at any time during the execution of a parametric program.

Arithmetic

Though different versions of parametric programming vary in terms of what arithmetic functions they allow, all offer at least basic functions like addition, subtraction, multiplication, and division. Most additionally allow trigonometry functions like sine, cosine, and tangent. Many also allow common functions like square root, exponential notation, absolute value, and rounding.

Logic

Almost all versions of parametric programming have program execution control, which provides a decision-making capability. Most allow basic logic statements like unconditional branching (the GOTO statement) and conditional branching (IF statement). Some allow more advanced control of program execution including AND/OR/ELSE/ENDIF capabilities within the conditional branching statements that are very similar to those of higher-level computer programming languages.

Looping

Like all computer programming languages, most versions of parametric programming offer ways to repeat a series of commands a specified number of times. Even if no special looping command is provided, looping still can be accomplished with conditional branching (IF) statements.

CNC-RELATED FEATURES

The computer-related features of parametric programming are impressive in themselves, but when combined with the CNC-related features that also are provided, parametric programming becomes an even more formidable—and available—programming tool. Most of the CNC-related features are obscure and require much explanation. Simply mentioning that they are available as we do here does not offer any implications of *when* they may be needed or what benefits they provide. We deal in depth with the implications of when you will use these features in Chapter 5.

Access to machine position

Most versions of parametric programming allow you to determine the machine's current position in several ways, including relative to the machine's reference position, relative to program zero location, and relative to the machine's most recent position.

Access to offsets

Most versions of parametric programming allow you to read and write to and from your offset tables. With most, *all* offsets are accessible in this manner, including tool-length and cutter-radius compensation offsets (on machining centers) and tool-wear and geometry offsets (on turning centers).

Access to the control's current modes

Most versions of parametric programming allow you to determine (from within a program) the current state of all modes (incremental/absolute, inch/metric, rapid/straight line/circular, and all other modes).

Alarm generation

Most versions of parametric programming actually let you command to go into alarm state if the parametric program determines that a mistake has been made or a tool/workpiece collision is imminent.

Creation of G-, M-, and T-words

At least one popular version of parametric programming allows the creation of G-, M-, and T-words. You can actually have the control execute a specified custom macro whenever the control reads the G-, M-, or T-code you create.

Access to control panel functions

Some versions of parametric programming provide the flexibility to let you disable certain control panel functions such as single block, feed hold, and feedrate override.

Timers

Many versions of parametric programming give you access to the control's clock and timer.

IMPRESSED?

As stated, if this is your first exposure to parametric programming, you are probably surprised at the new programming tools you will have available with parametric programming and the added capabilities they provide. *Yet we have barely scratched the surface.* In the following chapters we introduce the features of parametric programming in much more detail. With each new feature, you will see simple applications to illustrate how and when the feature can be used. Remember that these early examples are not intended to stress the power of parametric programming, but to familiarize you with the basic use of each feature. Once you understand the basics, we proceed to the more advanced implications of parametric programming.

Chapter 2

Introduction to Variables

Discussions in Chapter 1 are necessarily broad in nature. The points made can be applied to any version of parametric programming. Beginning with this chapter, we offer much more specific information. Discussions and specific examples are included for three popular versions of parametric programming: Fanuc's *custom macro B,* Okuma's *user task 2,* and Fadal's *macro.*

If you have programming experience with any computer language, you already know many of the computer-related functions of parametric programming. However, for those who haven't written computer programs, we begin in this chapter a thorough explanation of each computer-related feature. The presentations are in rather basic terms, so even computer programming novices will be able to follow.

We recommend once again that newcomers to computer programming obtain a beginner's book on BASIC. It will help you understand what can be done with the computer-related features of parametric programming. Such a book also will describe much of the *mentality* that goes into computer programming that is not included in this text, including how to *flowchart* the logic related to more complicated programs. Experienced computer programmers will still need to read the appropriate sections of Part One to learn the specific *syntax* of parametric programming.

WHAT ARE VARIABLES?

One of the most important features of parametric programming is the ability to use variables. Though the actual syntax and structure for variables change from one version of parametric programming to another, the general applications remain remarkably similar.

If you have been exposed to computer programming, you already know what variables are. The use and application of variables in parametric programming are much the same as they are in any computer programming language. However, most people who have no computer programming experience find it somewhat difficult to understand the concept of variables. For this reason, we include two simple analogies to illustrate what variables really are.

Variables are temporary storage locations for numerical values. A value can be placed into a variable for use later. In this sense, you can think of any *tool offset* as being a kind of variable. With a tool offset, a value is stored in the offset for use by the program. Depending on the compensation type, the offset value could represent (among other things) the length of a tool, a milling cutter radius, a wear amount, or a turning tool's nose radius.

With tool-length compensation, for example, the programmer includes a command in the program that points to the tool's tool-length compensation offset number. With one popular control, a G43 command specifies tool-length compensation as being instated and an H word actually points at the offset number. The command

N015 G43 H02 Z0.1

for example, tells this control to look in offset number two for the tool-length compensation value for this tool. Since the programmer will not know the exact length of each tool as the program is written, he or she can utilize each of the machine's tool offsets as a kind of variable. In effect, the programmer is creating a program that will function correctly regardless of how long the setup person makes each tool.

At the machine during setup, the setup person will measure the tool-length compensation value for tool number two and enter it into tool offset number two. By so doing, *the setup person is manually assigning the value of a variable that corresponds to tool number two's length-compensation value*. This value means nothing to the control until the tool-length compensation command (the G43 command in our case) is executed. Note that the value of any variable can be changed whenever it is necessary to do so, just as any tool offset value can be changed at any time. During the production run, for instance, the operator can change the offset value as needed to compensate for tool wear.

In brief, the similarity between offsets and variables is that variables are storage locations for values. And like offsets, a value (possibly the result of a calculation) can be stored in the variable. The control will not know the meaning or use of the variable until it is referenced in some way during the execution of the parametric program. Most control manufacturers can display the values of all variables right on the display screen. Some even allow certain variables to be entered and modified by the CNC operator through the keyboard and display screen.

However, variables allow *much more flexibility* than tool offsets. In most cases, a tool offset is used for only one form of compensation, and the compensation type is quite rigid in its use of offsets. With tool-length compensation, for example, the offset can only be used to represent the tool-length compensation value. If you do not like the way your tool-length compensation feature functions, there is nothing you can do about it, unless you have parametric programming. A parametric programming variable can be used to represent almost anything. Among the countless things variables can be used to represent are an axis position, an arc radius, a chamfer size, a spindle speed or feedrate, a tool diameter, and almost anything else. Indeed, your own imagination and ingenuity are the only limitations to what variables can represent.

Variables within parametric programs also can be thought of as memories in an electronic calculator. With a calculator, you can store a value that is needed in your calculations. This memory value can be quickly referenced as required by pressing one or two keys. Variables in parametric programs can be used to represent constant values (possibly the result of an arithmetic expression) and can be referenced as required within the parametric program. Some of the many applications for variables are:

- To get *arguments* into the parametric program and tell the program how to behave.
- To serve as temporary storage locations for values that will be used throughout the parametric program.
- To provide the programmer the ability to set and step counters.
- To provide the programmer the ability to set *flags*.
- To serve as constant values that seldom change.

Most programmers who prefer the inch mode, for example, use a value of 0.100 in. (2.54 mm) as their rapid approach distance. If programmed as a hard-and-fixed value, it can be quite difficult to change all approaches in the program to something smaller—say 0.050 in. (1.27 mm)—to reduce cycle time. However, if this approach distance is specified as a variable and referenced during every approach, the approach distance can be easily changed by changing one word in the program.

As is true with any computer programming language, different types of variables attend most versions of parametric programming. But though most versions of parametric programming do not require that you *declare* the variable type (as you must in some computer programming languages), the parametric programmer must *understand* the different types available with their particular version in order to write parametric programs.

VARIABLES IN FANUC'S CUSTOM MACRO B

There are five types of variables in custom macro B. With one exception, all are represented by a pound sign (#) and a number. The number next to the pound sign specifies the *variable number* and also implies its type.

ARGUMENTS

Arguments specify values that change from one execution of the custom macro to the next. They tell the custom macro program how to behave *right now*, and are especially helpful in user-created canned-cycle and family-of-parts applications. There are actually two ways in custom macro B to use arguments, based on which of these two application categories (user-created canned-cycle and family-of-parts) you are working with.

Arguments with user-created canned-cycle applications

With user-created canned cycles, it is almost always best to specify arguments in a G65 command, as briefly discussed in Chapter 1. Since user-created canned cycles only machine a small portion of the workpiece, and since they may have to be executed several times by the main program, G65 affords a very convenient way of getting data into the custom macro, since it resembles a canned-cycle command. Within the G65, a P word is included to specify the program number of the custom macro program. Then comes the list of arguments. The following is a list of argument letter addresses allowed in custom macro B.

A, B, C, D, E, F, H, I, J, K, M, Q, R, S, T, U, V, W, X, Y, and Z

Note that not all letters of the alphabet are permitted: G, L, N, O, and P are missing. You must choose from those letter addresses that are allowed by custom macro. For reasons we discuss later, I, J, and K also should be avoided.

Custom macro B gives the parametric programmer remarkable flexibility with argument naming, despite not having the entire alphabet at his or her disposal. In most cases, a very logical argument name can be chosen (A for angle, R for radius, T for tool, D for depth, and so on). When programming for user-created canned-cycle applications with custom macro B, you'll want to take advantage of this argument naming flexibility.

It is important in letter address arguments to include a decimal point with each argument value specified in the G65 command. Without the decimal point, the control will use the *fixed format* for the letter address word to interpret the argument value. As you probably know, this fixed format changes from one word type to another. The letter address *X*, for example, has a four-place decimal format in the inch mode. If the decimal point is left out, a value of X1 will be taken as X0.0001 on most controls. However, the letter address B has a three-place format. A value of B1 (without the decimal point) will be taken as B0.001. To avoid any confusion regarding the placement of the decimal point for each word, make it a point to specify the decimal point in all arguments included in the G65 command.

Figure 2-1 shows a very simple application. We limit our discussion at this point to argument name selection and will not (as yet) show the actual custom macro program. Notice that the right side of this workpiece must be milled. Based on his or her knowledge of the company's work, the programmer knows that this is just one of more than 300 workpieces that requires the right side to be milled. Rather than writing the redundant coding needed to mill the right side

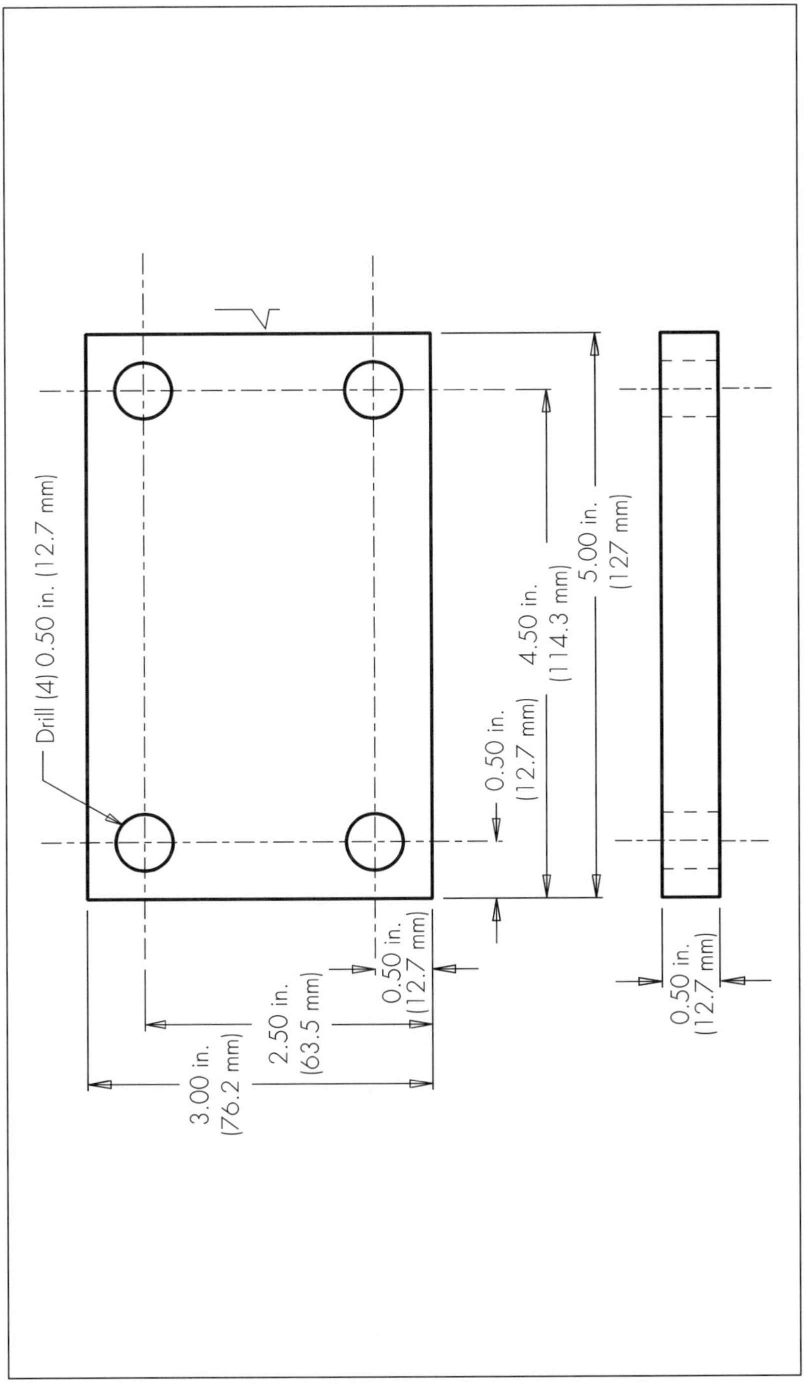

Figure 2-1. One of many workpieces that require a milling operation on the right end. Since in this example the right side of more than 300 workpieces must be machined, it is an ideal candidate for a user-defined canned cycle.

in over 300 programs, the programmer wishes to write a user-defined canned cycle to do the milling.

Figure 2-2 illustrates one very good way of documenting your parametric program, because anyone involved with this application will know the meaning of each argument being assigned. It shows the arguments this programmer has chosen to represent each value that changes from one workpiece to the next. Since every dimension relative to the right side of the workpiece changes from one workpiece to the next, each dimension requires an argument assignment. Note that *only* the dimensions related to the milling of the right side require arguments.

Note also that arguments are chosen in as logical a manner as possible. Arguments *X*, *Y*, and *Z* represent the corner position of this workpiece; W is the width of the workpiece and H is the height; T is the thickness; D is the tool diameter; and F is the cutting feedrate. Since this program will be used often, argument names chosen should be easy to remember. As the parametric programmer, *you* select the argument names to be used with your applications. Be sure to do so as logically as possible.

If the parametric program has already been written and verified, the following program will machine the entire workpiece (this includes a call to the *right-end-milling* custom macro).

Program:
O0001 (Program number)
N005 T01 M06 (Place 1-in. [25.4-mm] endmill in spindle)
N010 G54 G90 S400 M03 T02 (Select coordinate system, absolute mode, turn spindle on, and get next tool ready)
N015 G00 X5.0 Y0 (Rapid over to the workpiece)
N020 G43 H01 Z0.1 (Instate tool-length compensation, move up close to work surface)
N025 G65 P1000 X0. Y0. Z0. W5.0 H3.0 T0.5 D1.0 F5.0 (Mill right side)
N030 G91 G28 Z0 M19 (Move to tool-change position, orient spindle)
N035 M01 (Optional stop)
N040 T02 M06 (Place 1/2-in. [12.7-mm] drill two in spindle)
N045 G54 G90 S800 M03 T01 (Select coordinate system, absolute mode, start spindle, get tool number one ready)
N050 G00 X0.5 Y0.5 (Rapid to first hole position)
N055 G43 H02 Z0.1 (Instate tool-length compensation, rapid up close to work surface)
N060 G81 R0.1 Z– 0.85 F4.0 (Drill first hole)
N065 Y2.5 (Drill second hole)
N070 X4.5 (Drill third hole)
N075 Y0.5 (Drill fourth hole)
N080 G80 (Cancel cycle)
N085 G91 G28 Z0 M19 (Return to tool-change position, orient spindle)
N090 M30 (End of program)

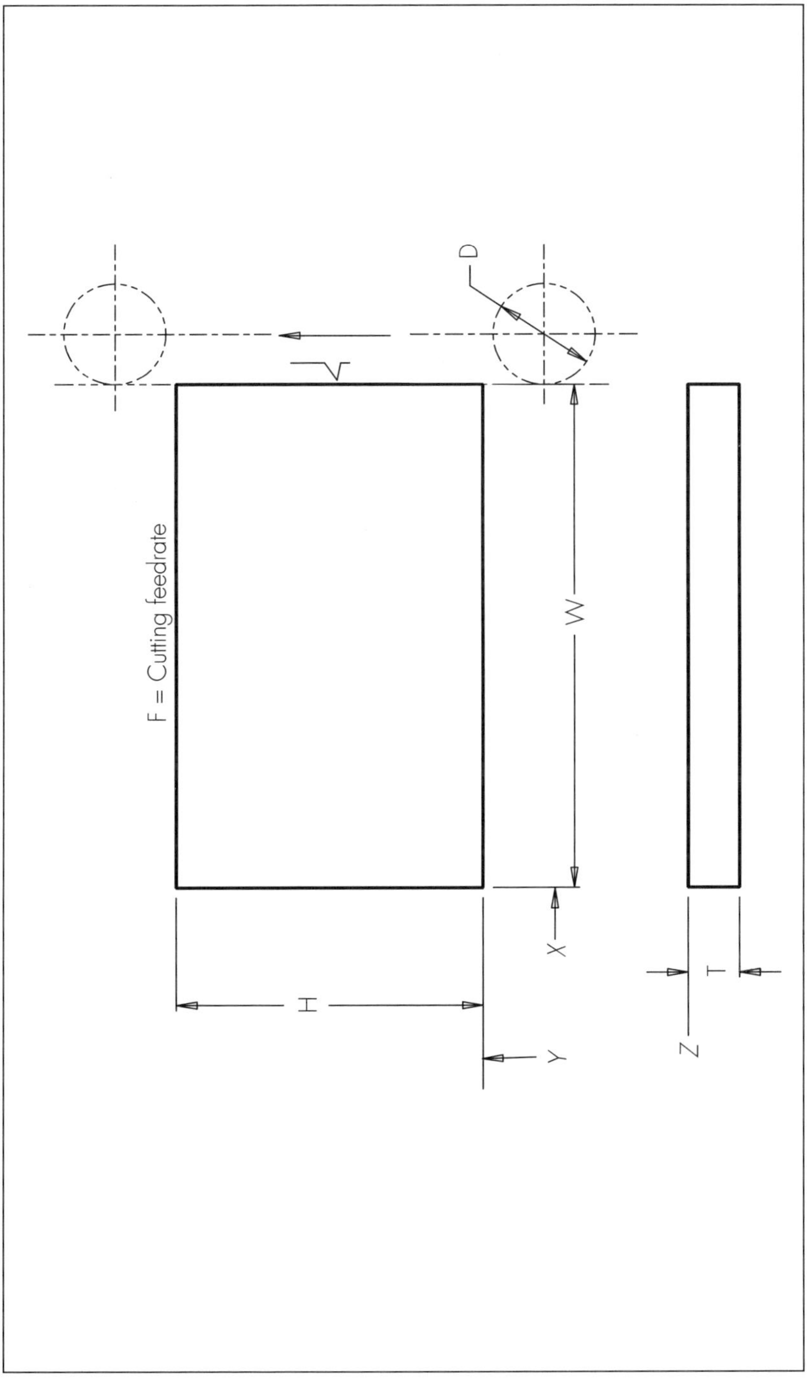

Figure 2-2. Argument selection for the workpiece illustrated in Figure 2-1. By documenting arguments in this way, anyone reading the program should have little difficulty understanding each argument's meaning.

Once again, note how little of this program actually deals with the milling of the right side. Only one command is involved (line N025). If more workpieces must be milled by this program, it is easy to simply specify more calling commands (G65) to program number O1000.

Before we can show the custom macro that actually does the machining for this application, you must be exposed to another variable type (local variables). These will be discussed a little later.

Arguments with family-of-parts applications

The G65 command is not always the most convenient way to get data into the custom macro, even though it is a very helpful command and is almost always the method of choice for user-created canned-cycle applications. When programming family-of-parts applications, for example, the entire workpiece is being machined by the parametric program. In this case, it is usually more convenient to state the values of all arguments at the beginning of the program rather than call the parametric program from another program. This involves using another of the custom macro variable types called *common variables.*

Common variables range in the #100 series. At least 50 common variables are available (more can be purchased as an option) and they range from #100 through (at least) #149. (We deal with common variables in more detail later.)

Though the naming of each argument must be well documented, common variables make better arguments with family-of-part applications. We will again use the workpiece in Figure 2-1 as the example program, but will change the criteria of the application. Instead of just milling the right end for a variety of different workpieces, say the workpiece shown in Figure 2-1 is but one of over 100 very similar workpieces. All workpieces have the four 0.5-in. (12.7-mm)-diameter holes drilled through the workpiece 0.5 in. from each edge. This new criteria (machining the entire workpiece) changes the application from a user-created canned-cycle application to a *family-of-parts* application. Figure 2-3 shows the documentation for this application, including the needed common variable arguments.

Notice that the argument names themselves (#100-#108) now mean nothing. There is no way to relate the 100 series number to the workpiece attribute it represents. For this reason, argument meanings must be clarified in the program (as shown in the following example program) as well as in any supporting documentation for the program. The custom macro program that machines all workpieces in this part family looks like this:

Program:
O0002 (Program number)
(Argument assignments)
#100 = 0 (Left side position in *X*)
#101 = 0 (Lower side position in *Y*)
#102 = 0 (Top surface of workpiece in *Z*)

#103 = 5.0 (Length of workpiece in *X*)
#104 = 3.0 (Width of workpiece in *Y*)
#105 = 0.5 (Workpiece thickness)
#106 = 1.0 (Milling cutter diameter)
#107 = 400 (rpm for milling cutter)
#108 = 5.0 (Feedrate of milling cutter in ipm)
(Machining portion of the program)
N005 T01 M06 (Place 1-in. [25.4-mm] endmill in spindle)
N010 G54 G90 **S#107** M03 T02 (Select coordinate system, absolute mode, turn spindle on, and get next tool ready)
N015 G00 **X[#100 + #103 + #106/2] Y[#101 – 0.1 – #106/2]** (Rapid over to the workpiece)
N020 G43 H01 **Z[#102 + 0.1]** (Instate tool-length compensation, move up close to work surface)
N025 G01 **Z[#102 – #105 – 0.1]** F50.0 (Fast feed past surface)
N030 **Y[#101 + #104 + 0.1 +#106/2] F#108** (Mill right side)
N035 G00 **Z[#102 + 0.1]** (Rapid to above top surface)
N035 G91 G28 Z0 M19 (Move to tool-change position, orient spindle)
N040 M01 (Optional stop)
N045 T02 M06 (Place 1/2-in. [12.7-mm] drill two in spindle)
N050 G54 G90 S800 M03 T01 (Select coordinate system, absolute mode, start spindle, get tool number one ready)
N055 G00 **X[#100 + 0.5] Y[#101 + 0.5]** (Rapid to first hole position)
N060 G43 H02 **Z[#102 + 0.1]** (Instate tool-length compensation, rapid up close to work surface)
N065 G81 **R[#102 + 0.1] Z[#102 – #105 – 0.18]** F4.0 (Drill first hole)
N070 **Y[#101 + #104 – 0.5]** (Drill second hole)
N075 **X[#100 + #103 – 0.5]** (Drill third hole)
N080 **Y[#101 + 0.5]** (Drill fourth hole)
N085 G80 (Cancel cycle)
N090 G91 G28 Z0 M19 (Return to tool-change position, orient spindle)
N095 M30 (End of program)

Since this is the first full parametric programming example we have shown, it bears further explanation. First, notice how well arguments #100 through #108 are documented in the program. A message in parentheses is placed next to each argument to tell anyone viewing the program exactly what each argument means. An operator or setup person could confidently modify one or more of these arguments during setup for the next workpiece in the family to be machined without fear of making mistakes.

Second, notice that *arguments are only required for those things that change* from one workpiece to the next. Any value that does not change (speed and feed for 1/2-in. [12.7-mm] drill, 0.5-in. [12.7-mm] hole position, 0.1-in. [2.54-mm]

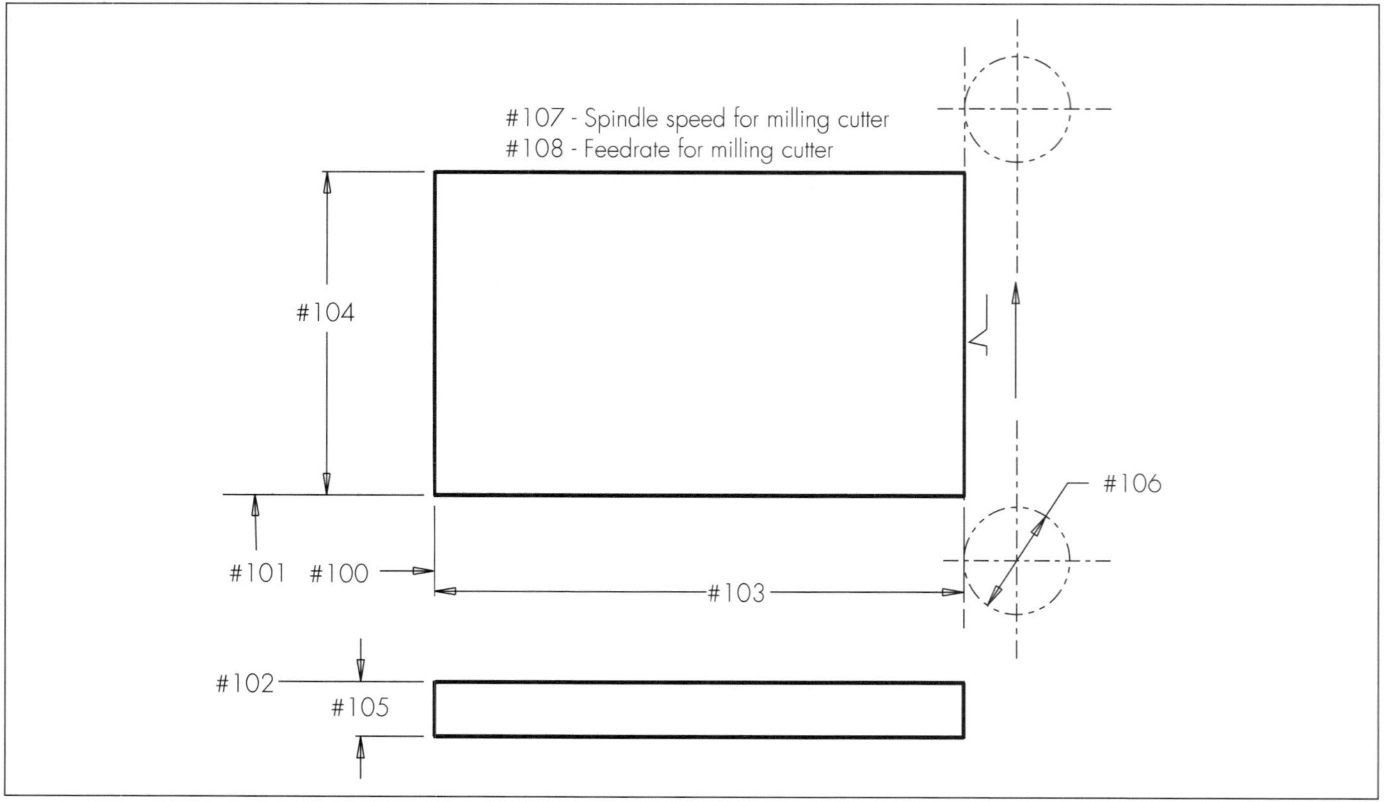

Figure 2-3. Argument selection based on family-of-parts application of the workpiece shown in Figure 2-1. Because the arguments using this type of coding are not as intuitive as those in Figure 2-2, their meanings must be included in the program.

approach distance, etc.) is treated as a *constant value*. Constant values, of course, cannot be changed by simply manipulating argument values.

Third, notice that *calculations* are being done in many of the program's commands to determine axes positions. It is almost impossible to come up with any good parametric programming example that does not require arithmetic functions. Look specifically at line N015. The *X* and *Y* positioning movements will be based on the result of a calculation. Based on the current values of the arguments set at the beginning of this program, the *X* value will be calculated as 5.5 (0 + 5.0 + 1.0/2). The *Y* value will be calculated as -0.6 (the negative result of 0.1 + 1.0/2). When executed, this command will appear to the control as "N015 G00 X5.5 Y – 0.6." Keep in mind, however, that if the values of #100, #103, or #106 are changed (as will be necessary for other workpieces in the family), the positioning movement in line N015 also will change.

Finally, notice the brackets ([]) that enclose the calculations being made to determine the value of a CNC word. These brackets are *not* parentheses. Custom macro B requires that brackets be used in this manner when making calculations to determine the value of CNC words. More discussion on brackets appears in the section on arithmetic functions.

While this is a relatively simple family-of-parts application, it represents well how arguments can be introduced to family-of-parts parametric programs. Again, we are just scratching the surface, but it should give you a sense of the power you have available with parametric programming.

Note once more the subtle distinction between argument assignment for user-created canned-cycle applications and argument assignment for family-of-parts applications. With part families, the custom macro is machining the entire workpiece, and assigning arguments in the main program is usually the *most convenient* method of argument assignment since only one program is required.

On the other hand, since user-created canned-cycle applications always require a separate program for the canned-cycle-like operation, and since this program may be required many times even to machine one workpiece, arguments *must* be set in the main program (within the G65 command). And it *must* be possible to manipulate the arguments from one call statement to the next.

LOCAL VARIABLES WITH ARGUMENT ASSIGNMENT ONE

When a custom macro is called with a G65 command, argument values are passed in the form of letter addresses. As stated earlier, valid letter address arguments include:

A, B, C, D, E, F, H, I, J, K, M, Q, R, S, T, U, V, W, X, Y, and Z

Though it may seem a little confusing at first, you are not allowed to directly reference an argument within the custom macro by its letter address. The control would confuse the argument name (A, B, C, X, Y, Z, etc.) with the true CNC meaning for the letter address. For this reason, you must reference arguments by their corresponding *local variables*. For illustration, the following list is presented to show all letter address arguments and their corresponding local variable numbers. This is the most popular form of argument assignment, called *argument assignment number one*.

A:#1, B:#2, C:#3, D:#7, E:#8, F:#9, H:#11, I:#4, J:#5, K:#6, M:#13, Q:#17, R:#18, S:#19, T:#20, U:#21, V:#22, W:#23, X:#24, Y:#25, Z:#26

When a G65 command is executed, two things happen. First, all local variables that have corresponding letter address arguments in the G65 command will be set. Second, the custom macro specified by the P word will be executed. During the execution of the custom macro, local variables will contain the values of arguments set in the G65 command. Within the custom macro, these local variables must be used to reference the arguments coming from the G65 calling command.

As an example, the command

N025 G65 P1000 X0. Y0. Z0. W5.0 H3.0 T0.5 D1.0 F5.0 (Mill right side)

first sets the local variables corresponding to *X*, *Y*, *Z*, W, H, T, D, and F. In this case, local variable #24 will be set to 0 (the *X* value), #25 to 0 (the *Y* value), #26 to 0 (the *Z* value), #23 to 5.0 (the W value), #11 to 3.0 (the H value), #20 to 0.5 (the T value), #7 to 1.0 (the D value), and #9 to 5.0 (the F value). Then the control executes program O1000. When the programmer wishes to reference the value of *X* within the custom macro program, #24 must be used.

It is also important to know that all *other* local variables (#1, #2, #3, etc.) will have no value when the custom macro is executed, meaning they will be left *vacant*. But zero doesn't mean vacant. Zero is a value. When a variable is vacant, it has no value. We will discuss some implications of vacant variables later.

Also note that, as the name implies, local variables only remain set during the execution of the custom macro. At the completion of the custom macro (the M99 command), they are all set back to vacant.

Here is a full example that uses the workpiece shown in Figure 2-1. Again, we are milling the right side of the workpiece, and the arguments used for this user-created canned-cycle application are shown in Figure 2-2. The main program that machines the workpiece in the same form shown earlier is

Program:
O0001 (Program number)
N005 T01 M06 (Place 1-in. [25.4-mm] endmill in spindle)
N010 G54 G90 S400 M03 T02 (Select coordinate system, absolute mode, turn spindle on, and get next tool ready)

N015 G00 X5.0 Y0 (Rapid over to the workpiece)
N020 G43 H01 Z0.1 (Instate tool-length compensation, move up close to work surface)
N025 G65 P1000 X0. Y0. Z0. W5.0 H3.0 T0.5 D1.0 F5.0 (Mill right side)
N030 G91 G28 Z0 M19 (Move to tool-change position, orient spindle)
N035 M01 (Optional stop)
N040 T02 M06 (Place 1/2-in. [12.7-mm] drill two in spindle)
N045 G54 G90 S800 M03 T01 (Select coordinate system, absolute mode, start spindle, get tool number one ready)
N050 G00 X0.5 Y0.5 (Rapid to first hole position)
N055 G43 H02 Z0.1 (Instate tool-length compensation, rapid up close to work surface)
N060 G81 R0.1 Z – 0.85 F4.0 (Drill first hole)
N065 Y2.5 (Drill second hole)
N070 X4.5 (Drill third hole)
N075 Y0.5 (Drill fourth hole)
N080 G80 (Cancel cycle)
N085 G91 G28 Z0 M19 (Return to tool change position, orient spindle)
N090 M30 (End of program)

Again, note that line N025 will perform the milling of the right side. Armed with your new knowledge of local variables, you should now be able to understand the custom macro program:

O1000 (Custom macro to mill right side)
N1 G00 **X[#24 + #23 + #7/2] Y[#25 - 0.1 – #7/2]** (Rapid to starting position)
N2 G01 **Z[#26 – #20 – 0.1]** F50. (Fast feed to work surface)
N3 **Y[#25 + #11 + #7/2] F#9** (Mill right side)
N4 G00 **Z[#26 + 0.1]** (Rapid to above workpiece)
N5 M99 (End of custom macro)

In line N1, with the call statement set as it is in line N025 of the main program, the *X* position will be calculated as X5.5 because argument *X* (referenced by #24) is set to 0, W (referenced by #23) is set to 5.0, and D (referenced by #7) is set to 1.0. The control will calculate this *X* position as 0 + 5.0 + 1.0/2 which results in a value of 5.5.

Keeping track of local variables

Admittedly, local variable numbers do not make much sense in argument assignment number one. Even for experienced custom macro programmers, it is difficult to remember which local variable corresponds to a given letter address argument. For this reason, you must keep close at hand the argument assignment list shown earlier as you write custom macros.

One helpful suggestion is to write the letter address argument (in a different color) next to the local variable number as you actually write the custom macro program. To illustrate: let's say you must reference the letter address argument D in the custom macro. As you look at the local variable assignment list, you find that #7 corresponds to D. As you write #7 in the custom macro, also write the letter D in a different color close by to remind you of which letter address argument #7 represents.

Running out of variables?

While local variables are used predominantly to represent letter address arguments coming from the G65 command, nothing prohibits you from using unused local variables for other purposes. Almost anything that can be done with a common variable also can be done with an unused local variable. The exception is any application that requires the value of the variable to be retained after the completion of the custom macro (after the M99); local variables are reset to vacant at M99.

More on vacancy

Remember that when a letter address argument is left out of the call statement, its corresponding local variable is set to vacant when the custom macro is called by the G65 command. The custom macro representation for vacancy is #0. Though we do not wish to dig in too deep yet, in certain instances it can be helpful to know about vacant local variables.

In the previous right-side-milling custom macro, for example, all arguments shown in Figure 2-2 must be included in the G65 command. (Arguments such as these that must be included in the call statement are called *mandatory arguments*. In this example, all arguments are mandatory arguments.) Based on the way the custom macro shown earlier is written, if the programmer using this custom macro forgets to include one or more of the arguments in the G65 command, it will be difficult to predict what will happen. Surely the custom macro will not behave as expected.

However, with the knowledge that a local variable is left set to vacant (#0) if it is not included in the call statement, and knowing how to perform tests in the custom macro, the custom macro programmer can test all local variables being assigned to determine if any are vacant. If some are, an alarm can be generated to halt the machine before any damage is done. In Chapter 4 we show how to perform tests and generate alarms. This is just one of several instances in which understanding variable vacancy can be helpful.

COMMON VARIABLES

Common variables were introduced during our discussion of arguments used in family-of-parts applications. Again, common variables range in the #100 series and at least 50 are available (from #100 through #149). More can be purchased as an option.

As the name implies, common variables remain active even after the execution of the custom macro program. Most CNC controls that have custom macro are set to retain the values of common variables (#100 series) until the power to the machine is turned off, at which time they are set back to vacant.

While the *global* nature of common variables has special implications, for now we concentrate on the more basic uses of them. Aside from being used as arguments in family-of-parts applications, common variables have a variety of applications, as described in the following paragraphs.

To minimize redundant calculations

In the right-end-milling custom macro shown earlier, the *diameter* of the endmill is passed to the custom macro (with D). Yet in the custom macro, it is necessary to utilize the endmill's radius. The redundant calculation #7/2 is done three times in this custom macro. While this is a simple calculation and only performed three times, imagine that the endmill is being used to machine several surfaces. In this case, it is likely that the expression #7/2 would be stated many times. These calculations are tedious to write and error prone, and force the control to perform unnecessary calculations. This can result in wasted cycle time.

Common variables can be used to minimize redundant calculations. The result of a redundant calculation can be stored into a common variable early in the custom macro. From that point, the variable can be used instead of the repeated calculation. Following is the custom macro program to mill the right side of a workpiece modified to take advantage of this technique.

O1000 (Custom macro to mill right side)
N1 **#100 = #7/2**
N2 G00 X[#24 + #23 + **#100**] Y[#25 – 0.1 – **#100**] (Rapid to starting position)
N3 G01 Z[#26 – #20 – 0.1] F50. (Fast feed to work surface)
N4 Y[#25 + #11 + **#100**] F#9 (Mill right side)
N5 G00 Z[#26 + 0.1] (Rapid to above workpiece)
N6 M99 (End of custom macro)

For general-purpose calculations

Common variables are typically the variable type of choice for general-purpose calculations that must be done in the custom macro. Another version of the right-side-milling custom macro is shown below that calculates all positioning movements up front, storing the results of all calculations in common variables. The movement commands simply reference the values previously calculated.

```
O1000 (Custom macro to mill right side)
N1 #100 = #24 + #23 + #7/2
N2 #101 = #25 – 0.1 – #7/2
N3 #102 = #26 – #20 – 0.1
N4 #103 = #25 + #11 + #7/2
```

```
N5 #104 = #26 + 0.1
N2 G00 X#100 Y#101 (Rapid to starting position)
N3 G01 Z#102 F50. (Fast feed to work surface)
N4 Y#103 F#9 (Mill right side)
N5 G00 Z#104 (Rapid to above workpiece)
N6 M99 (End of custom macro)
```

As you develop your own custom macros, there will be many opportunities for you to use common variables for general purpose calculations.

As counters

Looping functions often require that variables change each time the loop is executed. Common variables are used as the loop counter and for any other variable that must be initialized at the beginning of the loop and stepped each time through the loop.

As flags

It is sometimes necessary to set a *flag* that tells the custom macro how to behave. For example, you may wish to set up criteria in your right-side milling custom macro that allows the milling cutter to either climb or conventional mill. A special flag will specify which milling method you wish to use. In effect, you are *flagging* the custom macro as to how it should behave. Other examples of when flags are required (which we will discuss in greater detail in Chapter 4) include right-hand versus left-hand tooling, inch versus metric input, and whether or not coolant is to be used. While letter address arguments in the G65 command commonly set flags, the global nature of common variables makes it possible to maintain the values of flags even among several programs.

PERMANENT COMMON VARIABLES

Permanent common variables range in the #500 series. Ten are standard (from #500 through #509), and more permanent common variables can be purchased as an option. As the name implies, these variables are retained even after the power is turned off. In this sense, they are much like tool offsets and, like tool offsets, most controls even let you modify the values of permanent common variables through the keyboard and display screen.

Since there are so few permanent common variables in many controls, it is wise to reserve their use for those applications when it is necessary to retain data from one day to the next. A good example of this type of application is a utility custom macro for part counting. Suppose you wish to set up a part counter to let the machine run 750 workpieces and then stop. As you begin running production, your custom macro-based part counter will begin counting workpieces. It isn't very likely that you will finish all 750 workpieces before the end of the day or shift. More likely it will take several days to complete the run. In this case, you will need your part-counting custom macro program to remember where it left off from day to day. If you store the current part count in a #500 series permanent common variable, the value of the variable will not be lost when the power is turned off. It will continue to 750, whereupon it will turn off the machine.

Do you have a probing system?

Permanent common variables also are used to retain *system data* when certain accessory devices are used. If you have a probing system, for example, and if custom macro is used to program the probe, it is quite likely that at least some of your permanent common variables are being used to store calibration values for the probing system. Before you begin using any #500 series permanent common variables, you must confirm that they are not already being used for other purposes.

Setting your own system constants

Think of any constant hard-and-fixed value that you use throughout your CNC programming. One that may come to mind is your rapid approach and retract distance. When programming in the inch mode, for example, many programmers like to stay a constant 0.100 in. away from the surface being approached before entering a cutting mode. When cutting is completed, many programmers feed the tool away from the workpiece by this same 0.100-in. amount.

While this approach distance is quite safe, when high production quantities must be run, the time that each approach and retract at the cutting feedrate adds to cycle time may be excessive. However, if all approach and retract movements are programmed with hard-and-fixed values (calculated by the programmer), it can be quite difficult to change the approach and retract values without a massive overhaul of the CNC program.

If instead, a permanent common variable is used in which to store the rapid approach and retract distance (you could even use one permanent common variable for each), and if this variable is referenced in *every CNC program during every tool's approach and retract movement*, changing the approach/retract distance will be as simple as changing one permanent common variable. And as stated, permanent common variables can be entered and changed through the keyboard and display screen, just like tool offsets.

Consider the following program that drills a few holes and mills a slot. It assumes the rapid approach/retract distance is stored in permanent common variable #500.

```
O0003 (Program that allows easy approach/retract distance change)
N005 T01 M06 (Place 1/2-in. [12.7-mm] drill in spindle)
N010 G54 G90 S800 M03 T02 (Select coordinate system, absolute mode, turn
   spindle on, and get tool number two ready)
N015 G00 X.5 Y.5 (Rapid to first hole location)
```

N020 G43 H01 **Z#500** (Instate tool-length compensation, rapid up to workpiece)
N025 G81 **R#500** Z – 0.85 F3.5 (Drill first hole)
N030 Y3.5 (Drill second hole)
N035 X4.5 (Drill third hole)
N040 Y.5 (Drill fourth hole)
N045 G80 (Cancel cycle)
N050 G91 G28 Z0 (Return to tool-change position)
N055 M01 (Optional stop)
N060 T02 M06 (Place 3/8-in. [9.53-mm] drill in spindle)
N065 G54 G90 S950 M03 T03 (Select coordinate system, absolute mode, turn spindle on, and get tool three ready)
N075 G00 X3.0 Y2.0 (Rapid to hole location)
N080 G43 H02 **Z#500** (Instate tool-length compensation, rapid to approach position)
N085 G81 **R#500** Z – 0.55 F3.5 (Drill hole)
N090 G80 (Cancel cycle)
N095 G91 G28 Z0 (Return to tool-change position)
N100 M01 (Optional stop)
N105 T03 M06 (Place 1/4-in. [6.35-mm] endmill in spindle)
N110 G54 G90 S1000 M03 T01 (Select coordinate system, absolute mode, turn spindle on, and get tool one ready)
N115 G00 X3.75 **Y – [0.125 + #500]** (Rapid to slot approach position)
N120 G43 H03 Z – 0.125 (Instate tool-length compensation, rapid to work surface)
N125 G01 **Y[4.125 + #500]** F2.5 (Mill slot)
N130 G00 **Z#500** (Rapid to above work surface)
N135 G91 G28 Z0 (Return to tool-change position)
N140 M30 (End of program)

Notice that any time a tool in this program approaches or retracts (lines N020, N025, N080, N085, N115, N125, and N130), permanent common variable #500 is referenced. Notice also in lines N115 and N125, that the programmer can even perform a calculation that has the control figure out the correct approach/retract position based on the value of permanent common variable #500. With this program, changing the approach/retract distance will be as easy as changing #500.

SYSTEM VARIABLES

Many of the CNC-related features of custom macro are handled with system variables that range from #1000 through about #6000. They allow access to many CNC functions, including current machine position, offsets, alarm generation, and many more. These variables are covered at length in Chapter 5.

Summary of custom macro variables

As you have seen, there are five types of variables in custom macro B. *Arguments* are used in family-of-parts and user-created canned-cycle applications to get input data into the custom macro. *Local variables*, which get reset to vacant at the completion of the custom macro (at M99), are used to represent letter address arguments coming from the G65 command. *Common variables*, which also can be used as arguments (in family-of-parts applications), are used for a variety of purposes including general-purpose calculations, counters, and for minimizing redundant calculations. *Permanent common variables* are used for applications requiring the value of a variable to be retained from day to day. Finally, *system variables* allow access to many CNC-related features of custom macro. While these discussions should leave you with a pretty good idea of what custom macro variables can do, remember that all presentations to this point have been rather elementary. There are some advanced implications that we address in later chapters.

VARIABLES IN OKUMA'S USER TASK 2

There are three types of variables in user task 2. The same kinds of functions are available as in Fanuc's custom macro B; only the terminology and syntax change.

Actually, Okuma does name variables in user task 2 in the same manner as in custom macro B: local variables, common variables, and system variables. However, subtle differences exist in usage which may lead to confusion if you are working with both of these powerful versions of parametric programming.

LOCAL VARIABLES

As the name implies, local variables will remain active only during the execution of the user task program. At the completion of the program (or when the control is reset), local variables are cleared. A total of 127 local variables is available with most control models.

Local variable names are chosen by the user task programmer. While there are some strict rules about local variable naming, this ability to name your own local variables allows you to choose very logical names. Local variable names can be up to four characters long. The first two characters must be letters of the alphabet. The next two can be letters or numbers. The letters O, N, and V cannot be used in a local variable name. Examples of acceptable local variable names include DIA1, THRD, FLG1, XY, and AB11. You will want to take advantage of this flexible local variable naming feature as you develop user task programs to keep variable names as logical and easy to remember as possible.

Using local variables as arguments

Any variable that passes data into the user task program is called an *argument*. In most cases, local variables can be used for this purpose.

Local variable arguments for user-created canned-cycle applications. In most user-created canned-cycle applications, a CALL command will be used to pass arguments and activate the user task program. The CALL command includes the program name being called and a list of arguments (local variables) to be passed to the user task program.

Figure 2-4 illustrates a very simple application. As in our last example, the right side of this workpiece must be milled. Based on his or her knowledge of the company's work, the programmer knows that this is just one of more than 300 workpieces that requires the right side to be milled. Rather than writing the code needed to mill the right side in more than 300 programs, the programmer prefers to write a user-defined canned cycle to do the milling.

Figure 2-5 shows the local variable arguments one programmer has chosen to represent each value that changes from one workpiece to the next. Since every dimension relative to the right side of the workpiece changes from one workpiece to the next, each dimension requires an argument assignment. Note that *only* the dimensions related to the milling of the right side require arguments. (Figure 2-5, incidentally, illustrates one very good way of documenting your parametric program for everyone to see. Anyone involved with this application will know the meaning of each local variable argument being assigned.)

Note also that arguments are chosen in as logical a manner as possible. XZER, YZER, and ZZER represent the position of the corner of this workpiece. WIDT is the width of the workpiece and HGHT is the height. THK is the thickness. TDIA is the tool diameter and FEED is the cutting feedrate. Since this program will be used often, argument names selected should be easy to remember. As the parametric programmer, *you* select the argument names to be used with your applications. Be sure to do so as logically as possible.

Assuming the parametric program has already been written and verified, here is a program that will machine the entire workpiece that includes a call to the *right-end-milling* user task program.

Program:
```
O0001 (Program name)
N005 G15 H01 (Select coordinate system)
N010 T01 M06 (Place 1-in. [25.4-mm] endmill in spindle)
N015 G90 S400 M03 T02 (Select absolute mode, turn spindle on, and get next
   tool ready)
N020 G00 X5.0 Y0 (Rapid over to the workpiece)
N025 G56 H01 Z0.1 (Instate tool-length compensation, move up close to work
   surface)
N030 CALL O1000 XZER = 0  YZER = 0  ZZER = 0  WIDT = 5.0  HGHT = 3.0
   THK = 0.5  TDIA = 1.0  FEED = 5.0 (Mill right side)
N035 G15 H01 (Select coordinate system)
```

```
N040 G00 Z50.0 (Move to tool-change position)
N045 T02 M06 (Place 1/2-in. [12.7-mm] drill two in spindle)
N050 G90 S800 M03 T01 (Select absolute mode, start spindle, get tool number
   one ready)
N055 G00 X0.5 Y0.5 (Rapid to first hole position)
N060 G56 H02 Z0.1 (Instate tool-length compensation, rapid up close to work
   surface)
N065 G81 R0.1 Z – 0.85 F4.0 (Drill first hole)
N070 Y2.5 (Drill second hole)
N075 X4.5 (Drill third hole)
N080 Y0.5 (Drill fourth hole)
N085 G80 (Cancel cycle)
N090 G00 Z50. (Move away in Z)
N095 M02 (End of program)
```

Note again how little of this program deals with the milling of the right side. Only one command is involved (line N030). If more workpieces must be milled by this program, it is easy to simply specify more calling commands (with more CALL commands) to the program named O1000.

Here is the user task program that actually mills the right side. Notice how easy it is to directly reference the value of each argument (local variable) being set in the main program.

O1000 (User task program to mill right side)
N1 G00 **X = XZER + WIDT + TDIA/2 Y = YZER – 0.1 – TDIA/2** (Rapid to starting position)
N2 G01 **Z = ZZER – THK – 0.1** F50. (Fast feed to work surface)
N3 **Y = YZER + HGHT + TDIA/2 F = FEED** (Mill right side)
N4 G00 **Z = ZZER + 0.1** (Rapid to above workpiece)
N5 RTS (End of user task program)

While this user task program should be relatively easy to follow, there are a few points to make about its structure. First, each local variable argument is being referenced exactly as it is named in the CALL command from the main program. Second, each actual CNC word that relates to a local variable includes an equal sign (=) to set the word equal to the local variable or calculation result. Finally, arithmetic calculations are being done within several commands. While these simple arithmetic functions should be quite understandable, we discuss them in greater detail later.

Local variables for family-of-parts applications. While the CALL command is very helpful, and it is almost always the method of choice for user-created canned-cycle applications, it is not always the most convenient way to get data into the user task program. When programming family-of-parts applications, the entire workpiece is being machined by the parametric program. This being the

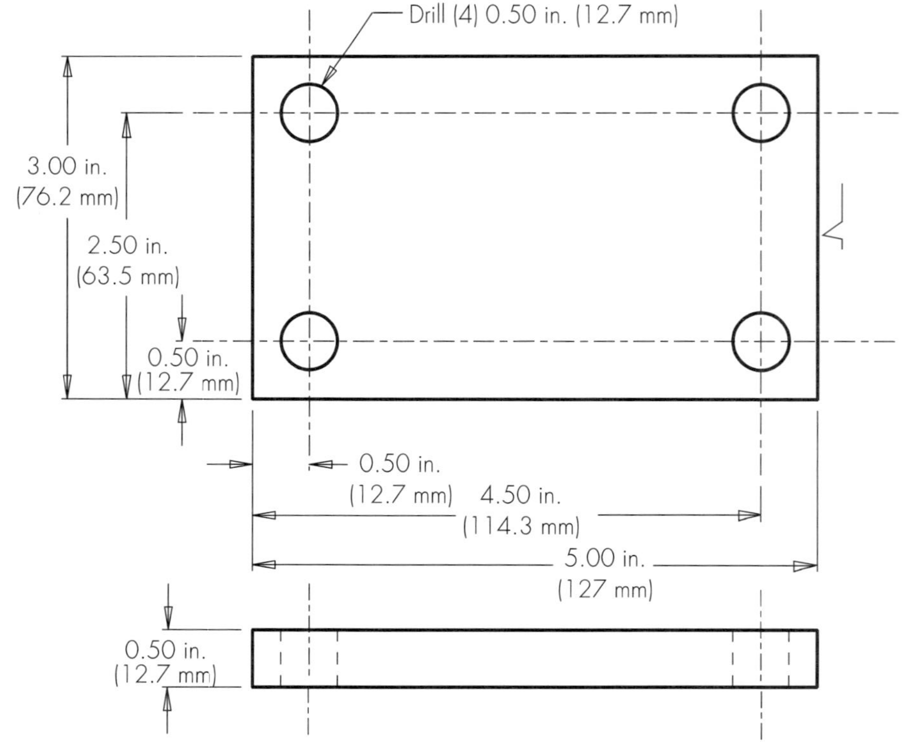

Figure 2-4. Another of many workpieces that require a milling operation on the right end. With more than 300 workpieces to be machined, the programmer in this example wisely opts to write a user-created canned cycle.

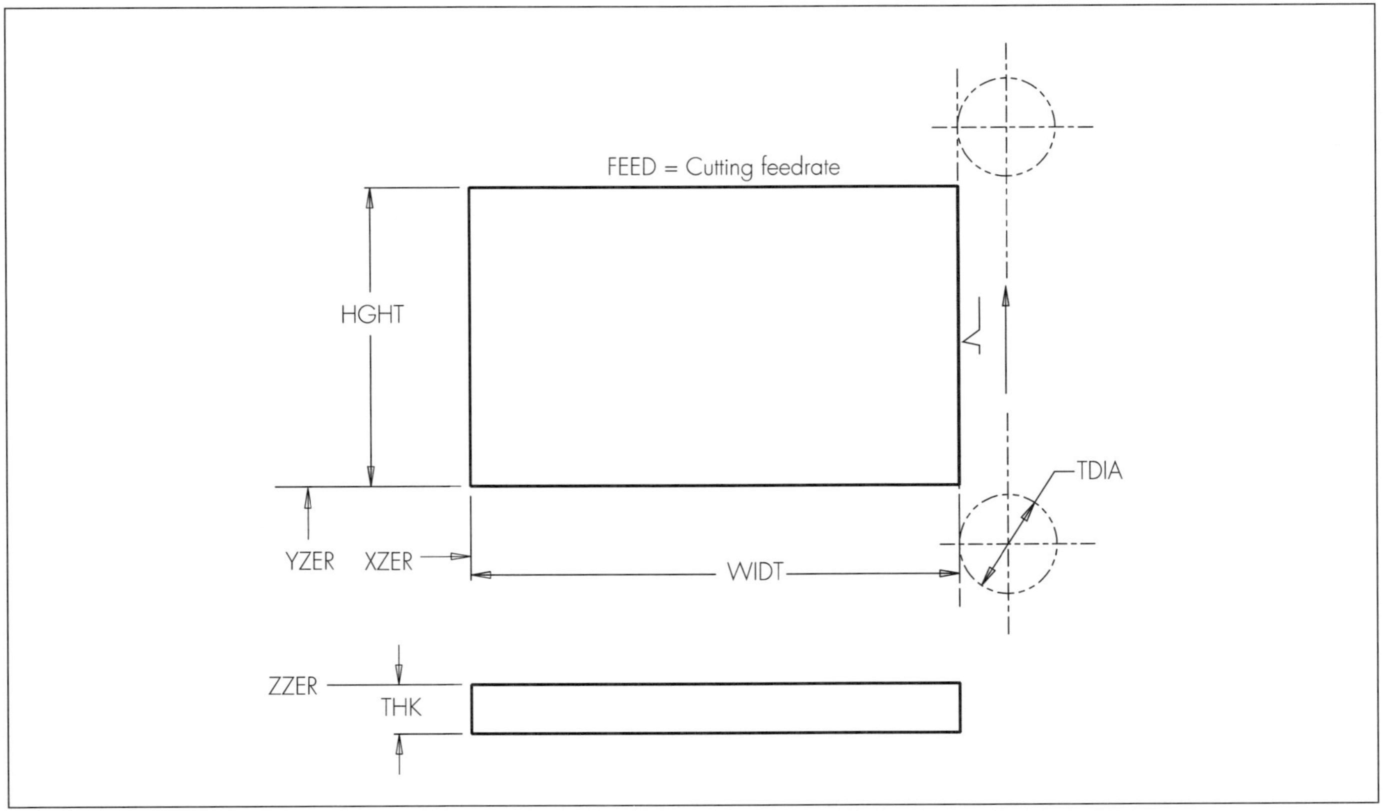

Figure 2-5. The local variable argument selection for a user-created canned cycle based on Figure 2-4. Only the dimensions related to machining the right side of the workpiece require arguments.

case, it is usually more convenient to state the values of all local variable arguments at the beginning of the program rather than to call the user task program from another program.

Again using the workpiece in Figure 2-4 for the example program, we change the criteria of the application: Instead of just milling the right side for a variety of different workpieces, let us say the workpiece shown in Figure 2-4 is one of more than 100 very similar workpieces. All workpieces have four 0.5-in. (12.7-mm)-diameter holes drilled 0.5 in. from each edge. This new criterion (machining the entire workpiece) changes the application from a user-created canned-cycle application to a *family-of-parts* application. Again, Figure 2-5 shows the local variable arguments related to this application. Since the hole locations are always 0.5 in. in diameter and 0.5 in. from the edges, no additional local variable arguments are required. Hence, the user task program to machine all workpieces in this part family is

Program:
O0002 (Program name)
(Argument assignments)
XZER = 0 (Left side position in *X*)
YZER = 0 (Lower side position in *Y*)
ZZER = 0 (Top surface of workpiece in *Z*)
WIDT = 5.0 (Length of workpiece in *X*)
HGHT = 3.0 (Width of workpiece in *Y*)
THK = 0.5 (Workpiece thickness)
TDIA = 1.0 (Milling cutter diameter)
FEED = 5.0 (Feedrate of milling cutter in ipm)
(Machining portion of the program)
N005 G15 H01 (Select coordinate system)
N010 T01 M06 (Place 1-in. [25.4-mm] endmill in spindle)
N015 G90 **S = 3.82 * 80/TDIA** M03 T02 (Select absolute mode, turn spindle on, and get next tool ready)
N020 G00 **X = XZER + WIDT + TDIA/2 Y = YZER – 0.1 – TDIA/2** (Rapid over to the workpiece)
N025 G56 H01 **Z = ZZER + 0.1** (Instate tool-length compensation, move up close to work surface)
N030 G01 **Z = ZZER – THK – 0.1** F50.0 (Fast feed past surface)
N035 **Y = YZER + HGHT + 0.1 + TDIA/2 F = FEED** (Mill right side)
N040 G00 **Z = ZZER + 0.1** (Rapid to above top surface)
N045 G00 Z50.0 (Rapid to tool-change position)
N050 G15 H01 (Select coordinate system)
N055 T02 M06 (Place 1/2-in. [12.7-mm] drill two in spindle)
N060 G90 S800 M03 T01 (Select absolute mode, start spindle, get tool number one ready)

N065 G00 **X = XZER + 0.5 Y = YZER + 0.5** (Rapid to first hole position)
N070 G56 H02 **Z = ZZER + 0.1** (Instate tool-length compensation, rapid up close to work surface)
N075 G81 **R = ZZER + 0.1 Z = ZZER – THK – 0.18** F4.0 (Drill first hole)
N080 **Y = YZER + HGHT – 0.5** (Drill second hole)
N085 **X = XZER + WIDT – 0.5** (Drill third hole)
N090 **Y = YZER + 0.5** (Drill fourth hole)
N095 G80 (Cancel cycle)
N100 G00 Z50.0 (Move to tool-change position)
N105 M02 (End of program)

Notice how well the local variable arguments are documented at the beginning of this program. While their names are quite clear and logical, the messages in parentheses stress the meaning of each argument in greater detail. An operator or setup person could confidently modify one or more of these arguments during the setup procedure for the next workpiece in the family to be machined.

Local variable arguments are only required for those things that change from one workpiece in the family to the next. Any value that does not change (speed and feed for 1/2-in. [12.7-mm] drill, 0.5 in. [12.7-mm] hole position, 0.1-in. [2.54-mm] approach distance, etc.) is treated as a *constant value*. Constant values, remember, cannot be changed by simply manipulating argument values.

While this is a relatively simple family-of-parts application, it clearly states how local variable arguments can be introduced to family-of-parts parametric programs. It also should provide a hint as to some of the power you have with parametric programming.

Note the subtle distinction between argument assignment for user-created canned-cycle applications and argument assignment for family-of-parts applications. With part families, the user task program is machining the entire workpiece, and assigning arguments in the main program is the most convenient method of argument assignment since only one program is required.

On the other hand, since user-created canned-cycle applications always require a separate program to perform the canned-cycle-like operation, and since this program may be required several times even to machine one workpiece, arguments must be set in the main program (within the CALL command). It must also be possible to manipulate the arguments from one call statement to the next.

Using local variables to minimize redundant calculations

Their use as arguments is just one of a host of applications for local variables. In the simple right-end-milling user-created canned-cycle application, for example, the *diameter* of the endmill is passed to the user task program (with TDIA). Yet during the user task program, it is necessary to utilize the endmill's *radius*. The redundant calculation TDIA/2 is done three times in the user task program. Admittedly, this is a rather simple calculation and it is only being done three times; however, imagine that this milling cutter must machine several surfaces. It is likely that the TDIA/2 calculation would be repeated numerous times. These calculations are tedious to write and error prone and force the control to perform unnecessary calculations. This can result in wasted execution (cycle) time.

Local variables can be used to minimize redundant calculations by storing the result of a redundant calculation in a local variable early on in the custom macro. From that point forward, the variable can be used instead of the repeated calculation. The user task program to mill the right side of a workpiece modified to take advantage of this technique looks like this:

O1000 (User task program to mill right side)
N1 **TRAD = TDIA/2**
N2 G00 **X = XZER + WIDT + TRAD Y = YZER – 0.1 – TRAD** (Rapid to starting position)
N3 G01 **Z = ZZER – THK – 0.1** F50. (Fast feed to work surface)
N4 **Y = YZER + HGHT + TRAD F = FEED** (Mill right side)
N5 G00 **Z = ZZER + 0.1** (Rapid to above workpiece)
N6 RTS (End of custom macro)

Using local variables for general-purpose calculations

Local variables are also the variable type of choice for general-purpose calculations that must be done in the user task program. Here is another version of the right-side-milling user task program that calculates all positioning movements up front, storing the results of all calculations in common variables. The movement commands simply reference the values previously calculated.

O1000 (User task program to mill right side)
N1 TRAD = TDIA/2
N2 XRAP = XZER + WIDT + TRAD
N3 YRAP = YZER – 0.1 – TRAD
N4 ZBTM = ZZER – THK - 0.1
N5 YCUT = YZER + HGHT + TRAD
N6 ZRAP = ZZER + 0.1
N1 G00 **X = XRAP Y = YRAP** (Rapid to starting position)
N2 G01 **Z = ZBTM** F50. (Fast feed to work surface)
N3 **Y = YCUT F = FEED** (Mill right side)
N4 G00 **Z = ZRAP** (Rapid to above workpiece)
N5 RTS (End of custom macro)

As you develop your own user task programs, many opportunities will arise for you to use local variables for general-purpose calculations.

As counters

Looping functions often require that variables change each time the loop is executed. Local variables can be used as the loop counter and for any other variable that must be initialized at the beginning of the loop then stepped each time through the loop.

As flags

It is sometimes necessary to set a *flag* that tells the user task program how to behave. For example, you may wish to set up criteria in your right-side milling user task program that allows the milling cutter to either climb or conventional mill. A special flag will specify which milling method you wish to use. In fact, you are *flagging* the user task program as to how it should behave. Other instances of when flags are required (discussed in greater detail in the section on conditional branching) include right-hand versus left-hand tooling, inch versus metric input, and in cases where coolant is or is not to be used. While local variable address arguments in the CALL command commonly set flags, the global nature *of common variables* makes it possible to maintain the values of flags even among several programs.

COMMON VARIABLES

Unlike local variables, a common variable will retain its value even after the user task program is completed. In fact, common variable values in user task will even be retained after the power has been turned off. In this sense, they are much like tool offsets. You have 32 common variables at your command in your user task programs. However, the naming of common variables is not nearly as flexible as it is with local variables. Common variable names are fixed, and range from V1 through V32.

Since there are a limited number of common variables, reserve their use for those applications that require them. Use them only when it is necessary to retain data among several programs or when you must retain data from one day to the next.

Retaining data among several programs

Many parametric programming applications require several programs working with each other. The more complicated the application, the more likely it is that multiple programs will be needed and the more likely that data will be shared. Common variables are used for any data that must be shared among two or more programs (this kind of data is commonly called *global* data).

As with any complicated problem, it is often helpful to divide the parametric programming application into smaller and easier-to-handle segments.

It may be helpful, for example, to have one user task program calculating speeds and feeds. Another may be setting up the tool station numbers; yet another doing the machining. In this case, speed, feed, and tool-station values must be available to the machining program. Common variables must be used to maintain the value of each feed, speed, and tool station number from one program to another.

Retaining data from one day to the next

In many cases it is necessary to retain data from day to day. Part counting is a good example. Using the same scenario as our last example, suppose you wish to set up a part counter to let the machine run 750 workpieces only and then stop; as you begin running production, the user task-based part counter will begin counting workpieces. Since it isn't very likely that you will finish all 750 workpieces before the end of the day or shift, you will need your part-counting user task program to remember where it left off from day to day. If you store the current part count in a common variable, the value of the variable will not be lost when the power is turned off.

Setting your own system constants

One of the many constant hard-and-fixed values that you use regularly throughout your CNC programming is your rapid approach and retract distance. Many programmers, when working in the inch mode, like to stay a constant 0.100 in. away from the surface being approached before entering a cutting mode. When cutting is completed, they feed the tool away from the workpiece by this same 0.100-in. amount.

Though this approach distance is quite safe, when high production quantities are run, the time it takes to complete each approach and retract at the cutting feedrate adds to cycle time and may be excessive. However, if all approach and retract movements are programmed with hard-and-fixed values (calculated by the programmer), it can be quite difficult to change the approach and retract values without a massive overhaul of the CNC program.

If instead, a common variable is used in which to store the rapid approach and retract distance, and if this variable is referenced in every CNC program during every tool's approach and retract movement, changing the approach/retract distance will be as simple as changing one common variable. And, as stated, common variables can be entered and changed through the keyboard and display screen, just like tool offsets.

Consider the following program that drills a few holes and mills a slot. It assumes the rapid approach/retract distance is stored in common variable V1.

O0003 (Program that allows easy approach/retract distance change)
N005 G15 H01 (Select coordinate system)
N010 T01 M06 (Place 1/2-in. [12.7-mm] drill in spindle)
N015 G90 S800 M03 T02 (Select absolute mode, turn spindle on, and get tool number two ready)
N020 G00 X.5 Y.5 (Rapid to first hole location)
N025 G56 H01 **Z =V1** (Instate tool-length compensation, rapid up to workpiece)

N030 G81 **R = V1** Z – 0.85 F3.5 (Drill first hole)
N035 Y3.5 (Drill second hole)
N040 X4.5 (Drill third hole)
N045 Y.5 (Drill fourth hole)
N050 G80 (Cancel cycle)
N055 G00 Z50.0 (Rapid to tool-change position)
N060 G15 H01 (Select coordinate system)
N065 T02 M06 (Place 3/8-in. [9.53-mm] drill in spindle)
N070 G90 S950 M03 T03 (Select absolute mode, turn spindle on, and get tool three ready)
N075 G00 X3.0 Y2.0 (Rapid to hole location)
N080 G56 H02 **Z = V1** (Instate tool-length compensation, rapid to approach position)
N085 G81 **R = V1** Z – 0.55 F3.5 (Drill hole)
N090 G80 (Cancel cycle)
N095 G00 Z50.0 (Rapid to tool-change position)
N100 G15 H01 (Select coordinate system)
N105 T03 M06 (Place 1/4-in. [6.35-mm] endmill in spindle)
N110 G90 S1000 M03 T01 (Select absolute mode, turn spindle on, and get tool one ready)
N115 G00 X3.75 **Y = –[0.125 + V1]** (Rapid to slot approach position)
N120 G56 H03 Z – 0.125 (Instate tool-length compensation, rapid to work surface)
N125 G01 **Y = 4.125 + V1** F2.5 (Mill slot)
N130 G00 **Z = V1** (Rapid to above work surface)
N135 G00 Z50.0 (Move to tool-change position)
N140 M02 (End of program)

Notice that any time a tool in this program approaches or retracts (lines N025, N030, N080, N085, N115, N125, and N130), common variable V1 is referenced. In lines N115 and N125, notice how the programmer can even perform a calculation that has the control figure out the correct approach/retract position based on the value of common variable V1. With this program, changing the approach/ retract distance will be as easy as changing V1.

SYSTEM VARIABLES

Many of the CNC-related features of user task 2 are handled with system variables. These variables have special names (chosen by Okuma) and allow the user to access many CNC functions, including current machine position, offsets, alarm generation, and many more. These system variables are covered in ample detail in Chapter 5.

Summary of user task variables

As we have seen, there are three types of variables in user task 2; *local variables*, which can be used for a variety of purposes, including as arguments, to minimize redundant calculations, for general-purpose calculations, and as counters; *common variables*, used for applications that require the value of a variable to be retained from one program to another or from day to day; and *system variables*, which allow access to many CNC-related features of user task 2. While you should have a pretty good idea of what variables do, bear in mind that discussions to this point are intentionally broad in scope. There are some advanced implications which we will address later.

VARIABLES IN FADAL'S MACRO

Fadal distinguishes between *CNC* commands and *macro* commands. While CNC commands require no special consideration, *all macro commands must begin with a pound sign (#).* There is only one exception to this rule, which is explained following presentation of the next sample program.

Just about every variable function possible with Fanuc's custom macro B and Okuma's user task 2 is possible with Fadal's macro, which incorporates three types of variables: the *R-series, V-series,* and *system.*

R-SERIES VARIABLES IN CNC COMMANDS

The R-series variables are used primarily to bring data into CNC words. That is, every value must be stored in an R-series variable before it can be referenced in a CNC word. They range from R0 through R9, meaning you have 10 R-series variables with which to work.

While the small number of R series variables may sound limiting, bear in mind that for complicated applications, the same R-series variables can be used over and over again. While R-series variables also can be used for other purposes (e.g., general-purpose calculations and as arguments), another series of variables (V1-V100) are more commonly used for these purposes.

A simple example highlighting the use of R-series variables follows. As you read through it, consider the commands.

```
.
.
.
N050 R0 + 5.0 (Set variable R0 to 5.0)
N055 R1 + 4.0 (Set variable R1 to 4.0)
N060 R2 + 0.6 (Set variable R2 to 0.6)
N065 G00 X + R0  Y – R2 (Rapid into position)
N070 G01 Z – 0.5 F50. (Fast feed to work surface)
```

```
N075 Y + R1 F5.0 (Machine right side of workpiece)
N080 G00 Z0.1 (Rapid away in Z)
.
.
.
```

In lines N050, N055, and N060, variables R0, R1, and R2 are being assigned. Note that the actual format for assigning variables within CNC commands includes a plus sign (it also could be a minus sign). Also note that when R-series variables are referenced within CNC commands (N065 and N075), you must specify the plus (or minus) sign. Note, too, that lines N050, N055, and N060 illustrate the one time that a pound sign (#) is not required at the beginning of a macro command. If you are simply assigning the value of an R-series variable (setting it to a constant numerical value), the pound sign is not required.

R-SERIES VARIABLES IN MACRO COMMANDS

While assigning and using R-series variables in CNC commands as shown in the previous program can sometimes be helpful, most applications require the values of variables to be manipulated using some form of arithmetic operations (addition, subtraction, multiplication, etc.). Whenever arithmetic operations must be performed, they must be done in a macro command that begins with the pound sign (#). Consider these commands:

```
.
.
.
N060 R0 + 5.0 (Assign R0 as 5.0)
N065 R1 + 4.0 (Assign R1 as 4.0)
N070 R2 + 1.0 (Assign R2 as 1.0)
N075 #R3 = R0 + R2/2 'Assign R3 as 5 + 1/2 (5.5)
N080 #R4 = 0.1 + R2/2 'Assign R4 as 0.1 + 1/2 (0.6)
N085 #R5 = R1 + R2/2 'Assign R5 as 4 + 1/2 4.5)
N090 G00 X + R3 Y – R4 (Rapid into position)
N095 G01 Z – 0.5 F50.0 (Fast feed to work surface)
N100 Y + R5 F5.0 (Mill right side)
N105 G00 Z0.1 (Move away from surface)
.
.
.
```

In lines N075, N080, and N085, the pound sign specifies that this is a macro command. Macro commands allow all kinds of arithmetic operations to be performed (as well as many other functions). Take special note of the fact that, before the value of any variable can be referenced within the CNC program, it must be transferred into an R-series variable. Also note that since parentheses are used for other purposes in macro commands, an apostrophe (') must be used to designate comments. Everything following an apostrophe is a comment.

V-SERIES VARIABLES

The programmer has at his disposal 100 V-series variables in Fadal's macro, ranging from V1 through V100. These variables are global, meaning their values will be available to any active program. As stated, V-series variables cannot be directly referenced in CNC commands—only macro commands (beginning with the pound sign) can reference V-series variables. Since there are so many of them, and since they are more flexible in their use than R-series variables, V-series variables make the best choice for most variable applications. The programmer using V-series variables is able to reserve the R-series variables for transferring values into the CNC program.

Using V-series variables for arguments

We have stated that any variable that gets data into the parametric program is called an *argument*. V-series variables can be used as arguments.

V-series variables as arguments for user-created canned-cycle applications. In most user-created canned-cycle applications, a simple subprogram call command (M98) is used to activate the parametric program. The M98 command includes a P word that specifies the program number for the macro program. Prior to the M98 command, one or more command lines can be used to set the values of V-series variable arguments to be passed. Since V-series variables are *global*, their values will remain intact during the execution of the macro program.

Figure 2-6 shows an example of a very simple application. Notice that the right side of this workpiece must be milled.

The programmer knows that more than 300 workpieces of this type will be machined, all requiring the right side to be milled. Rather than writing code to mill the right side in over 300 programs, the programmer chooses to write a user-created canned cycle to do this milling.

Figure 2-7 shows the V-series arguments one programmer has chosen to represent the values that change from one workpiece to the next. Since every dimension relative to the right side of the workpiece changes with every workpiece, each dimension requires an argument assignment. Note that *only* the dimensions related to the milling of the right side require arguments. Note also that the argument names themselves (V1 through V8) do not make much sense. Only through good documentation will everyone know the meaning of each argument name. (Figure 2-7, incidentally, illustrates one very good way of documenting your parametric program. Anyone involved with this application will know the meaning of each V-series variable argument being assigned.)

Assuming the parametric program has already been written and verified, the following program will machine the entire workpiece shown in Figure 2-6 (which includes calling commands to the *right-end-milling* macro program).

Program:
O0001 (Program name)
N005 T01 M06 (Place 1-in. [25.4-mm] endmill in spindle)
N010 G54 H01 G90 S400 M03 T02 (Select coordinate system, absolute mode, turn spindle on, and get next tool ready)
N015 G00 X5.0 Y0 (Rapid over to the workpiece)
N020 G43 H01 Z0.1 (Instate tool-length compensation, move up close to work surface)
N025 #V1 = 0 'Assign argument V1
N030 #V2 = 0 'Assign argument V2
N035 #V3 = 0 'Assign argument V3
N040 #V4 = 5.0 'Assign argument V4
N045 #V5 = 3.0 'Assign argument V5
N050 #V6 = 0.5 'Assign argument V6
N055 #V7 = 1.0 'Assign argument V7
N060 #V8 = 5.0 'Assign argument V8
N065 M98 P1000 (Mill right side)
N070 G91 G28 Z0 M19 (Move to tool-change position, orient spindle)
N075 M01 (Optional stop)
N080 T02 M06 (Place 1/2-in. [12.7-mm] drill two in spindle)
N085 G54 G90 S800 M03 T01 (Select coordinate system, absolute mode, start spindle, get tool number one ready)
N090 G00 X0.5 Y0.5 (Rapid to first hole position)
N095 G43 H02 Z0.1 (Instate tool-length compensation, rapid up close to work surface)
N100 G81 R0.1 Z-0.85 F4.0 (Drill first hole)
N105 Y2.5 (Drill second hole)
N110 X4.5 (Drill third hole)
N115 Y0.5 (Drill fourth hole)
N120 G80 (Cancel cycle)
N125 G91 G28 Z0 M19 (Return to tool-change position, orient spindle)
N130 M02 (End of program)

Note again how little of this program deals with the milling of the right side. The series of commands from N025 through N060 simply set the V-series variable arguments. Line N065 calls the macro program. If more workpieces must be milled by this program, it is easy to simply specify more calling commands (after changing the related V-series variable arguments) to the program named O1000. Note from this example that V-series variables must be designated one per com-

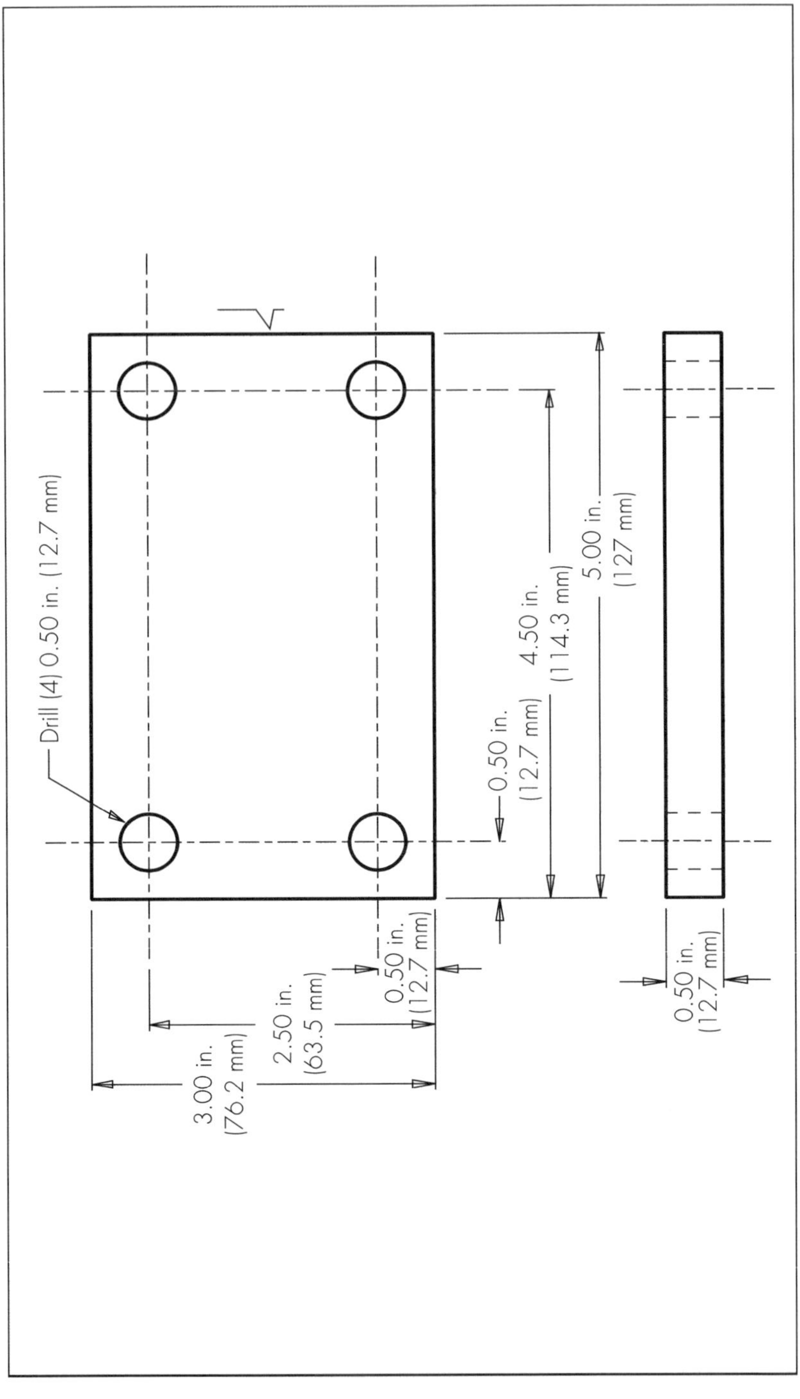

Figure 2-6. Again, a user-created canned cycle to machine the right side of this workpiece relieves the programmer of writing code for each of the hundreds of workpieces to be milled.

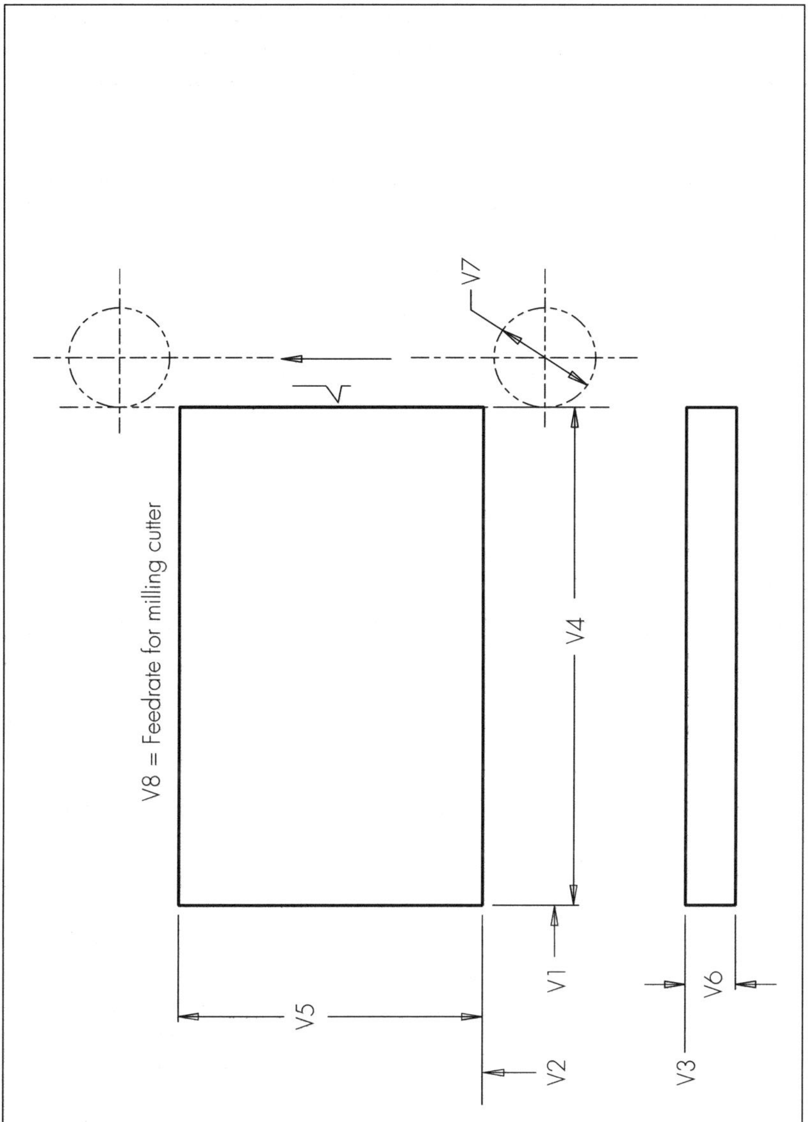

Figure 2-7. The V-series argument selection for the workpiece in Figure 2-6. Since the coding of these arguments does not provide direct insight into the meaning of the arguments, thorough documentation is needed to avoid any confusion.

mand. Though this may at first seem a little cumbersome, keep in mind that this requirement can actually be a benefit when it comes to multiple call statements. Only those V-series variables that change must be specified from one call statement to the next.

For actually milling the right side, the following macro is used. Notice that each value needed within the CNC program must first be transferred to an R-series variable. Also notice that each argument is referenced directly by the V-series variable being set in the main program.

O1000 (Macro program to mill right side)
N1 **#R0 = V1 + V4 + V7/2** 'Calculate *X* position for cutter
N2 **#R1 = V2 – 0.1 – V7/2** 'Calculate *Y* starting position
N3 **#R2 = V3 – V6 – 0.1** 'Calculate work surface position in *Z*
N4 **#R3 = V2 + V5 + 0.1 + V7/2** 'Calculate *Y* ending position
N5 **#R4 = V8** 'Transfer feedrate into R4
N6 **#R5 = V3 + 0.1** 'Calculate clearance position in *Z*
N7 G00 **X + R0 Y + R1** (Rapid to starting position)
N8 G01 **Z + R2** F50.0 (Fast feed to work surface)
N9 **Y + R3 F + R4** (Mill right side)
N10 G00 **Z + R5** (Rapid to above workpiece)
N11 M99 (End of parametric program)

While this macro program should be relatively easy to follow, there are a few points that are important to understand about its structure. First, calculations must be done to determine the positions needed during the milling cutter's motions. Calculations require macro commands (as opposed to CNC commands), and all macro commands must begin with the pound sign (lines N1 through N6).

Second, values needed during CNC words must first be stored in R-series variables before they can be referenced in CNC commands. Even the simple feedrate value that does not require a calculation (in line N5) must first be placed into an R-series variable. This line also is considered a macro command (it must begin with a pound sign) since it references data stored in a V-series variable.

Finally, arithmetic calculations are being done within several commands. While these simple arithmetic functions are quite understandable, we detail the arithmetic functions of macro later.

Running out of R-series variables? Even with this rather simple example, you may have noticed that we used 6 of the 10 available R-series variables. This raises the question of whether there will be enough R-series variables available for more complicated applications. While this is an understandable concern, remember that R-series variables can be used over and over again, even within the same program. Once an R-series variable's value is no longer needed, it can be reassigned and reused. Consider this second version of the macro program that only requires *two* R-series variables.

```
O1000 (Macro program to mill right side)
N1 #R0 = V1 + V4 + V7/2 'Calculate X position for cutter
N2 #R1 = V2 – 0.1 – V7/2 'Calculate Y starting position
N3 G00 X + R0 Y + R1 (Rapid to starting position)
N4 #R0 = V3 – V6 – 0.1 'Calculate work surface position in Z
N5 G01 Z + R0 F50. (Fast feed to work surface)
N6 #R0 = V2 + V5 + 0.1 + V7/2 'Calculate Y ending position
N7 #R1 = V8 'Transfer feedrate into R4
N8 Y + R0 F + R1 (Mill right side)
N9 #R0 = V3 + 0.1 'Calculate clearance position in Z
N10 G00 Z + R0 (Rapid to above workpiece)
N11 M99 (End of parametric program)
```

Notice that as soon as an R-series variable is no longer needed, it is reassigned and reused. This dramatically minimizes the number of R-series variables needed in your macro programs and also reinforces the reason why you should only use R-series variables for the purpose of getting data into CNC words.

V-series variables as arguments for family-of-parts applications. While use of the M98 command and corresponding V-series variable argument assignments is the best way to pass data to a user-created canned-cycle macro program, it is not always the most convenient way to get data into the macro program. When programming family-of-parts applications, the entire workpiece is being machined by the parametric program. In this case, it is usually more convenient to state the values of all V-series variable arguments at the very beginning of the program rather than to call the macro program from another program.

Again using the workpiece in Figure 2-6, we demonstrate a new program to accommodate changes in the criteria of the application. Instead of just milling the right side for a variety of different workpieces, we now stipulate that the workpiece in Figure 2-6 is just one of more than 100 very similar workpieces. All workpieces have four 0.5-in. (12.7-mm)-diameter holes drilled 0.5 in. from each edge. This new criterion (machining the entire workpiece) changes the application from a user-created canned-cycle application to a *family-of-parts* application. Again, Figure 2-7 shows the V-series variable arguments related to this application. Since the hole locations are always 0.5 in. in diameter and 0.5 in. from the edges, no additional V-series variable arguments are required. The following macro program machines all workpieces in this part family.

Program:

```
O0002 (Program name)
(Argument assignments)
N005 #V1 = 0 'Left side position in X
N010 #V2 = 0 'Lower side position in Y
N015 #V3 = 0 'Top surface of workpiece in Z
```

N020 **#V4 = 5.0** 'Length of workpiece in *X*
N025 **#V5 = 3.0** 'Width of workpiece in *Y*
N030 **#V6 = 0.5** 'Workpiece thickness
N035 **#V7 = 1.0** 'Milling cutter diameter
N040 **#V8 = 5.0** 'Feedrate of milling cutter in ipm
(Machining portion of the program)
N045 T01 M06 (Place 1-in. [25.4-mm] endmill in spindle)
N050 #R0 = 3.82 * 80/V7 'Calculate spindle speed in rpm
N055 G54 G90 **S + R0** M03 T02 (Select coordinate system, absolute mode, turn spindle on, and get next tool ready)
N060 #R0 = V1 + V4 + V7/2 'Calculate *X* position for milling cutter
N065 #R1 = V2 – 0.1 – V7/2 'Calculate starting position in *Y*
N070 G00 **X + R0 Y + R1** (Rapid over to the workpiece)
N075 #R0 = V3 + 0.1 'Calculate rapid approach position in *Z*
N080 G43 H01 **Z + R0** (Instate tool-length compensation, move up close to work surface)
N085 #R0 = V3 – V6 – 0.1 'Calculate work surface position in *Z*
N090 G01 **Z + R0** F50.0 (Fast feed past surface)
N095 #R0 = V2 + V5 + 0.1 + V7/2 'Calculate *Y* ending position
N100 **#R1 = V8** 'Transfer feedrate value to R1
N105 **Y + R0 F + R1** (Mill right side)
N110 **#R0 = V3 + 0.1** 'Calculate clearance position in *Z*
N115 G00 **Z + R0** (Rapid to above top surface)
N120 G91 G28 Z0 M19 (Move to tool-change position, orient spindle)
N125 M01 (Optional stop)
N130 T02 M06 (Place 1/2-in. [12.7-mm] drill two in spindle)
N135 G54 G90 S800 M03 T01 (Select coordinate system, absolute mode, start spindle, get tool number one ready)
N140 #R0 = V1 + 0.5 'Calculate first hole *X* position
N145 #R1 = V2 + 0.5 'Calculate first hole *Y* position
N150 G00 **X + R0 Y + R1** (Rapid to first hole position)
N155 #R3 = V3 + 0.1 'Calculate rapid plane for hole machining
N160 G43 H02 **Z + R3** (Instate tool-length compensation, rapid up close to work surface)
N165 #R0 = V3 – V6 – 0.18 'Calculate hole bottom position
N170 G81 **R0 + R3 Z + R0** F4.0 (Drill first hole)
N175 #R0 = V2 + V5 – 0.5 'Calculate second hole *Y* position
N180 **Y + R0** (Drill second hole)
N185 #R0 = V1 + V4 – 0.5 'Calculate third hole *X* position
N190 **X + R0** (Drill third hole)
N195 #R0 = V2 + 0.5
N200 **Y + R0** (Drill fourth hole)

```
N205 G80 (Cancel cycle)
N210 G91 G28 Z0 M19 (Return to tool-change position, orient spindle)
N215 M02 (End of program)
```

Notice how well the V-series variable arguments are documented at the beginning of this program. Since their names are not recognizable, the explanatory comments to the right of the apostrophe stress the meaning of each. An operator or setup person could confidently modify any one these arguments during setup for the next workpiece in the family to be machined.

Also notice that arguments are only required for those things that change from one workpiece in the family to the next. Any workpiece value that does not change (speed and feed for 1/2-in. [12.7-mm] drill, 0.5-in. [12.7-mm] hole position, 0.1-in. [2.54-mm] approach distance, etc.) is treated as a *constant value*. And constant values, of course, cannot be changed by simply manipulating argument values.

While this is a relatively simple family-of-parts application, it clearly demonstrates how local variable arguments can be introduced to family-of-parts parametric programs. It should also provide at least an inclination of the power available to you with parametric programming.

Note again the subtle distinction between argument assignment for user-created canned-cycle applications and argument assignment for family-of-parts applications. With part families, the macro program is machining the entire workpiece, and assigning arguments in the main program is the most convenient method of argument assignment since only one program is required.

On the other hand, user-created canned-cycle applications always require a separate program to perform the canned-cycle-like operation. Because this program may be required several times even to machine one workpiece, and because arguments must be manipulable from one execution of the parametric program to the next, arguments must be set in the main program.

Getting input from the operator. A unique feature of Fadal's macro is that of enabling operator input during execution of the CNC program. The macro program can query the operator for arguments to be used in the macro program. While you should always test the input data for errors (we'll show how later), this feature provides an effective way to get arguments assigned for family-of-parts applications.

Here is a modified version of the family-of-parts application for the workpiece shown in Figure 2-7 calling to mill the right side of the workpiece and drill four holes. Notice the use of the PRINT and INPUT statements that ask for data from the operator.

Program:

```
O0003 (Program name)
(Argument assignments)
N005 #PRINT "ENTER LEFT SIDE POSITION IN X"
```

```
N010 #INPUT V1
N015 #PRINT "ENTER THE LOWER SURFACE POSITION IN Y"
N020 #INPUT V2
N025 #PRINT "ENTER TOP SURFACE POSITION IN Z"
N030 #INPUT V3
N035 #PRINT "ENTER LENGTH OF WORKPIECE IN X"
N040 #INPUT V4
N045 #PRINT "ENTER WIDTH OF WORKPIECE IN Y"
N050 #INPUT V5
N055 #PRINT "ENTER WORKPIECE THICKNESS"
N060 #INPUT V6
N065 #PRINT "ENTER MILLING CUTTER DIAMETER"
N070 #INPUT V7
N075 #PRINT "ENTER MILLING FEEDRATE"
N080 #INPUT V8
(Machining portion of the program)
N085 T01 M06 (Place 1-in. [25.4-mm] endmill in spindle)
N090 #R0 = 3.82 * 80/V7 'Calculate spindle speed in rpm
N095 G54 G90 S + R0 M03 T02 (Select coordinate system, absolute mode, turn
   spindle on, and get next tool ready)
N100 #R0 = V1 + V4 + V7/2 'Calculate X position for milling cutter
N105 #R1 = V2 = 0.1 – V7/2 'Calculate starting position in Y
N110 G00 X + R0 Y + R1 (Rapid over to the workpiece)
N115 #R0 = V3 + 0.1 'Calculate rapid approach position in Z
N120 G43 H01 Z + R0 (Instate tool-length compensation, move up close to work
   surface)
N125 #R0 = V3 – V6 – 0.1 'Calculate work surface position in Z
N130 G01 Z + R0 F50.0 (Fast feed past surface)
N135 #R0 = V2 + V5 + 0.1 + V7/2 'Calculate Y ending position
N140 #R1 = V8 'Transfer feedrate value to R1
N145 Y + R0 F + R1 (Mill right side)
N150 #R0 = V3 + 0.1 'Calculate clearance position in Z
N155 G00 Z + R0 (Rapid to above top surface)
N160 G91 G28 Z0 M19 (Move to tool-change position, orient spindle)
N165 M01 (Optional stop)
N170 T02 M06 (Place 1/2-in. [12.7-mm] drill two in spindle)
N175 G54 G90 S800 M03 T01 (Select coordinate system, absolute mode, start
   spindle, get tool number one ready)
N180 #R0 = V1 + 0.5 'Calculate first hole X position
N185 #R1 = V2 + 0.5 'Calculate first hole Y position
N190 G00 X + R0 Y + R1 (Rapid to first hole position)
N195 #R3 = V3 + 0.1 'Calculate rapid plane for hole machining
```

```
N200 G43 H02 Z + R3 (Instate tool-length compensation, rapid up close to work
   surface)
N205 #R0 = V3 – V6 – 0.18 'Calculate hole bottom position
N210 G81 R0 + R3 Z + R0 F4.0 (Drill first hole)
N215 #R0 = V2 + V5 – 0.5 'Calculate second hole Y position
N220 Y + R0 (Drill second hole)
N225 #R0 = V1 + V4 – 0.5 'Calculate third hole X position
N230 X + R0 (Drill third hole)
N235 #R0 = V2 + 0.5
N240 Y + R0 (Drill fourth hole)
N245 G80 (Cancel cycle)
N250 G91 G28 Z0 M19 (Return to tool-change position, orient spindle)
N255 M02 (End of program)
```

As mentioned, this program requires refinement on at least two points. First, nothing guards against operator mistakes when inputting data. The control will execute the motions of this program regardless of input error, which could be disastrous. (We discuss how to test for mistakes during argument assignment in Chapter 4.) Second, note that with this version of the program, the PRINT and INPUT commands will be executed *every time* the program is run. Unless your operator is making only one workpiece, entering so much data for each workpiece is not practical. (We also show how to handle this problem later in this text.)

Using V-series variables as counters

Looping functions often require that variables change each time the loop is executed. V-series variables are used as the loop counter and for any other variable that must be initialized at the beginning of the loop and stepped each time through the loop.

Using V-series variables as flags

It is sometimes necessary to set a *flag* that tells the macro program how to behave. If, for example, you set up criteria in your right-side milling user task program that allow the milling cutter to either climb or conventional mill, a special flag will specify the milling method you wish to use. You have *flagged* the macro program, telling it how to behave. Other examples of when flags are required include right-hand versus left-hand tooling, inch versus metric input, and whether or not coolant is to be used. While V-series variables prior to the M98 calling command commonly set flags, the *global* nature of V-series variables makes it possible to maintain the values of flags even among several programs.

Retaining data among several programs

Many parametric programming applications require several programs to work together. The more complicated the application, the more likely it is that multiple programs will be needed and that data will be shared. V-series variables should be used for any data that must be shared among two or more programs (this kind of data is commonly called *global* data). As with any complicated problem, it is often helpful to divide the parametric programming application into smaller and easier-to-handle segments.

It may be helpful, for example, to have one macro program calculating speeds and feeds. Another may be setting up the tool station numbers, and yet another actually doing the machining. In this case, speed, feed, and tool station values must be available to the machining program. V-series variables should be used to maintain the value of each feed, speed, and tool station number from one program to another.

Retaining data from one day to the next

When it is necessary to retain data from day to day, such as in part counting, a macro program with part count stored in a V-series variable is needed. If you set up a part counter to let the machine run 750 workpieces and then stop the machine, a macro-based part counter will begin counting workpieces as you begin running production. Since it isn't likely that you will finish all 750 workpieces before the end of the day/shift, your part counting macro program will have to remember where it left off from day to day. If you store the current part count in a V-series variable, the value of the variable will not be lost when the power is turned off. Like tool offsets, V-series variables are retained after power is turned off, and are even available for modification through the control's display screen and keyboard.

Setting your own system constants

Cutting cycle time is an ever-present objective in the CNC environment, and programmers have become quite adept at ferreting out areas for improvement. Constant hard-and-fixed values used throughout CNC programming seemed impregnable before parametric programming. Again selecting one of these values at random—rapid approach and retract distance—we find that when programming in the inch mode many programmers like to stay a constant 0.100 in. away from the surface being approached before entering a cutting mode. When cutting is completed, these programmers feed the tool away from the workpiece by this same amount.

Though this approach distance is quite safe, when high production quantities are run, the time it takes to complete each approach and retract at the cutting feedrate adds to cycle time and may be excessive. However, if all approach and retract movements are programmed with hard-and-fixed values (calculated by the programmer), it can be quite difficult to change these values without a massive overhaul of the CNC program.

If, instead, a V-series variable is used to store the rapid approach and retract distance, and if this variable is referenced in every CNC program during every tool's approach and retract movement, changing the approach/retract distance will be as simple as changing one V-series variable. Like tool offsets, V-series variables can even be entered and changed through the keyboard and display screen, making it very easy to change the rapid approach/retract distance.

Consider the following program that drills a few holes and mills a slot. It assumes the rapid approach/retract distance is stored in V-series variable V1.

O0003 (Program that allows easy approach/retract distance change)
N005 T01 M06 (Place 1/2-in. [12.7-mm] drill in spindle)
N010 G54 G90 S800 M03 T02 (Select coordinate system, absolute mode, turn spindle on, and get tool number two ready)
N015 G00 X.5 Y.5 (Rapid to first hole location)
N017 **#R1 = V1** 'Set Approach distance to R1
N020 G43 H01 **Z + R1** (Instate tool-length compensation, rapid up to workpiece)
N025 G81 **R0 + R1** Z – 0.85 F3.5 (Drill first hole)
N030 Y3.5 (Drill second hole)
N035 X4.5 (Drill third hole)
N040 Y0.5 (Drill fourth hole)
N045 G80 (Cancel cycle)
N050 G91 G28 Z0 (Return to tool-change position)
N055 M01 (Optional stop)
N060 T02 M06 (Place 3/8-in. [9.53-mm] drill in spindle)
N065 G54 G90 S950 M03 T03 (Select coordinate system, absolute mode, turn spindle on, and get tool three ready)
N075 G00 X3.0 Y2.0 (Rapid to hole location)
N080 G43 H01 **Z + R1** (Instate tool-length compensation, rapid to approach position)
N085 G81 **R0 + R1** Z – 0.55 F3.5 (Drill hole)
N090 G80 (Cancel cycle)
N095 G91 G28 Z0 (Return to tool-change position)
N100 M01 (Optional stop)
N105 T03 M06 (Place 1/4-in. [6.35-mm] endmill in spindle)
N110 G54 G90 S1000 M03 T01 (Select coordinate system, absolute mode, turn spindle on, and get tool one ready)
N113 **#R2 = 0.1 + V1** 'Set R2 to *Y* approach position
N115 G00 X3.75 **Y – R2** (Rapid to slot approach position)
N120 G43 H01 Z – 0.125 (Instate tool-length compensation, rapid to work surface)
N123 **#R2 = 4.125 + V1** 'Set R2 to *Y* plus side of slot
N125 G01 **Y + R2** F2.5 (Mill slot)
N130 G00 **Z + R1** (Rapid to above work surface)
N135 G91 G28 Z0 (Return to tool change position)
N140 M02 (End of program)

Notice that any time a tool in this program approaches or retracts (lines N020, N025, N080, N085, N115, N125, and N130), the value requiring the proper approach distance stored in V1 is referenced. In lines N113 and N123, the programmer can even perform a calculation that has the control figure out the correct approach/retract position based on the current value of common variable V1. With this program, changing the approach/retract distance will be as easy as changing variable V1.

SYSTEM VARIABLES

Many of the CNC-related features of Fadal's macro are handled with *system* variables. These variables have special names (chosen by Fadal) and allow access to many CNC functions, including current machine position, offsets, alarm generation, and many others. These variables are detailed later in the book.

Summary of macro variables

There are three types of variables in macro: *R-series,* which are used to get data into a form by which it can be used in a CNC word; *V-series*, which can be used for a variety of purposes, including as arguments, to minimize redundant calculations, for general-purpose calculations, and as counters; and *system* which allow access to many CNC-related features of *macro*.

In this chapter our intent is to provide an overview of macro variables, what they do, and when they can be used. Each of the variables has some advanced implications and these are addressed in upcoming chapters.

SAMPLE PROGRAMS

Though we have barely scratched the surface of what can be done with parametric programming, you should be ready to study some sample programs that exemplify its features. We are introducing nothing new, so you should have no trouble understanding these programs. For this reason, explanations are kept to a minimum. (In the interest of reader continuity, we have not interrupted the written programs by placing figures as they are called out; the figures appear after the sample programs.)

MILLING A STRAIGHT SLOT

This simple user-created canned-cycle example mills a straight slot along the *X* axis, using an endmill with the same diameter as the slot width (only one pass is made). The parametric program will plunge to depth at the left side (at one-half the specified feedrate) and mill to the right side of the slot. Figure 2-8 (page 76) shows a print that requires our parametric program to be called twice.

Fanuc's custom macro B format

Figure 2-9 (page 77) shows the letter address arguments used with the straight slot milling application.

The beginning of the main program calls the custom macro B program twice to machine the two slots.

```
O0010 (Main program)
N005 T01 M06 (Place 1/2-in. [12.7-mm]-diameter endmill in spindle)
N010 G54 G90 S800 M03 T02 (Select coordinate system, absolute mode, start
   spindle, get next tool ready)
N015 G00 X1.0 Y0.75 (Rapid over to first slot)
N020 G43 H01 Z0.1 (Instate tool-length compensation, rapid up close to work
   surface)
N025 G65 P1010 X1.0 Y0.75 U4.0 Z0 D0.25 F5.0 (Mill lower Y slot)
N030 G65 P1010 X1.0 Y3.25 U4.0 Z0 D0.25 F5.0 (Mill upper Y slot)
.
.
.
```

The custom macro B program:

```
O1010 (Slot milling custom macro program)
G00 X#24 Y#25 (Move to left side of slot in X)
Z[#26 + 0.1] (Move to approach position in Z)
G01 Z[#26 – #7] F[#9/2] (Plunge into slot at 1/2 feedrate)
X[#24 + #21] F#9 (Mill slot)
G00 Z[#26 + 0.1] (Rapid out of slot)
M99 (End of custom macro)
```

Okuma's user task 2 format

Figure 2-10 (page 78) shows the local variable arguments used with the slot milling application.

At the beginning of the main program the user task 2 program is called twice to machine the two slots.

```
O0010 (Main program)
N005 G15 H01 (Select coordinate system)
N010 T01 M06 (Place 1/2-in. [12.7-mm]-diameter endmill in spindle)
N015 G90 S800 M03 T02 (Select absolute mode, start spindle, get next tool ready)
N020 G00 X1.0 Y0.75 (Rapid over to first slot)
N025 G56 H01 Z0.1 (Instate tool-length compensation, rapid up close to work
   surface)
N030 CALL O1010 XLFT = 1.0 YCTR = 0.75 LGTH = 4.0 ZSFR = 0 DPTH =
   0.25 FEED = 5.0 (Mill lower Y slot)
```

```
N035 CALL O1010 XLFT = 1.0 YCTR = 3.25 LGTH = 4.0 ZSFR = 0 DPTH =
   0.25 FEED = 5.0 (Mill upper Y slot)
.
.
.
```

The user task 2 program:

```
O1010 (Slot milling user task program)
G00 X = XLFT Y = YCTR (Move to left side of slot in X)
Z = ZSRF + 0.1 (Move to approach position in Z)
G01 Z = ZSRF – DPTH F = FEED/2 (Plunge into slot at 1/2 feedrate)
X = XLFT + LGTH F = FEED (Mill slot)
G00 Z = ZSRF + 0.1 (Rapid out of slot)
RTS (End of user task program)
```

Fadal's macro format

Figure 2-11 (page 79) shows the V-series variable arguments used with the slot-milling application.

Again the beginning of the main program calls the macro program twice to machine the two slots.

```
O0010 (Main program)
N005 T01 M06 (Place 1/2-in. [12.7-mm]-diameter endmill in spindle)
N010 G54 G90 S800 M03 T02 (Select coordinate system, absolute mode, start
   spindle, get next tool ready)
N015 G00 X1.0 Y0.75 (Rapid over to first slot)
N020 G43 H01 Z0.1 (Instate tool-length compensation, rapid up close to work
   surface)
N025 #V1 = 1.0 (Left side of slot in X)
N030 #V2 = 0.75 (Y position of lower slot)
N035 #V3 = 4.0 (Length of slot)
N040 #V4 = 0 (Z surface of slot top)
N045 #V5 = 0.25 (Slot depth)
N050 #V7 = 5.0 (Feedrate)
N055 M98 P1010 (Call macro, machine lower slot)
N060 #V2 = 3.25 (Y position of upper slot)
N065 M98 P1010 (Call macro, machine upper slot)
.
.
.
```

The macro program:

```
O1010 (Slot milling user task program)
#R1 = V1 'Set R1 to V1
#R2 = V2 'Set R2 to V2
G00 X + R1 Y + R2 (Move to left side of slot in X)
#R9 = V4 + 0.1 'Set R9 to approach position in Z
Z + R9 (Move to approach position in Z)
#R1 = V4 – V5 'Set R1 to slot bottom
#R2 = V7/2 'Set R2 to 1/2 of feedrate
G01 Z + R1 F + R2 (Plunge into slot at 1/2 feedrate)
#R1 = V1 + V3 'Set R1 to right side X position
#R2 = V7 'Set R2 to feedrate
X + R1 F + R2 (Mill slot)
G00 Z + R9 (Rapid out of slot)
M99 (End of macro program)
```

TAPPING A HOLE ON A TURNING CENTER

If your turning center does not have a tapping cycle, you can easily create one. With this example, we assume that the hole has already been drilled and is ready for tapping. Additionally, we assume that the command used for synchronized motion (as is needed when thread chasing) is G32. With most controls, the G32 command will disable feedrate override, forcing the programmed feedrate to be used. The only limitation of this tapping cycle over control manufacturer-based tapping cycles is that the single block function must be turned off for this cycle to work (even this limitation can be overcome with some versions of parametric programming).

Fanuc's custom macro B format

Figure 2-12 (page 80) shows the letter address arguments that we will use in this example.

Here is a portion of a main program that uses the user-created canned cycle for tapping. Note that the custom macro program will include the rapid approach movement up to the workpiece.

```
O0011 (Main program)
.
. (Facing and drilling operations)
.
N090 T0303 (Index turret to tap)
N095 G97 S500 M03 (Start spindle)
N100 G65 P1011 R0.2 Z0.75 F0.0625 (Tap hole)
```

```
N105 G00 X6.0 Z5.0 (Rapid to tool change position)
.
.
.
```

The custom macro B program:

```
O1011 (Custom macro for tapping)
G00 X0 Z#18 (Rapid into position)
G32 Z – #26 F#9 (Feed into hole)
M04 (Reverse spindle direction)
G32 Z#18 (Feed out of hole)
M99 (End of custom macro)
```

Note that the spindle reversal at the hole bottom is handled differently among machine tool builders based on how M codes are executed. Before running this cycle in an actual workpiece, you would have to confirm that the spindle reversal occurs in conjunction with the movement out of the hole.

Okuma's user task 2 format

Figure 2-13 (page 81) shows the local variable arguments for this example.

The following portion of a main program uses the user-created canned cycle for tapping. Note that the user task program includes the rapid approach movement up to the workpiece.

```
O0011 (Main program)
.
. (Facing and drilling operations)
.
N090 T0303 (Index turret to tap)
N095 G97 S500 M03 (Start spindle)
N100 CALL O1011 RPT = 0.2 DPTH = 0.75 PICH = 0.0625 (Tap hole)
N105 G00 X6.0 Z5.0 (Rapid to tool-change position)
.
.
.
```

The user task 2 program:

```
O1011 (Custom macro for tapping)
G00 X0 Z = RPT (Rapid into position)
G32 Z = [– DPTH] F = PICH (Feed into hole)
M04 (Reverse spindle direction)
```

```
G32 Z = RPT (Feed out of hole)
RTS (End of user task program)
```

Note in this example, too, that the spindle reversal at the hole bottom is handled differently among machine tool builders based on how M codes are executed. Before running this cycle in an actual workpiece, you would have to confirm that the spindle reversal occurs in conjunction with the movement out of the hole.

Fadal's macro format

Figure 2-14 (page 82) shows the macro V-series variable arguments that we use for this example.

This is a part of a user-created canned-cycle main program for tapping. Note that this format includes the rapid approach movement up to the workpiece.

```
O0011 (Main program)
.
. (Facing and drilling operations)
.
N090 T0303 (Index turret to tap)
N095 G97 S500 M03 (Start spindle)
N100 #V1 = 0.25 (Set V1 to rapid approach position)
N105 #V2 = 0.75 (Set V2 to tap depth)
N110 #V3 = 0.0625 (Set V3 to thread pitch)
N115 M98 P1011 (Tap hole)
N120 G00 X6.0 Z5.0 (Rapid to tool change position)
.
.
.
```

The macro program:

```
O1011 (Macro program for tapping)
#R9 = V1 'Set R9 to V1
G00 X0 Z + R9 (Rapid into position)
#R1 = V2 'Set R1 to V2
#R2 = V3 'Set R2 to V3
G32 Z – R1 F + R2 (Feed into hole)
M04 (Reverse spindle direction)
G32 Z + R9 (Feed out of hole)
M99 (End of macro)
```

Again, before running this cycle in an actual workpiece, confirm that the spindle reversal occurs in conjunction with the movement out of the hole.

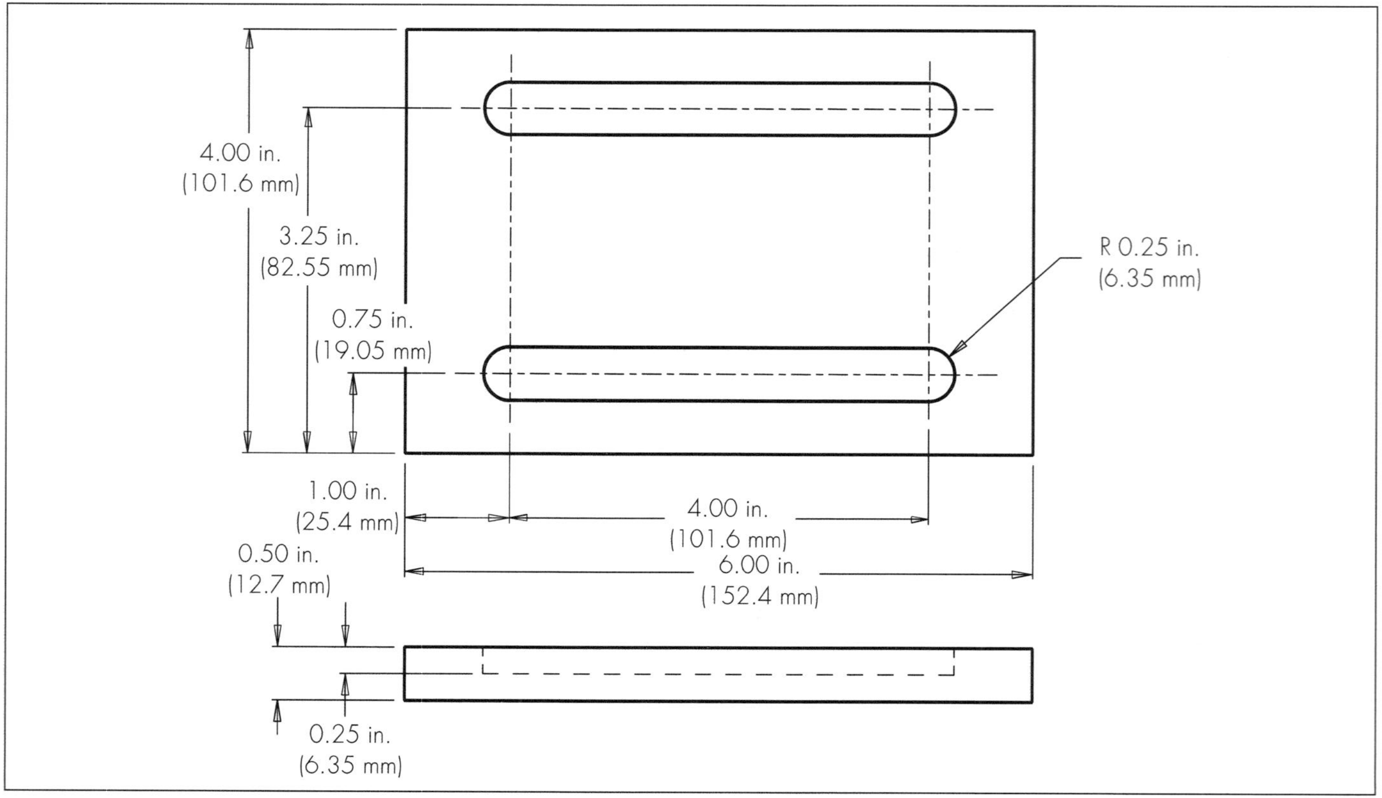

Figure 2-8. Milling a straight slot along the X axis. The tool will plunge to depth at the left side of the slot and mill to the right. The user-created canned cycle for this operation requires that the parametric program be called twice.

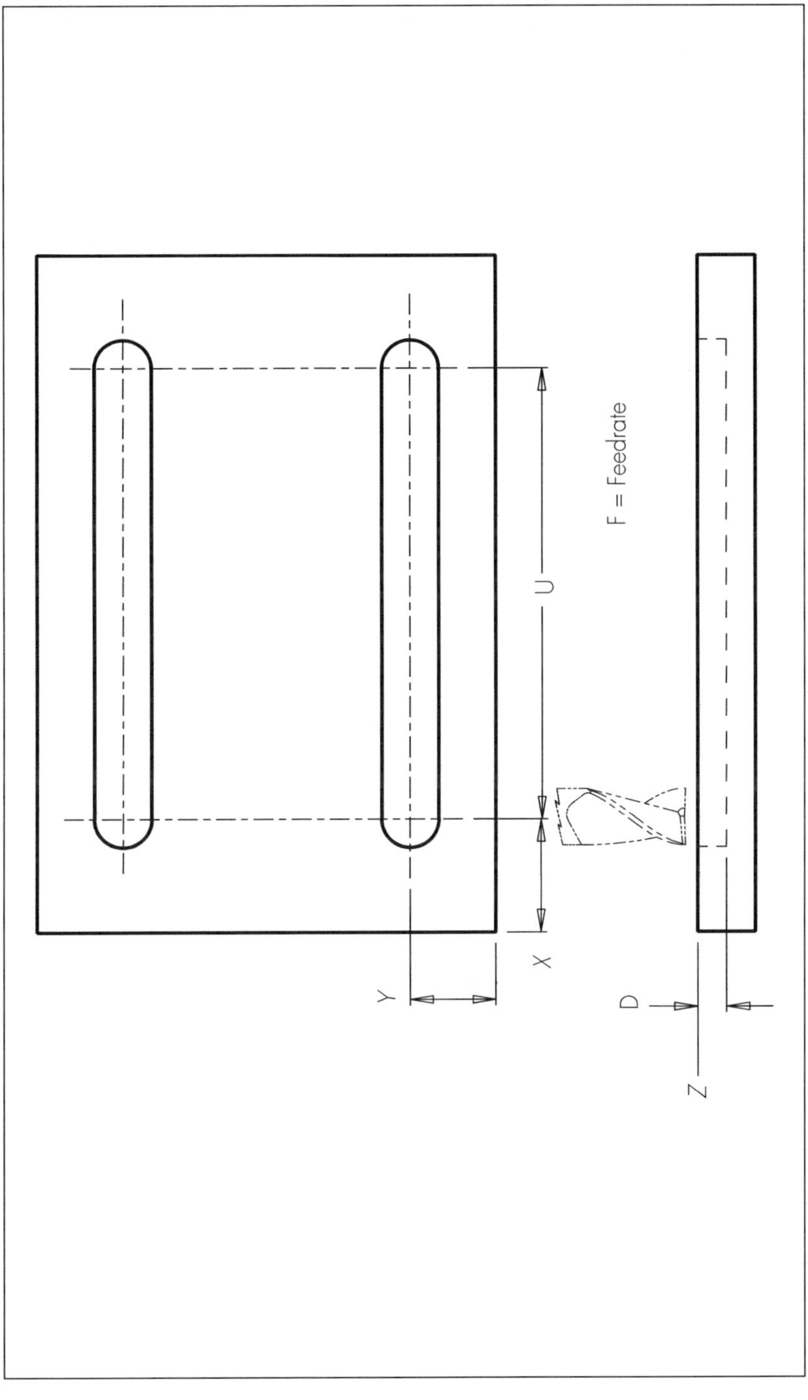

Figure 2-9. Letter address arguments are used with the custom macro B version of the slot-milling application. Most of these argument assignments in this version of parametric programming are initialisms of the plain-language words they represent.

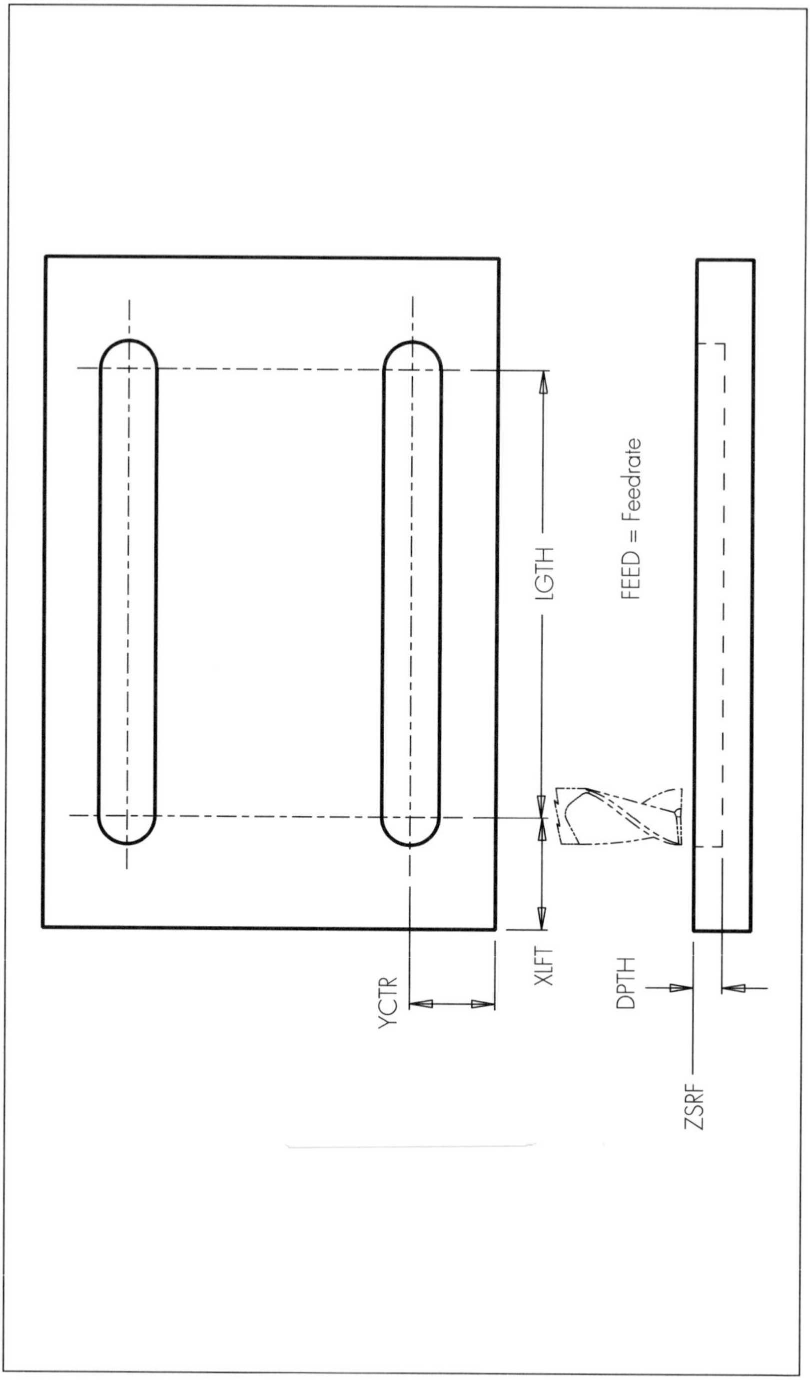

Figure 2-10. The user task 2 version of the slot-milling application is driven by local variable arguments. Again in this version, the argument-assignment codes are easy to understand.

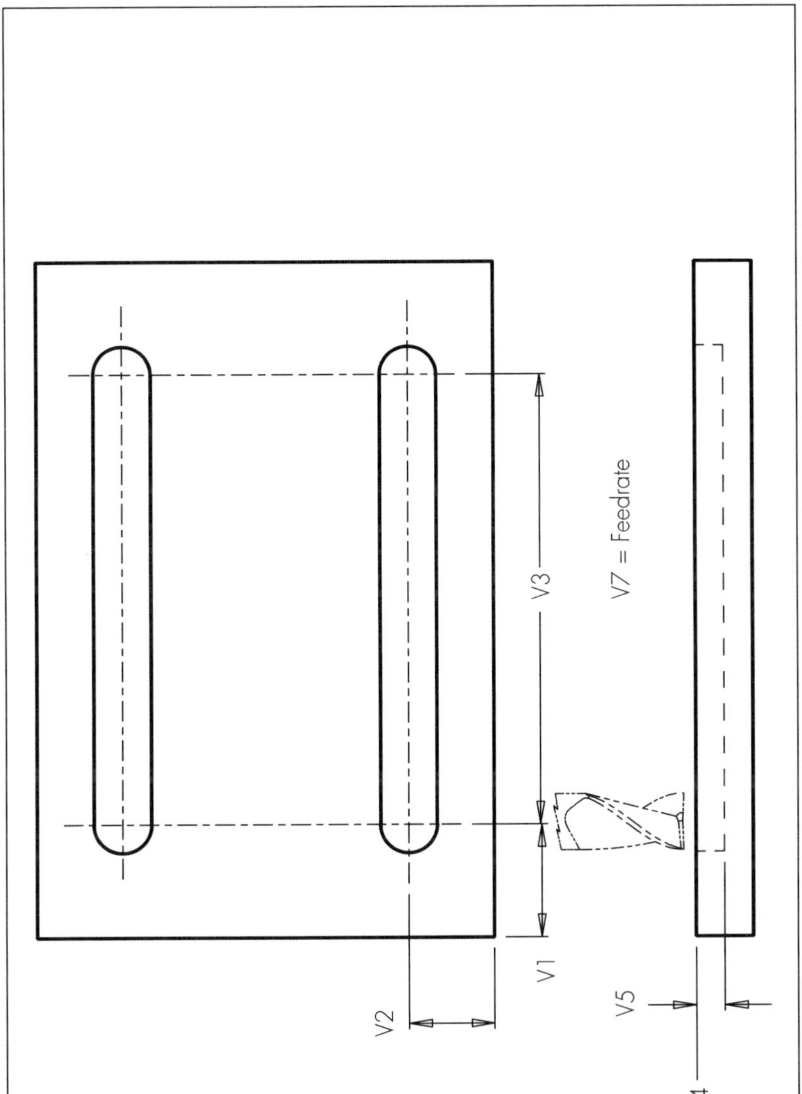

Figure 2-11. V-series variable arguments are used with Fadal's macro version of the slot-milling application. Though simple and consistent, these argument codes require thorough documentation to avoid confusion by readers of the program.

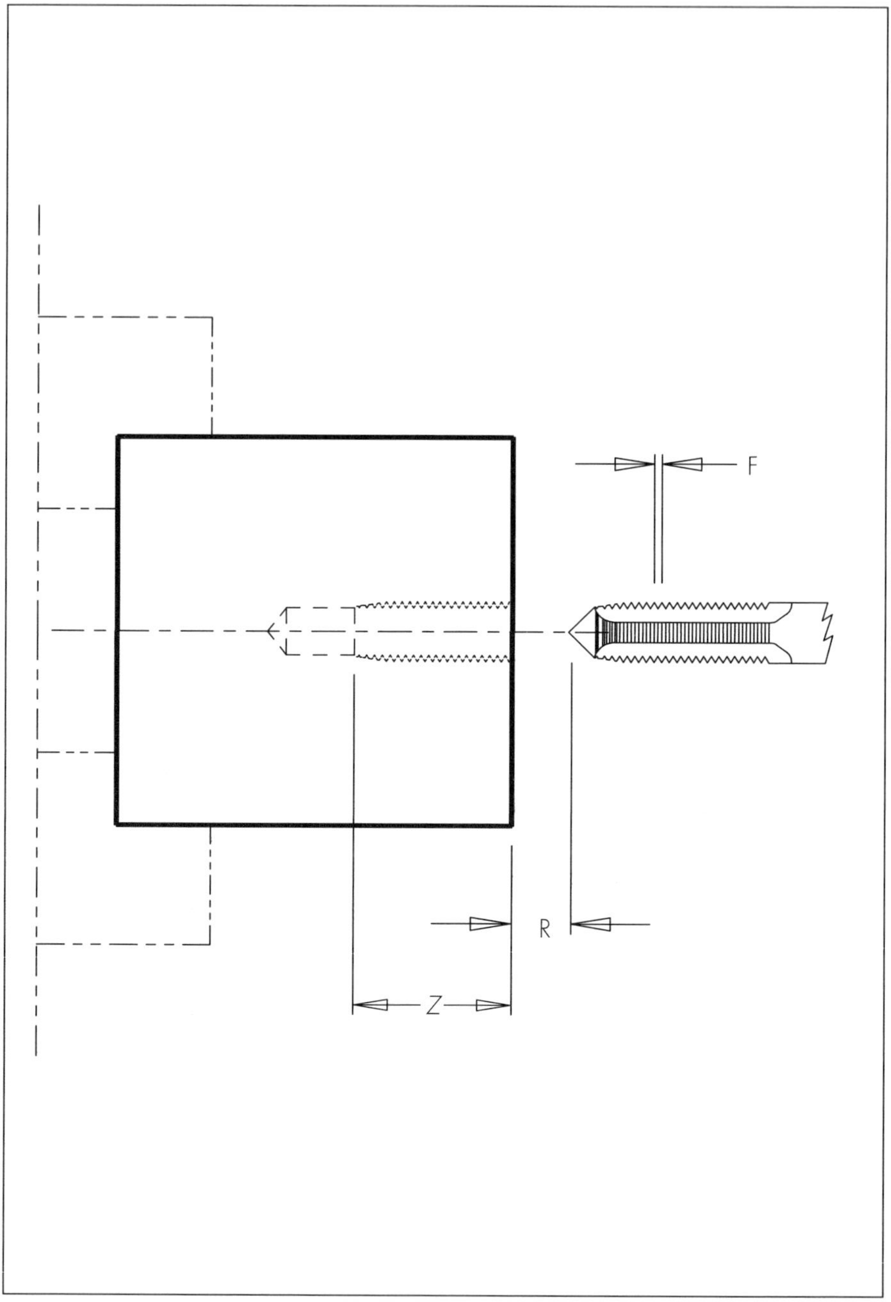

Figure 2-12. For tapping, letter address arguments direct the execution of the example program using Fanuc's custom macro B.

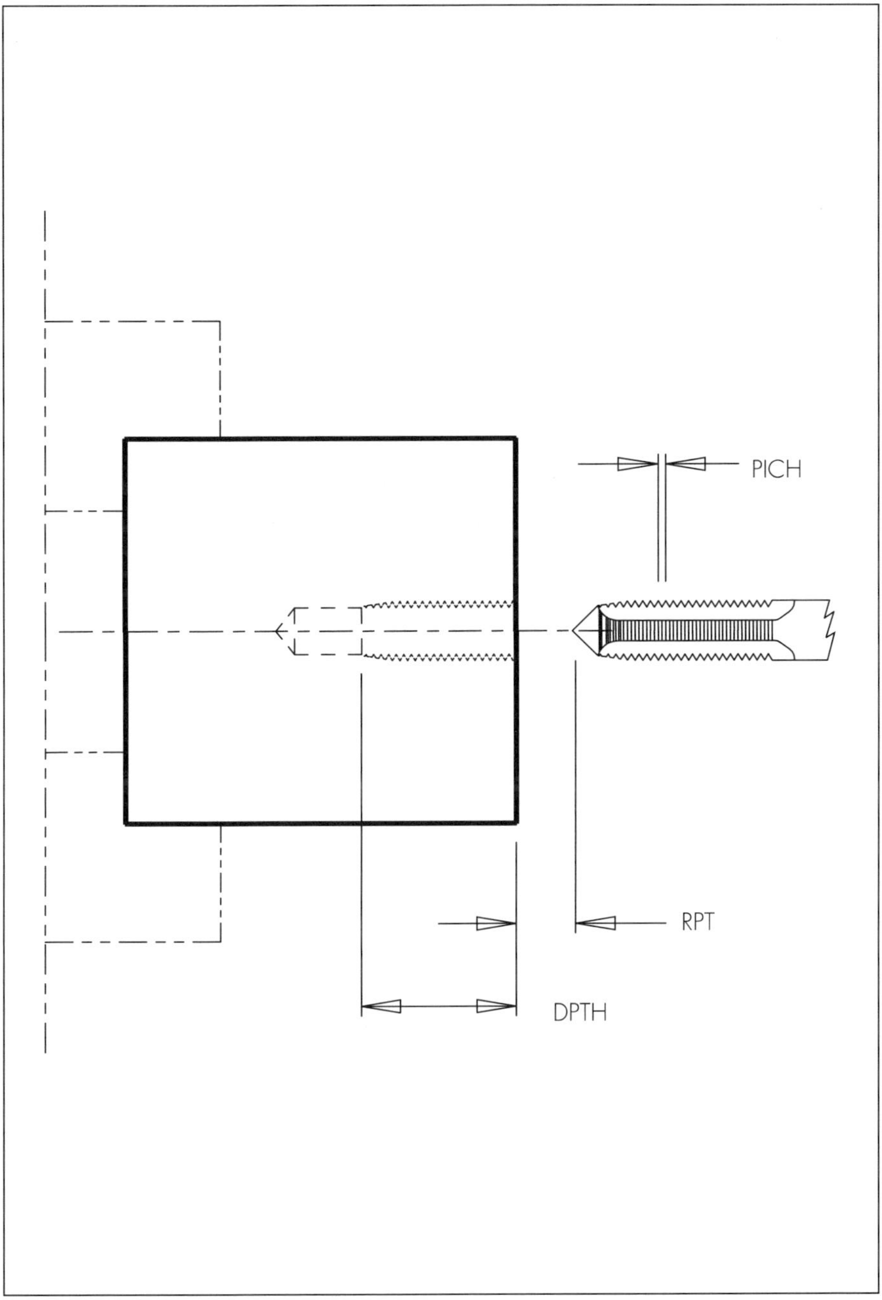

Figure 2-13. To execute the same tapping operation portrayed in Figure 2-12 using Okuma's user task 2, local variable arguments reflect even more closely their true meanings.

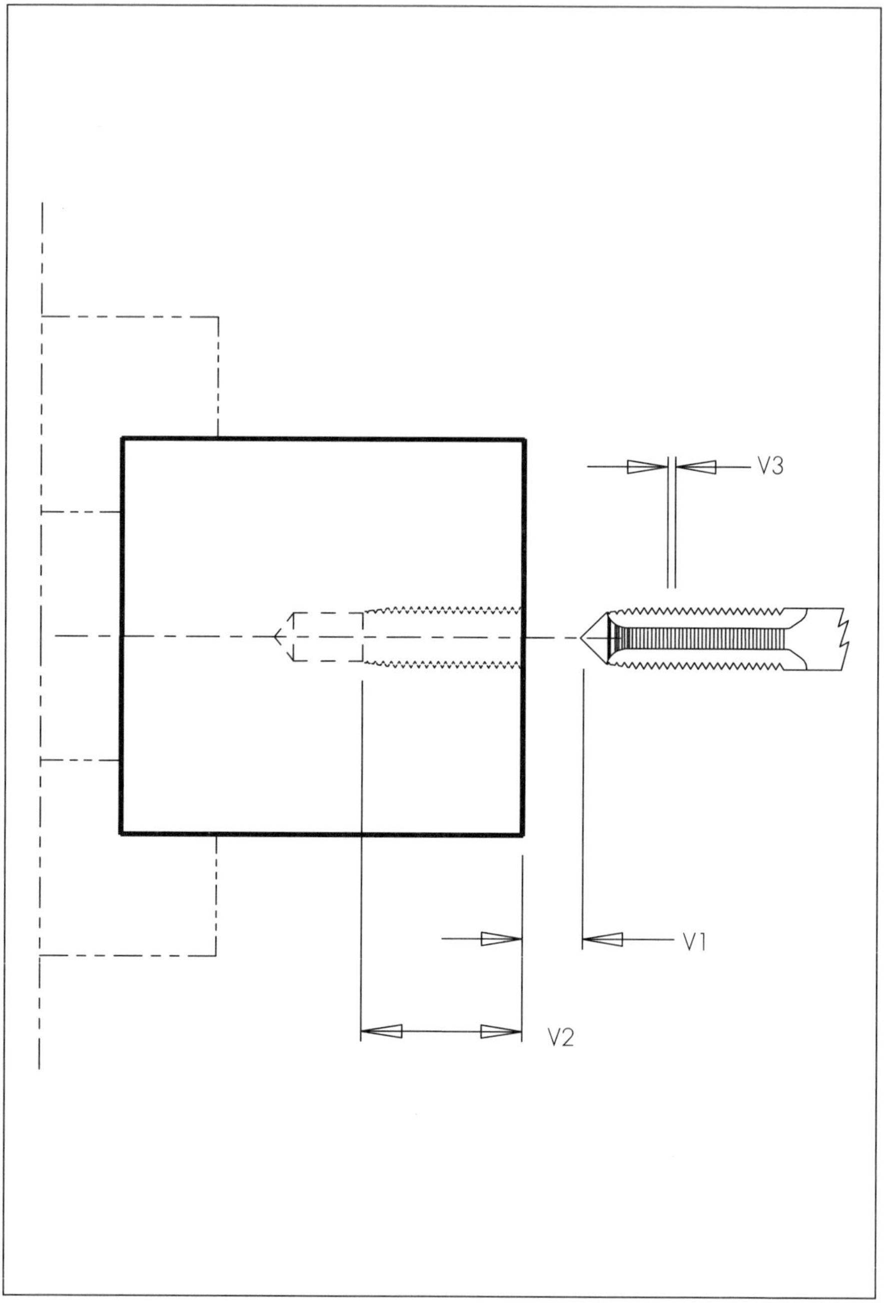

Figure 2-14. The tapping-related V-series variable arguments needed for the example program using Fadal's macro must be well documented.

Chapter 3

Arithmetic Capabilities

All versions of parametric programming allow arithmetic functions to be performed from within the parametric program. In fact, just about any thing that can be done on a scientific calculator can be done with most versions of parametric programming. This opens the door to handling a wide range of calculation-based applications.

In Chapter 2 some basic arithmetic functions (equality, addition, subtraction, multiplication, and division) were introduced as part of our discussion of variable techniques. Almost all versions of parametric programming allow you to go much farther. Most even allow you to combine arithmetic operations in an expression, minimizing the number of calculation commands needed in your parametric program. If you have computer programming experience, you will likely notice that most versions of parametric programming even follow the same priority rules (order of arithmetic operations) found in any computer programming language.

While the reasoning behind basic arithmetic operations like addition, subtraction, multiplication, and division are quite easy to visualize and understand, there are certain arithmetic functions that may seem more obscure, and you may question the reason why these functions are available. In this chapter, we not only introduce each of these functions, but we identify those instances when each function is required.

BASIC FUNCTIONS

As stated, all versions of parametric programming allow addition, subtraction, multiplication, and division. All will use the equal sign (=) for equality, the plus sign (+) for addition, the minus sign (–) for subtraction, the asterisk (*) for multiplication, and the slash code (/) for division. The use of these functions should be quite obvious, and several applications for their use are shown in Chapter 2.

COMBINING OPERATIONS TO FORM AN EXPRESSION

As you begin combining these basic functions to form expressions, you must be aware of the *priority* by which operations are executed within an expression. Misunderstandings will lead to incorrect calculations.

Consider this command, given in custom macro B format.

#101 = 3 + 4 * 3

We are setting common variable #101 to three plus four times three. What is the result?

You *must know* that multiplication has a higher priority than addition. First, the control will multiply 4 times 3 (Result: 12). Then the control will add 3 to 12 (Result: 15). If, on the other hand, the control first added 3 to 4 (Result: 7) and then multiplied it times 3, the result would be 21. Given the fact that calculation results are commonly used to determine axis position, feedrate, spindle speed, and other CNC functions, misunderstandings about the way arithmetic operations are performed can lead to serious problems.

All versions of parametric programming allow you to *force* the control to execute an expression in your desired manner. Custom macro B and user task 2 use brackets with which to force your desired execution order. Fadal's macro uses parentheses for the same purpose. Consider the following identical commands written in each format:

#101 = [3 + 4] * 3 (Fanuc's custom macro B)
TDIA = [3 + 4] * 3 (Okuma's user task 2)
#V1 = (3 + 4) * 3 'Fadal's macro

Note that the brackets (or parentheses) force the control to do the addition *before* the multiplication. If you are in doubt about the order of arithmetic operations, you can simply include the brackets or parentheses, regardless of whether they are needed. Consider these commands:

#101 = 3 + [4 * 3] (Fanuc's custom macro B)
TDIA = 3 + [4 * 3] (Okuma's user task 2)
#V1 = 3 + (4 * 3) 'Fadal's macro

In each of these cases, we are simply forcing the control to perform the arithmetic expression in exactly the same way it would have done without the brackets or parentheses.

You must keep in mind, however, that most versions of parametric programming have a limitation related to how deep you can *nest* operations with brackets or parentheses. By nesting, we mean placing brackets within brackets. While Fadal's macro has virtually no limitation in this regard, some versions of parametric programming are rather limited in terms of how deep you can nest brackets. Fanuc's custom macro B, for example, only allows you to nest operations in brackets four deep. If you do not fully understand the priority of arithmetic operations and include brackets simply because you are not sure which operation will be executed first, you will likely waste nest levels. Consider this command given in custom macro B format:

#101 = [3 + [[4 * 3] + 2]]

While this is an acceptable custom macro command, it wastes bracket nest levels. The brackets surrounding 4 * 3 are in the deepest nest. The control would have done this operation first anyway. The brackets surrounding [4 * 3] + 2 make the second nest level, and the brackets around the entire expression make the third nest level. This expression is within one nest level of the allowed number (four) in custom macro B. For this expression, the control would come up with the same answer (17) *with or without the brackets*. To keep from wasting nest levels, you must understand the rules of arithmetic priority. (The complete set of priority rules will be shown after we introduce all arithmetic functions.)

REFERENCING A PREVIOUSLY ASSIGNED VARIABLE

While this may seem rather obvious if you read Chapter 2, you can of course, reference variables within arithmetic expressions. Consider these commands given in custom macro B format:

N1 #101 = 1
N2 #102 = 5
N3 #103 = #101 + #102

In line N3, the control will add the value of common variable #101 (1) to the value of common variable #102 (5), and assign the result (6) to common variable #103. It will be just as if line N3 reads #103 = 1 + 5.

REFERENCING A VARIABLE'S VALUE WITHIN ITS OWN ASSIGNING COMMAND

It is often necessary to *step* a variable's value, especially in what are called looping applications. This involves utilizing the variable's *current* value in an arithmetic expression that calculates its *new* value. Consider these commands given in user task 2 format.

N015 PASS = 1 (Set pass counter to one)
.
.
N065 PASS = PASS + 1 (Step pass counter by one)

When the control reads line N015, it sets the value of local variable PASS to one. The first time it reads line N065, it steps the current value of PASS by one, setting its new value to two. If it reads line N065 again (as would be common in a *loop*), the value of PASS will become three.

In most loops, not only the loop counter (PASS in our case) will be initialized and stepped. It is quite common for other loop criteria to be initialized and stepped. Consider the application shown in Chapter 1 to mill the right end of a workpiece. Our application assumes that only one milling pass is required. If there is a great deal of rough stock on the workpiece, it may be necessary to make two or more passes. In this potential loop application, not only will the pass number variable be required, but a current *X* position variable will be needed as well. Consider this series of commands given in user task 2 format.

```
N005 PASS = 1 (Set current pass number)
N010 DEEP = 0.125 (Set depth of cut per pass)
N015 CURX = 5.75 (Set first milling pass X position)
.
.
.
.
.
N065 PASS = PASS + 1 (Step PASS counter)
N070 CURX = CURX – DEEP (Step current X position for pass)
.
.
.
```

It is helpful also to reference a variable's value in an expression that calculates its new value when an argument included in the call statement does not match the method by which the argument is required in the parametric program. While we may be getting a little ahead of ourselves, the parametric programmer should always try to make the arguments being passed to the parametric program match the method by which the print is dimensioned. This keeps the person using the parametric program from having to perform calculations to determine the values of arguments, making it easy to use the parametric program.

In the right-side milling application shown in Chapter 1, for example, the milling cutter's *diameter* is being passed to the parametric program. Yet in the parametric program, the cutter's *radius* is required for motion commands. Consider this command given in Fadal's macro format.

```
N005 #V7 = V7/2 'Divide the current value of V7 by two
```

Once this command is executed one time (at the beginning of the macro program), the current value of V-series variable V7 will represent the cutter radius.

ADVANCED ARITHMETIC FUNCTIONS

Most versions of parametric programming allow much more than simple equality, addition, subtraction, multiplication, and division. Recognizing that more advanced arithmetic functions are needed to effectively employ parametric programming, most manufacturers have provided higher mathematical capabilities, ranging from trigonometry to square root.

TRIGONOMETRY FUNCTIONS

While versions of parametric programming vary with regard to which specific trigonometry functions they support, most allow enough to handle even the most complex trigonometry problems. At the very least, common functions like sine, cosine, tangent, and arc tangent are available.

Trigonometry is required (even without parametric programming), for instance, with bolt-hole circles. Most bolt-hole circles are dimensioned with techniques similar to those shown in Figure 3-1.

These dimensioning techniques require trigonometry to calculate hole center positions. Figure 3-2 shows the arithmetic expressions needed for calculating *X* and *Y* positions relative to the center of the bolt-hole circle.

For each of the three versions of parametric programming discussed in this text the expressions are stated as follows:

In Fanuc's custom macro B (Angle A is represented by local variable #1, radius R by #18):

#100 = COS[#1] * #18 (*X* position of hole)
#101 = SIN[#1] * #18 (*Y* position of hole)

In Okuma's user task 2 (Angle A is represented by local variable AGL1, radius R by RAD1):

XP1 = COS[AGL1] * RAD1 (*X* position of hole)
YP1 = SIN[AGL1] * RAD1 (*Y* position of hole)

In Fadal's macro (Angle A is represented by V-series variable V1; radius R is represented by V2):

#R0 = COS(V1) * V2 '*X* position of hole
#R1 = SIN(V1) * V2 '*Y* position of hole

As stated, versions of parametric programming vary with regard to how many trigonometry functions they allow. We show most trigonometry functions, as well as their syntax, for custom macro B, user task 2, and macro.

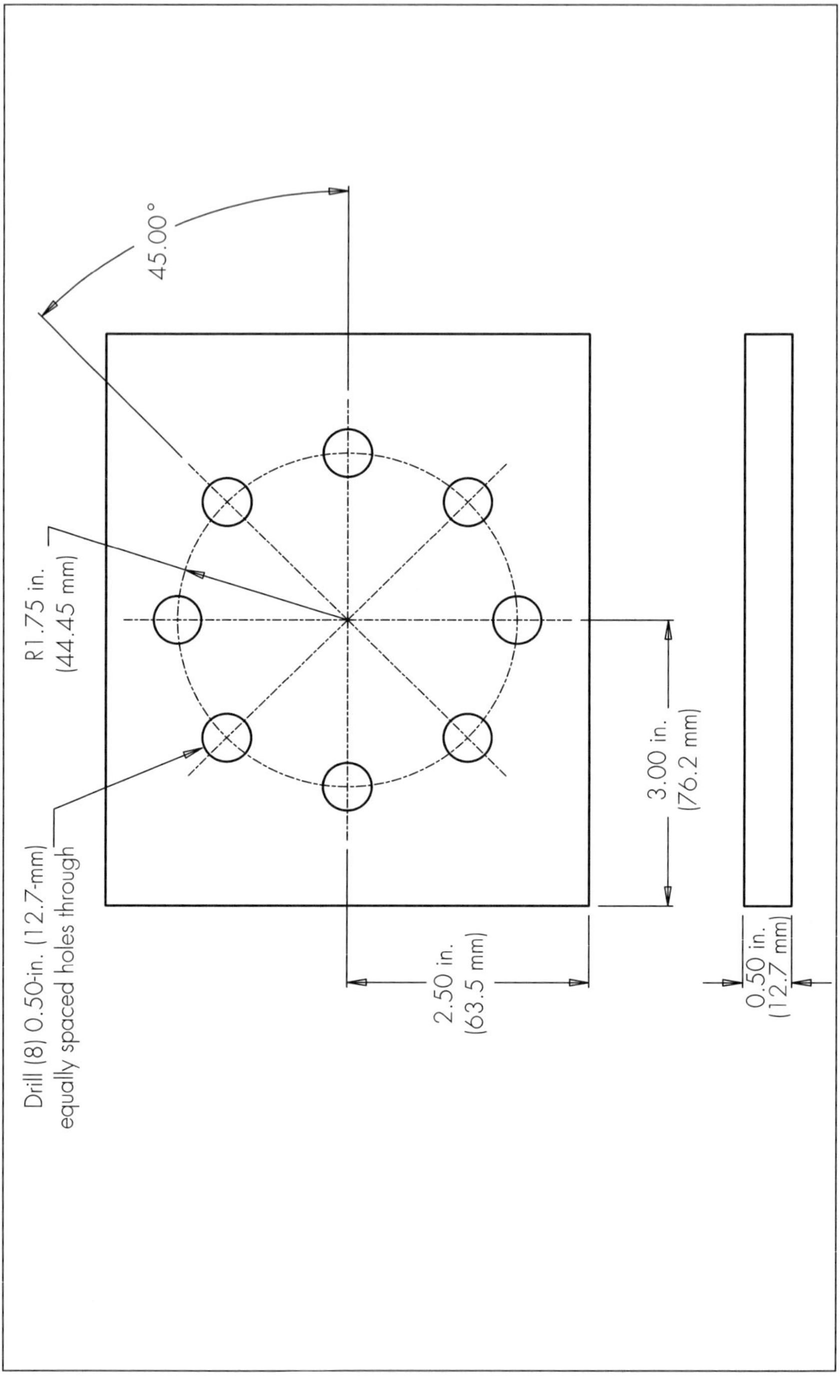

Figure 3-1. Trigonometric functions are essential to calculating certain workpiece positions, such as the bolt-hole circle shown here.

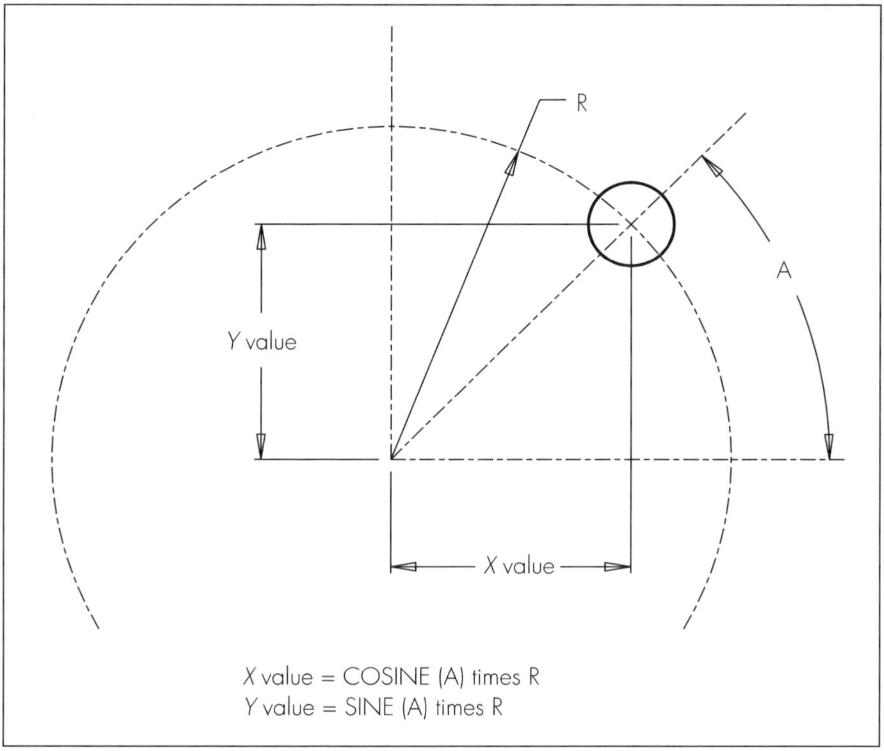

Figure 3-2. Positions of X and Y coordinates are calculated by arithmetic functions.

ROUNDING FUNCTIONS

Almost all versions of parametric programming allow rounding. At the very least, you will be allowed to round a number to the next closest integer. Some versions also allow you to specify the direction of rounding (to the next *higher or lower* integer). Some even allow you to specify rounding to a specified number of decimal places.

While the syntax for specific rounding functions varies, the applications for rounding remain remarkably similar. Whenever you need to convert a value to the closest integer or specified decimal place, rounding will be required.

Applications for rounding

At first glance, you may wonder why rounding functions are required in parametric programming. In truth, rounding functions are *often* required, especially when developing looping applications to determine the number of times the loop must be executed.

In some looping applications, the number of loop executions will be obvious. If developing a bolt-hole circle loop, for example, the number of times the loop must be executed will be equal to the number of holes to be machined. For machining eight holes, the loop must be executed eight times.

With most looping applications, however, the number of loop executions will not be so obvious. For example, if you wish to develop a deep-hole peck drilling cycle, this user-created canned-cycle must, of course, be able to handle any hole and peck depth combination specified.

If the hole depth is specified as 3.0 in. (76.2 mm), and the depth of peck is specified as 1.0 in. (25.4 mm), three passes (loop executions) will be required. In this specific example, it happens that the hole depth is evenly divisible by the depth of each peck. If, on the other hand, the hole depth is 3.045 in. (77.343 mm) and the depth of each peck is specified as 1.0 in. (25.4 mm), how many passes will you want the tool to make? Most standard deep-hole peck drilling canned cycles (normally specified with G83) will make four passes in this case. Three of the passes will be 1.0 in., and the fourth (rather wasteful) pass will be 0.045 in. (1.143 mm). Given this specific example, most programmers will disagree with the way their CNC control behaves. Most would rather have the control make three even passes (of 1.015 in. [25.781 mm]).

To improve upon the standard deep-hole drilling cycle, your user-created canned-cycle for deep-hole peck drilling must take advantage of the rounding function. To determine the number of pecks, you can round the value calculated when dividing the depth of the hole by the depth of each peck. Then you can recalculate the depth of each peck by dividing the hole depth by the number of pecks just calculated. This is how it looks in user task 2 format:

```
HDEP = 3.045 (Hole depth)
PKDP = 1.0 (Depth of each peck)
PASS = ROUND[HDEP/PKDP] (Number of pecks)
PKDP = HDEP/PASS (Recalculate peck depth)
```

Notice that PASS is calculated in this case as a value of three. The last command in this example recalculates the depth of each peck by dividing the hole depth by the number of pecks.

When using the rounding function in this manner, you must remember that the depth of each pass could increase or decrease. In our example, the depth of each pass happened to increase slightly. But consider another specific example.

```
HDEP = 3.75 (Hole depth)
PKDP = 1.0 (Depth of each peck)
PASS = ROUND[HDEP/PKDP] (Number of pecks)
PKDP = HDEP/PASS (Recalculate peck depth)
```

In this case, the number of passes (PASS) will be calculated as four. When 3.75 is divided by four, the depth of each pass (PKDP) will be recalculated as 0.9375 in. (23.8125 mm). In this case, the depth of each peck will be slightly less than the

1.0 in. (25.4 mm) that was specified. Most programmers will agree that this presents no problems when peck drilling. *Whenever the programmer is unconcerned with whether the amount per pass will be increased or decreased, the rounding function can be used.* (The word ROUND is used to specify rounding in user task 2 and custom macro B. Fadal's macro uses the abbreviation RND.)

You must remember, however, that certain looping applications will not be so flexible. Suppose, for example, you are developing a pocket milling parametric program and wish to utilize a loop to determine the number of *Z* passes to be made to get to the bottom of the pocket. You may be using a milling cutter that can go no deeper than 0.25 in. (6.35 mm) per pass. In this case, you will be very concerned not to let the depth of each pass be larger than the specified value. When you must specify a maximum (or minimum) depth of pass, you will need to use other rounding functions.

Custom macro B and user task 2 have two additional rounding functions called FIX and FUP. FIX will round a number down, while FUP will round it up. FUP is used when you wish to specify a *maximum* depth of cut, while FIX is used when you wish to specify a *minimum* depth of cut. Consider these commands given in user task 2 format.

```
PDEP  = 0.625 (Pocket depth)
PSDP  = 0.25 (Maximum depth of each Z pass)
PASS  = FUP[PDEP/PSDP] (Number of Z passes)
PSDP  = PDEP/PASS (Recalculate peck depth)
```

In this example, the depth of each pass (PSDP) will *never* be more than 0.25 in. (6.35 mm), regardless of the values of PDEP and the original PSDP. In this case, PSDP is setting a *maximum* depth of cut. Next consider these commands in user task 2 format.

```
PDEP  = 0.625 (Pocket depth)
PSDP  = 0.25 (Minimum depth of each Z pass)
PASS  = FIX[PDEP/PSDP] (Number of Z passes)
PSDP  = PDEP/PASS (Recalculate peck depth)
```

With these commands, the recalculated PSDP will never be *less* than its original value. The FIX function forces PSDP to be a *minimum* depth of cut.

Since we have not discussed looping in detail yet, a little clarification is perhaps in order regarding the rationale behind ROUND, FIX, and FUP. Just remember that rounding functions are helpful when you are developing parametric programs that require multiple passes, and when you need to have your parametric program calculate the number of passes. ROUND (RND with Fadal's macro) is used if you are not concerned with whether your specified pass depth is increased or decreased. You must round up (FUP with custom macro B and user task 2) when it is necessary to specify a maximum depth of cut. You must round down (FIX with custom macro B and user task 2, INT with macro) when you need to specify a minimum depth of cut.

OTHER COMMON ARITHMETIC FUNCTIONS

Other advanced arithmetic functions that are commonly available in most parametric programming versions help the programmer complete specific tasks.

Square root

This function (specified with SQRT with custom macro B and user task 2, and SQR with Fadal's macro) returns the square root of the specified value or expression. Consider this example given in custom macro B.

```
#100 = 5.0 (Short side of triangle)
#101 = 8.0 (Long side of triangle)
#102 = SQRT[[#100 * #100] + [#101 * #101]] (Hypotenuse of triangle)
```

Note that all higher level functions (including square root) allow arithmetic expressions to be performed internal to the function itself. In this case, everything in brackets will be done prior to the square root function.

Absolute value

This function, commonly specified ABS, returns the positive value of any value or expression specified. It can be very helpful when you are unsure of whether a person using your parametric program will enter a positive or negative value for an argument. Figure 3-3 illustrates a case when this kind of confusion could exist.

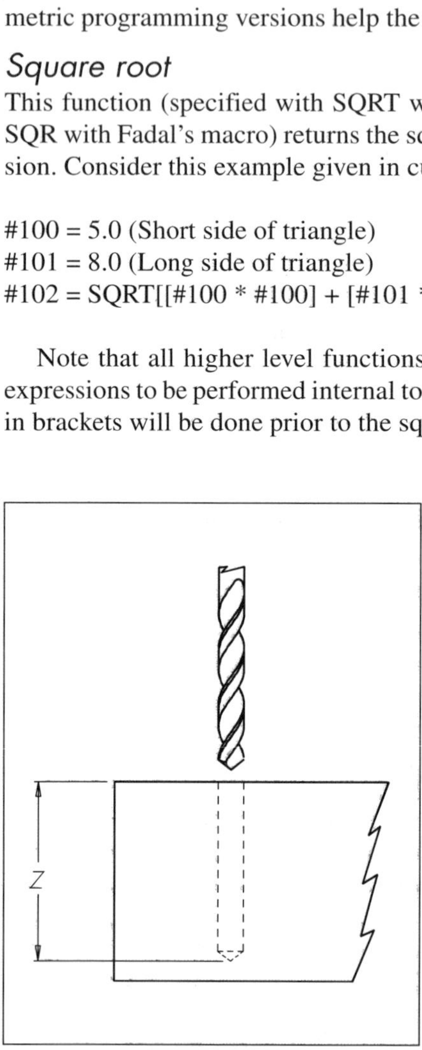

Figure 3-3. The user of your parametric program may not be sure of whether Z must be entered as a positive or negative value.

In this case, a programmer has developed a deep-hole peck-drilling parametric program. Using custom macro B, the programmer elected to use the letter address *Z* as the argument to represent the depth of the hole. The programmer in this application expects the user of the parametric program to pass *Z* as a positive value and the custom macro is written accordingly.

However, the user of this parametric program may not understand that *Z* is supposed to be passed as a positive value. Since this application so closely resembles a true canned cycle (G83), the user may incorrectly pass *Z* as a negative value. To guard against this possibility, the programmer can use the absolute value function. Consider this command written in custom macro B format:

```
G01 Z – [ABS[#26]] F5.0 (Feed to hole bottom)
```

This command will send the tool tip to the correct position regardless of whether *Z* is included in the call statement (G65) as a positive or negative value.

WHAT IF AN ARITHMETIC FUNCTION IS NOT AVAILABLE?

Most versions of parametric programming do not offer much more in the way of arithmetic functions. Yet there may be times when you need a function that is not included in your version of parametric programming. In such cases, it is reassuring to know that virtually all circumstances are covered because *you can develop your own parametric programming arithmetic functions.*

From what we have shown so far, for example, there is no function available to square a number (though some versions of parametric programming do have this function). But just because there is no squaring *function* does not mean you cannot square a number. As long as you know that to square a number simply means to multiply the number by itself, you can easily perform the multiplication operation, as this command given in Fadal's macro format shows.

```
#V2 = V1 * V1 'Perform square function
```

Admittedly, the square function is a simple application. Other arithmetic functions may not be so easy to figure out. Consider, for example, the FUP function (to round to the next higher integer). This function is not available in Fadal's version of parametric programming. Only the INT (for integer) function is available. This function is identical to the FIX function in custom macro B and user task 2. It disregards any decimal portion of the value. The INT result of 9.6598, for example, will be 9.

With the arithmetic tools you have available, if you can figure out how the FUP function works, you *can write your own function* to round to the next higher integer (the FUP function). Consider this short program:

```
O6000 (Program to perform FUP function)
(Pass variable V1 to this program to be rounded up)
N005 #V1 = INT(V1) 'Extract integer value of V1
N010 #V1 = V1 + 1 'Add one to the integer value to get FUP value
N015 M99 'End of macro
```

The person requiring a value to be rounded up will simply set V-series variable V1 to the value to be rounded up and call program O6000. From this point on, the new value of V1 will be the next higher integer.

There are other ways to arrive at the same solution and you may come up with a better way of handling this specific arithmetic problem, but the important point is that you should *always* be able to come up with a way of creating *any* arithmetic function you need. With a little ingenuity, almost anything is possible.

PRIORITY OF ARITHMETIC OPERATIONS

As mentioned earlier, you must understand the order by which arithmetic operations are performed in order to develop expressions that combine operations. And most versions of parametric programming follow the same priority as any computer programming language. Now that we have discussed the arithmetic *functions*, we can show the complete priority of arithmetic *operations*.

1. Operations in brackets (parentheses with Fadal's macro)
2. Higher level functions (SIN, COS, TAN, SQRT, etc.)
3. Multiplication then division
4. Addition then subtraction

While it should not have any bearing on the outcome of the expression, the control will always work through the expression at a given level (1-4) from left to right. If performing level 3 (multiplication then division), for example, it will perform multiplication from the left side of the expression to the right side. Then it will perform division from left to right.

If you are in doubt about operation execution order, you can always use brackets (or parentheses with Fadal's macro) to force the order you need. Remember that when you do so, however, you may be wasting nest levels.

SUMMARY OF ARITHMETIC CAPABILITIES

The material in this section will give you the ability to quickly reference the correct syntax for arithmetic functions in the three versions of parametric programming addressed. We simply list the function type, give its correct name (or symbol), and show a brief example of its use. Each example is concerned only with the single function being discussed (no combined expressions). Of course, you can combine operations as shown earlier if you follow the rules of arithmetic priority.

FANUC'S CUSTOM MACRO B

In each example we set the value of common variable #100.

Basic functions

Equality (=): **#100 = 2.5** (Result: #100 equals 2.5)
Addition (+): **#100 = 2 + 3** (Result: #100 = 5)

Subtraction (–): **#100 = 8 – 3** (Result: #100 = 5)
Multiplication (*): **#100 = 6 * 4** (Result: #100 = 24)
Division (/): **#100 = 12/3** (Result: #100 = 4)

Advanced functions

Sine (SIN): **#100 = SIN[30]** (Result: #100 = 0.5)
Cosine (COS): **#100 = COS[30]** (Result: #100 = 0.8660)
Arc cosine (ACOS): **#100 = ACOS[0.8660]** (Result: #100 = 30)
Tangent (TAN): **#100 = TAN[30]** (Result: #100 = 0.5773)
Arc tangent (ATAN): **#100 = ATAN[0.5][0.8660]** (Result: #100 = 30)

Note about arc tangent function: This function returns an *angle.* The value in the left set of brackets is the side of the triangle opposite the angle, and the value in the right set of brackets is the side adjacent.

Logarithm (LN): **#100 = LN[2]** (Result: #100 = 0.3010)
Square root (SQRT): **#100 = SQRT[9]** (Result: #100 = 3)
Absolute value (ABS): **#100 = ABS[2 – 5]** (Result: #100 = 3)
Rounding (ROUND): **#100 = ROUND[4.3992]** (Result: #100 = 4)
Round down (FIX): **#100 = FIX[3.8038]** (Result: #100 = 3)
Round up (FUP): **#100 = FUP[2.3221]** (Result: #100 = 3)

OKUMA'S USER TASK 2

In each example, we will set the value of local variable EXPL.

Basic functions

Equality (=): **EXPL = 2.5** (Result: EXPL equals 2.5)
Addition (+): **EXPL = 2 + 3** (Result: EXPL = 5)
Subtraction (–): **EXPL = 8 – 3** (Result: EXPL = 5)
Multiplication (*): **EXPL = 6 * 4** (Result: EXPL = 24)
Division (/): **EXPL = 12/3** (Result: EXPL = 4)

Advanced functions

Sine (SIN): **EXPL = SIN[30]** (Result: EXPL = 0.5)
Cosine (COS): **EXPL = COS[30]** (Result: EXPL = 0.8660)
Tangent (TAN): **EXPL = TAN[30]** (Result: EXPL = 0.5773)
Arc tangent (ATAN): **EXPL = ATAN[0.5773]** (Result: EXPL = 30)
Arc tangent form 2 (ATAN2): **EXPL = ATAN2[0.5, 0.8660]** (Result: EXPL = 30)

Note about second arc tangent function: This function also returns an *angle.* The value to the left of the comma is the side of the triangle opposite the angle, while to the right of the comma the value is the side adjacent.

Square root (SQRT): **EXPL = SQRT[9]** (Result: EXPL = 3)
Absolute value (ABS): **EXPL = ABS[2 – 5]** (Result: EXPL = 3)

Rounding (ROUND): **EXPL = ROUND[4.3992]** (Result: EXPL = 4)
Round down (FIX): **EXPL = FIX[3.8038]** (Result: EXPL = 3)
Round up (FUP): **EXPL = FUP[2.3221]** (Result: EXPL = 3)

Rounding to four places in the inch mode and three places in metric:
Round with decimal (DROUND): **EXPL = DROUND[1/12]** (Result: EXPL = 0.0833 [inch mode])
Round down at last decimal position (DFIX): **EXPL = DFIX[1/12]** (Result: EXPL = 0.0833 [inch mode])
Round up at last decimal position (DFUP): **EXPL = DFUP[1/12]** (Result: EXPL = 0.0834 [inch mode])

FADAL'S MACRO

In each example we set the value of V-series variable V1.

Basic functions

Equality (=): **#V1 = 2.5** 'Result: V1 equals 2.5
Addition (+): **#V1 = 2 + 3** 'Result: V1 = 5
Subtraction (–): **#V1 = 8 - 3** 'Result: V1 = 5
Multiplication (*): **#V1 = 6 * 4** 'Result: V1 = 24
Division (/): **#V1 = 12/3** 'Result: V1 = 4

Advanced functions

Sine (SIN): **#V1 = SIN(30)** 'Result: V1 = 0.5
Cosine (COS): **#V1 = COS(30)** 'Result: V1 = 0.8660

There is *no* tangent function in Fadal's macro. However, you can easily calculate the tangent value, as shown in the following example:

Tangent (not available): **#V1 = SIN(30)/COS(30)** 'Result: V1 = 0.5773, the tangent of 30 degrees
Arc tangent (ATN): **#V1 = ATN(0.5773]** 'Result: V1 = 30
Square root (SQR): **#V1 = SQR(9)** 'Result: V1 = 3
Absolute value (ABS): **#V1 = ABS(2 – 5)** 'Result: V1 = 3
Rounding (SET RND and RND):
#SET RND4 'Sets rounding to four places after decimal point
#V1 = RND(1/12) 'Result: V1 = 0.0833
#SET RND0 'Sets rounding for whole number
#V1 = RND(3.2235) 'Result: V1 = 3
Round down (INT): **#V1 = INT(3.8038)** 'Result: V1 = 3
Return the sign (SGN):
#V1 = SGN(10 – 3) 'V1 = +1
#V1 = SGN(3 – 10) 'V1 = –1

EXAMPLE PROGRAMS

With a firm understanding of parametric programming variables and arithmetic, you should be ready to tackle certain applications. While there is still more to understand, variables and arithmetic give you what you need to approach applications that do not require *logic* commands. These applications include many helpful user-created canned cycles that simply drive a cutting tool through a series of movements. At the very least, you should be able to understand these example programs with relative ease, since they show nothing more than what has previously been discussed.

With each example, we first briefly discuss the application. Then we show the specific programs in each version of parametric programming. Note that since these are examples, and we do not present any new information, we minimize our explanations of each parametric program. You may have to study them to fully understand. Also note that there are many ways to develop each application. We show just one way; our intent is to make it easy to understand the programs. With a little effort, you can surely improve the programs we show. (Again, in the interest of reader continuity, we have not interrupted the written programs by placing figures as they are called out; the figures follow the sample programs.)

MILLING AN OUTSIDE RECTANGULAR SHAPE

Figure 3-4 (see page 110) presents the drawing for this example, a rectangular shape with a radius in each corner. Machining this operation requires one milling pass around the shape. Suppose this is just one of more than 100 workpieces your company machines that requires the outside rectangular shape to be milled. While other machining operations are to be performed, there is not enough similarity among other workpieces requiring this milling operation to create a family-of-parts parametric program.

Instead of writing the longhand commands necessary to machine the rectangular shape in more than 100 programs, you elect to develop a user-created canned cycle. Every dimension shown on the drawing varies from one workpiece to another, meaning all dimensions related to the shape will require arguments.

Fanuc's custom macro B

Since only the milling of the rectangular shape is to be machined (all other operations will be handled separately), this is a user-created canned-cycle application. For this reason, we will use the G65 command and pass letter address arguments to the custom macro that machines the rectangular shape. Figure 3-5 (page 111) shows our selection of arguments.

Here is the beginning of a program using the custom macro.

```
O0004 (Program number)
N005 T01 M06 (Place 1-in. [25.4-mm] endmill in spindle)
```

```
N010 G90 G54 S300 M03 T02 (Select absolute mode, coordinate system, start
   spindle, and get next tool ready)
N015 G00 X4.0 Y0 (Rapid up close in XY)
N020 G43 H01 Z0.1 (Rapid up close in Z)
N025 G65 P1004 X4.0 Y3.0 D0.25 W0.25 R0.25 T1.0 F5.0 (Call custom macro
   with arguments set to machine the workpiece shown in Figure 3-4)
N030 G91 G28 Z0 M19 (Return to tool-change position in Z, orient spindle)
N035 M01 (Optional stop)
N040 . . .
.
.
.
```

The custom macro B program is:

```
O1004 (Program number)
#20 = #20/2 (Reset #20 as tool radius instead of diameter)
G00 X[#24 – #23 + #20] Y-[0.1 + #20] (Rapid to XY position)
Z- #7 (Rapid to work surface in Z)
G01 Y[#25 – #23 – #18] F#9 (Mill right side)
G03 X[#24 – #23 – #18] Y[#25 – #23 + #20] R[#18 + #20] (Machine upper right
   radius)
G01 X[#23 + #18] (Mill upper side)
G03 X[#23 – #20] Y[#25 – #23 – #18] R[#18 + #20] (Mill upper left radius)
G01 Y[#23 + #18] (Mill left side)
G03 X[#23 + #18] Y[#23 – #20] R[#18 + #20] (Mill lower left radius)
G01 X[#24 – #23 – #18] (Mill lower side)
G03 X[#24 – #23 + #20] Y[#23 + #18] R[#18 + #20] (Mill lower right radius)
G00 Z0.1 (Rapid to clearance position in Z)
M99 (End of custom macro)
```

Note that this custom macro has no sequence numbers. They are omitted for two reasons: first, since many user-created canned-cycle programs must be kept permanently in memory, we do not want to waste memory space with unneeded sequence numbers; second, we wish to reserve the use of sequence numbers for a special purpose. Custom macro B uses sequence numbers as *statement labels*.

Okuma's user task 2

Again, since only the milling of the rectangular shape is to be done, this is a user-created canned-cycle application. For this reason, we use the CALL command to pass local variable arguments to the user task program that machines the rectangular shape. Figure 3-6 (page 112) shows our selection of arguments.

Here is the beginning of a program using user task 2:

O0004 (Program number)
N005 G15 H01 (Select coordinate system)
N010 T01 M06 (Place 1-in. [25.4-mm] endmill in spindle)
N015 G90 S300 M03 T02 (Select absolute mode, start spindle, and get next tool ready)
N020 G00 X4.0 Y0 (Rapid up close in *XY*)
N025 G56 H01 Z0.1 (Rapid up close in *Z*)
N030 CALL O1004 XWID = 4.0YWID = 3.0 ZDEP = 0.25 STEP = 0.25 CRAD = 0.25 TDIA = 1.0 FEED = 5.0 (Call user task program with arguments set to machine the workpiece shown in Figure 3-4)
N035 . . .
.
.
.

The user task 2 program is:

O1004 (Program name)
TRAD = TDIA/2 (Set TRAD to tool radius)
G00 X = XWID – STEP + TRAD Y = – [0.1 + TRAD] (Rapid to *XY* position)
Z = – [ZDEP] (Rapid to work surface in *Z*)
G01 Y = YWID – STEP – CRAD F = FEED (Mill right side)
G03 X = XWID – STEP – CRAD Y = YWID – STEP + TRAD R = CRAD + TRAD (Machine upper right radius)
G01 X = STEP + CRAD (Mill upper side)
G03 X = STEP – TRAD Y = YWID – STEP – CRAD R = CRAD + TRAD (Mill upper left radius)
G01 Y = STEP + CRAD (Mill left side)
G03 X = STEP + CRAD Y = STEP – TRAD R = CRAD + TRAD (Mill lower left radius)
G01 X = XWID – STEP – CRAD (Mill lower side)
G03 X = XWID – STEP + TRAD Y = STEP + CRAD R = CRAD + TRAD (Mill lower right radius)
G00 Z0.1 (Rapid to clearance position in *Z*)
RTS (End of user task program)

This user task program has no sequence numbers. Many user-created canned-cycle application programs must be kept permanently in memory, and we do not want to waste memory space with unneeded sequence numbers.

Fadal's macro

Since this is a user-created canned-cycle application (milling only the rectangular shape of the workpiece), we set all V-series variable arguments in the main program and then call the macro program. Figure 3-7 (page 113) shows our selection of arguments.

Here is the beginning of a program using macro:

```
O0004 (Program number)
N005 T01 M06 (Place 1-in. [25.4-mm] endmill in spindle)
N010 G90 G54 S300 M03 T02 (Select absolute mode, coordinate system, start
   spindle, and get next tool ready)
N015 G00 X4.0 Y0 (Rapid up close in XY)
N020 G43 H01 Z0.1 (Rapid up close in Z)
N025 #V1 = 4.0 'Assign argument V1
N030 #V2 = 3.0 'Assign argument V2
N035 #V3 = 0.25 'Assign argument V3
N040 #V4 = 0.25 'Assign argument V4
N045 #V5 = 1.0 'Assign argument V5
N050 #V6 = 0.25 'Assign argument V6
N055 #V7 = 5.0 'Assign argument V7
N060 M98 P1004 (Call macro program)
N065 G91 G28 Z0 M19 (Return to tool-change position in Z, orient spindle)
N070 M01 (Optional stop)
N075 . . .
.
.
.
```

The macro program is:

```
O1004 (Program name)
#V5 = V5/2 (Set V5 to tool radius instead of diameter)
#R1 = V1 + V4 + V5 'Calculate X approach position
#R2 = 0.1 + V5 'Calculate Y approach position
G00 X + R1 Y – R2 (Rapid to XY position)
#R1 = V3 'Store depth variable in R1
Z – R1 (Rapid to work surface in Z)
#R1 = V2 – V4 – V6 'Calculate beginning Y position for radius
#R2 = V7 'Store feedrate variable in R2
G01 Y + R1 F + R2 (Mill right side)
#R1 = V1 – V4 – V6 'Calculate X end of upper right radius
#R2 = V2 – V4 + V5 'Calculate Y end of upper right radius
#R3 = V6 + V5 'Calculate tool path radius
G03 X + R1 Y + R2 R + R3 (Machine upper right radius)
```

```
#R1 = V4 + V6 'Calculate X beginning of upper left radius
G01 X + R1 (Mill upper side)
#R1 = V4 – V5 'Calculate X end of upper left radius
#R2 = V2 – V4 – V6 'Calculate Y end of upper left radius
G03 X + R1 Y + R2 R + R3 (Mill upper left radius)
#R1 = V4 + V6 'Calculate Y beginning of lower left radius
G01 Y + R1 (Mill left side)
#R1 = V4 + V6 'Calculate X end of lower left radius
#R2 = V4 – V5 'Calculate Y end of lower left radius
G03 X + R1 Y + R2 R + R3 (Mill lower left radius)
#R1 = V1 – V4 – V6 'Calculate X beginning of lower right radius
G01 X + R1 (Mill lower side)
#R1 = V1 – V4 + V5 'Calculate X end of lower right radius
#R2 = V4 + V6 'Calculate Y end of lower right radius
G03 X + R1 Y + R2 R + R3 (Mill lower right radius)
G00 Z0.1 (Rapid to clearance position in Z)
M99 (End of macro program)
```

Similar to the other versions, this macro program has no sequence numbers. Since many user-created canned-cycle application programs must be kept permanently in memory, it is wise that we not waste memory space with sequence numbers that we do not need.

A ONE-PLUNGE GROOVING CYCLE

Though we have been heavily favoring machining center applications to this point, keep in mind that there are equally as many excellent turning center applications. Many turning centers, for example, do not have an adequate grooving cycle. The following example shows a very helpful user-created one-plunge canned cycle for outside diameter grooving. (By one-plunge, we mean it assumes the tool width is equal to the groove width.) Figure 3-8 (page 114) shows one of more than 300 workpieces a company machines that requires this grooving operation. Notice that all dimensions related to the groove, including the chamfer size, vary from one workpiece to another.

Fanuc's custom macro B

This is a user-created canned-cycle application, so we'll use a G65 command to invoke the custom macro from the main program. Figure 3-9 (page 115) shows the letter address arguments to be included in the G65 command. Note that only the dimensions related to the groove are shown. In this example, the grooving tool width is equal to the groove width, so the groove width dimension does not need an argument assignment.

On the following page is a portion of one main program that calls the custom macro and machines the groove shown in Figure 3-8.

```
O0005 (Main program)
.
.
(Turning operations—getting workpiece ready to groove)
.
.
N150 T0303 (Index to 0.125-in. [3.175-mm] wide grooving tool)
N155 G96 S400 M03 (Select constant surface speed mode, start spindle)
N160 G00 X4.0 Z0 (Rapid up close to workpiece)
N165 G65 P1005 B2.0 S1.625 Z1.0 C0.031 F0.005 (Call custom macro to machine groove)
N170 G00 X6.0 Z5.0 (Return to tool-change position)
N175 M30 (End of program)
```

The custom macro B program is:

```
O1005 (Program number)
G00 X[#2 + 0.2] Z – #26 (Rapid to approach position)
G01 X#19 F#9 (Plunge groove)
G04 P500 (Pause for 1/2 second)
G00 X[#2 + 0.2] (Rapid out of groove)
Z – [#26 + #3] (Rapid to left side chamfer)
G01 X#2 (Feed flush with diameter)
X[#2 – 2 * #3] Z – #26 (Form left side chamfer)
G00 X[#2 + 0.2] (Rapid out of groove)
Z – [#26 – #3] (Rapid to right side chamfer)
G01 X#2 (Feed flush with diameter)
X[#2 – 2 * #3] Z – #26 (Form right side chamfer)
G00 X[#2 + 0.2] (Rapid out of groove)
M99 (End of custom macro)
```

Okuma's user task 2

This is a user-created canned-cycle application, so we'll use the CALL command to invoke the user task program from the main program. Figure 3-10 (page 116) shows the local variable arguments to be included in the CALL command. Note that only the dimensions related to the groove are shown. Since the grooving tool width equals the groove width, the groove width is not needed as an argument.

Here is a portion of one main program that calls the user task program and machines the groove shown in Figure 3-8.

```
O0005 (Main program)
.
.
```

(Turning operations—getting workpiece ready to groove)

.

.

N150 T0303 (Index to 0.125-in. [3.175-mm] wide grooving tool)
N155 G96 S400 M03 (Select constant surface speed mode, start spindle)
N160 G00 X4.0 Z0 (Rapid up close to workpiece)
N165 CALL O1005 BDIA = 2.0 SMLD = 1.625 ZPLC = 1.0 CHAM = 0.031 FEED = 0.005 (Call user task program to machine groove)
N170 G00 X6.0 Z5.0 (Return to tool-change position)
N175 M02 (End of program)

The user task 2 program is:

O1005 (Program number)
G00 X = BDIA + 0.2 Z = – ZPLC (Rapid to approach position)
G01 X = SDIA F = FEED (Plunge groove)
G04 P500 (Pause for 1/2 second)
G00 X = BDIA + 0.2 (Rapid out of groove)
Z = – [ZPLC+CHAM] (Rapid to left side chamfer)
G01 X = BDIA (Feed flush with diameter)
X = BDIA – 2 * CHAM Z = – ZPLC (Form left side chamfer)
G00 X = BDIA + 0.2 (Rapid out of groove)
Z = – [ZPLC – CHAM] (Rapid to right side chamfer)
G01 X = BDIA (Feed flush with diameter)
X = BDIA – 2 * CHAM Z = – ZPLC (Form right side chamfer)
G00 X = BDIA + 0.2 (Rapid out of groove)
RTS (End of user task program)

Fadal's macro program

This, too, is a user-created canned cycle application, so we include V-series variables in the main program and call the macro program with an M98 command. Figure 3-11 (page 117) shows the V-series variable arguments to be included in the main program command. (Again, only the dimensions related to the groove are shown. The grooving tool width equals the groove width, so the groove width is not needed as an argument.)

This is a portion of one main program that calls the macro program and machines the groove shown in Figure 3-8:

O0005 (Main program)

.

.

(Turning operations—getting workpiece ready to groove)

.

.

```
N150 T0303 (Index to 0.125-in. [3.175-mm]-wide grooving tool)
N155 G96 S400 M03 (Select constant surface speed mode, start spindle)
N160 G00 X4.0 Z0 (Rapid up close to workpiece)
N165 #V1 = 2.0 'Assign argument V1
N170 #V2 = 1.625 'Assign argument V2
N175 #V3 = 1.0 'Assign argument V3
N180 #V4 = 0.031 'Assign argument V4
N185 #V5 = 0.005 'Assign argument V5
N190 M98 P1005 (Call macro program to machine groove)
N195 G00 X6.0 Z5.0 (Return to tool-change position)
N200 M02 (End of program)
```

The macro program is:

```
O1005 (Program number)
#R5 = V1 + 0.2 'Calculate clearance position in X
#R6 = V3 'Store Z position in R6
#R7 = V1 'Store big diameter in R7
#R8 = V1 – 2 * V4 'Calculate bottom of chamfer in X
G00 X + R5 Z – R6 (Rapid to approach position)
#R1 = V2 'Store small diameter in R1
#R2 = V5 'Store feedrate in R2
G01 X + R1 F + R2 (Plunge groove)
G04 P500 (Pause for 1/2 second)
G00 X + R5 (Rapid out of groove)
#R1 = V3 + V4 'Calculate Z of left side of chamfer
Z – R1 (Rapid to left side chamfer)
G01 X + R7 (Feed flush with diameter)
X + R8 Z – R6 (Form left side chamfer)
G00 X + R5 (Rapid out of groove)
#R1 = V3 – V4 'Calculate right side of chamfer in Z
Z = – R1 (Rapid to right side chamfer)
#R1 = V1
G01 X + R1 (Feed flush with diameter)
X + R8 Z – R6 (Form right side chamfer)
G00 X + R5 (Rapid out of groove)
M99 (End of macro program)
```

AN INTERNAL THREAD-MILLING CYCLE

This internal thread-milling example utilizes a thread-milling cutter that can machine the entire thickness of the workpiece in one pass around the hole. It drives the milling cutter in a clockwise (conventional-milling) direction from top to bottom to machine the thread. To keep it simple, the approach and retract arc are one quarter of a circle (90° arcs). Figure 3-12 (page 118) shows the workpiece and the milling cutter's tool path.

While it is not the intention of this text to teach thread milling, to fully understand this application you must know that most CNC controls utilize a simple G02 or G03 for helical motion. The control can easily determine the difference between helical motion and circular motion. If a *Z*-axis departure is included in the G02 or G03 command, the control assumes that helical interpolation is being performed. The actual value of the *Z* departure is directly related to the portion of a full circle being machined in XY. For a quarter circle (90°), as is needed during the approach and retract, the *Z* departure must be one quarter of the pitch.

Fanuc's custom macro B

Since this is yet another user-created canned-cycle application, we will use the G65 command in the main program to pass letter address arguments to the thread milling custom macro. Figure 3-13 (page 119) shows the arguments.

Following is a portion of the program that includes a call statement to the thread milling custom macro:

```
O0006 (Program number)
.
.
(Machine hole to get ready for thread milling)
.
.
N150 T04 M06 (Place thread milling cutter in spindle)
N155 G90 G54 S400 M03 T05 (Select absolute mode, coordinate system, start
   spindle, get next tool ready)
N160 G00 X2.0 Y2.0 (Rapid over to center of hole)
N165 G43 H04 Z.1 (Rapid up to workpiece in Z)
N170 G65 P1006 X2.0 Y2.0 Z0.75 Q0.125 T1.0 M3.0 A1.0 F5.0 (Call custom
   macro and mill thread)
N175 G91 G28 Z0 M19 (Return to tool-change position)
N180 M01 (Optional stop)
.
.
.
```

The custom macro B program to mill threads is:

```
O1006 (Program number)
#20 = #20/2 (Set #20 to tool radius instead of diameter)
#101 = – [#26 + 0.1] (Initial Z position)
```

G00 X#24 Y[#25 + #13/2 – #1] (Rapid to center of approach radius, point 1)
Z#101 (Rapid to beginning *Z* position)
#101 = #101 – #17/4 (Calculate *Z* position at end of approach radius)
G01 X[#24 – #1 + #20] F[#9 * 5] (Fast feed to beginning of approach radius, point 2)
G02 X#24 Y[#25 + #13/2 – #20] Z#101 R[#1 – #20] F#9 (Circular approach to beginning of thread, point 3)
#101 = #101 – #17/2 (Calculate *Z* position half way around thread)
G02 Y[#25 – #13/2 + #20] Z#101 R[#13/2 – #20] (Mill half way around thread, point 4)
#101 = #101 – #17/2 (Calculate *Z* position at end of thread)
G02Y[#25 + #13/2 – #20] Z#101 R[#13/2 – #20] (Mill balance of thread, point 5)
#101 = #101 – #17/4 (Calculate *Z* position at end of retract radius)
G02 X[#24 + #1 – #20] Y[#25 + #13/2 – #1] Z#101 R[#1 – #20] (Retract to point 6)
G00 X#24 (Return to center of approach radius)
Z0.1 (Rapid to clearance position above workpiece)
M99 (end of custom macro)

Okuma's user task 2

Since this is yet another user-created canned cycle application, we will use the CALL command in the main program to pass local variable arguments to the thread-milling user task program. Figure 3-14 (page 120) shows the arguments.

Here is a portion of the program that includes a call statement to the thread-milling user task program.

O0006 (Program number)
.
.
(Machine hole to get ready for thread milling)
.
.
N145 G15 H01 (Select coordinate system)
N150 T04 M06 (Place thread-milling cutter in spindle)
N155 G90 S400 M03 T05 (Select absolute mode, start spindle, get next tool ready)
N160 G00 X2.0 Y2.0 (Rapid over to center of hole)
N165 G56 H04 Z.1 (Rapid up to workpiece in *Z*)
N170 CALL O1006 XCTR = 2.0 YCTR = 2.0 THK = 0.75 PICH = 0.125 TDIA = 1.0 MDIA = 3.0 ARAD = 1.0 FEED = 5.0 (Call user task program and mill thread)
.
.
.

The user task 2 program to mill threads is:

```
O1006 (Program number)
TRAD = TDIA/2 (Set TRAD to the radius of the cutter)
CURZ = – [THK + 0.1] (Initial Z position)
G00 X = XCTR Y = YCTR + MAJR/2 – ARAD (Rapid to center of approach
   radius, point 1)
Z = CURZ (Rapid to beginning Z position)
CURZ = CURZ – PICH/4 (Calculate Z position at end of approach radius)
G01 X = XCTR – ARAD + TRAD F = FEED * 5 (Fast feed to beginning of
   approach radius, point 2)
G02 X = XCTR Y = YCTR + MAJR/2 – TRAD Z = CURZ R = ARAD – TRAD
   F = FEED (Circular approach to beginning of thread, point 3)
CURZ = CURZ – PICH/2 (Calculate Z position half way around thread)
G02 Y = YCTR – MAJR/2 + TRAD Z = CURZ R = MAJR/2 – TRAD (Mill half
   way around thread, point 4)
CURZ = CURZ – PICH/2 (Calculate Z position at end of thread)
G02 Y = YCTR + MAJR/2 – TRAD Z = CURZ R = MAJR/2 – TRAD (Mill
   balance of thread, point 5)
CURZ = CURZ – PICH/4 (Calculate Z position at end of retract radius)
G02 X = XCTR + ARAD – TRAD Y = YCTR + MAJR/2 – ARAD Z = CURZ R
   = ARAD – TRAD (Retract to point 6)
G00 X = XCTR (Return to center of approach radius)
Z0.1 (Rapid to clearance position above workpiece)
RTS (end of user task program)
```

Fadal's macro program

This being yet another user-created canned-cycle application, we set the V-series variable arguments in the main program and call the macro with an M98 command. Figure 3-15 (page 121) shows the arguments.

Here is a portion of the program that includes a call to the thread-milling macro program.

```
O0006 (Program number)
.
.
(Machine hole to get ready for thread milling)
.
.
N150 T04 M06 (Place thread-milling cutter in spindle)
N155 G90 G54 S400 M03 T05 (Select absolute mode, coordinate system, start
   spindle, get next tool ready)
N160 G00 X2.0 Y2.0 (Rapid over to center of hole)
```

```
N165 G43 H04 Z.1 (Rapid up to workpiece in Z)
N170 #V1 = 2.0 'Assign argument V1
N175 #V2 = 2.0 'Assign argument V2
N180 #V3 = 0.75 'Assign argument V3
N185 #V4 = 0.125 'Assign argument V4
N190 #V5 = 1.0 'Assign argument V5
N195 #V6 = 3.0 'Assign argument V6
N200 #V7 = 1.0 'Assign argument V7
N205 #V8 = 5.0 'Assign argument V8
N210 M98 P1006 (Call macro program and mill thread)
N215 G91 G28 Z0 M19 (Return to tool-change position)
N220 M01 (Optional stop)
.
.
.
```

The macro program to mill threads is:

```
O1006 (Program number)
#V5 = #V5/2 'Set V5 to the radius of the cutter instead of its diameter
#R9 = V3 + 0.1 'Initial Z position
#R1 = V1 'Set R1 to X center of hole
#R2 = V2 + V6/2 – V7 'Set R2 to Y center of approach radius
G00 X + R1 Y + R2 (Rapid to center of approach radius, point 1)
Z – R9 (Rapid to beginning Z position)
#R9 = R9 + V4/4 (Calculate Z position at end of approach radius)
#R1 = V1 – V7 + V5 'Set R1 to X position approach radius
#R2 = V8 * 5 'Set R2 to fast moving feedrate
G01 X + R1 F + R2 (Fast feed to beginning of approach radius, point 2)
#R1 = V1 'Set R1 to X center of hole
#R2 = V2 + V6/2 – V5 'Set R2 to Y end point of approach radius
#R3 = V7 – V5 'Set R3 to tool path radius for circular motion
#R4 = V8 'Set R4 to milling feedrate
G02 X + R1 Y + R2 Z – R9 R + R3 F + R4 (Circular approach to beginning of
   thread, point 3)
#R9 = R9 + V4/2 'Calculate Z position half way around thread
#R1 = V2 – V6/2 + V5 'Set R1 to Y position half way around thread
#R2 = V6/2 – V5 'Calculate radius for circular move
G02 Y + R1 Z – R9 R + R2 (Mill half way around thread, point 4)
#R9 = R9 + V4/2 'Calculate Z position at end of thread
#R1 = V2 + V6/2 – V5 'Set R1 to Y end of circular move
G02 Y + R1 Z – R9 R + R2 (Mill balance of thread, point 5)
#R9 = R9 + V4/4 'Calculate Z position at end of retract radius
```

```
#R1 = V1 + V7 – V5 'Calculate X end point of retract radius
#R2 = V2 + V6/2 – V7 'Calculate Y end point of retract radius
#R3 = V7 – V5 'Calculate radius of retract
G02 X + R1 Y + R2 Z – R9 R + R3 (Retract to point 6)
#R1 = V1 'Store X center position in R1
G00 X + R1 (Return to center of approach radius)
Z0.1 (Rapid to clearance position above workpiece)
M99 (End of macro program)
```

WARNING ABOUT THE EXAMPLE PROGRAMS

While these examples should work nicely to reinforce your understanding of what has been presented thus far, keep in mind that these programs are not as fail-safe as they could be. We were very careful *not* to use the features of parametric programming that have yet to be introduced. Some of these features will improve safety, ease of use, and, in general, make parametric programming more flexible.

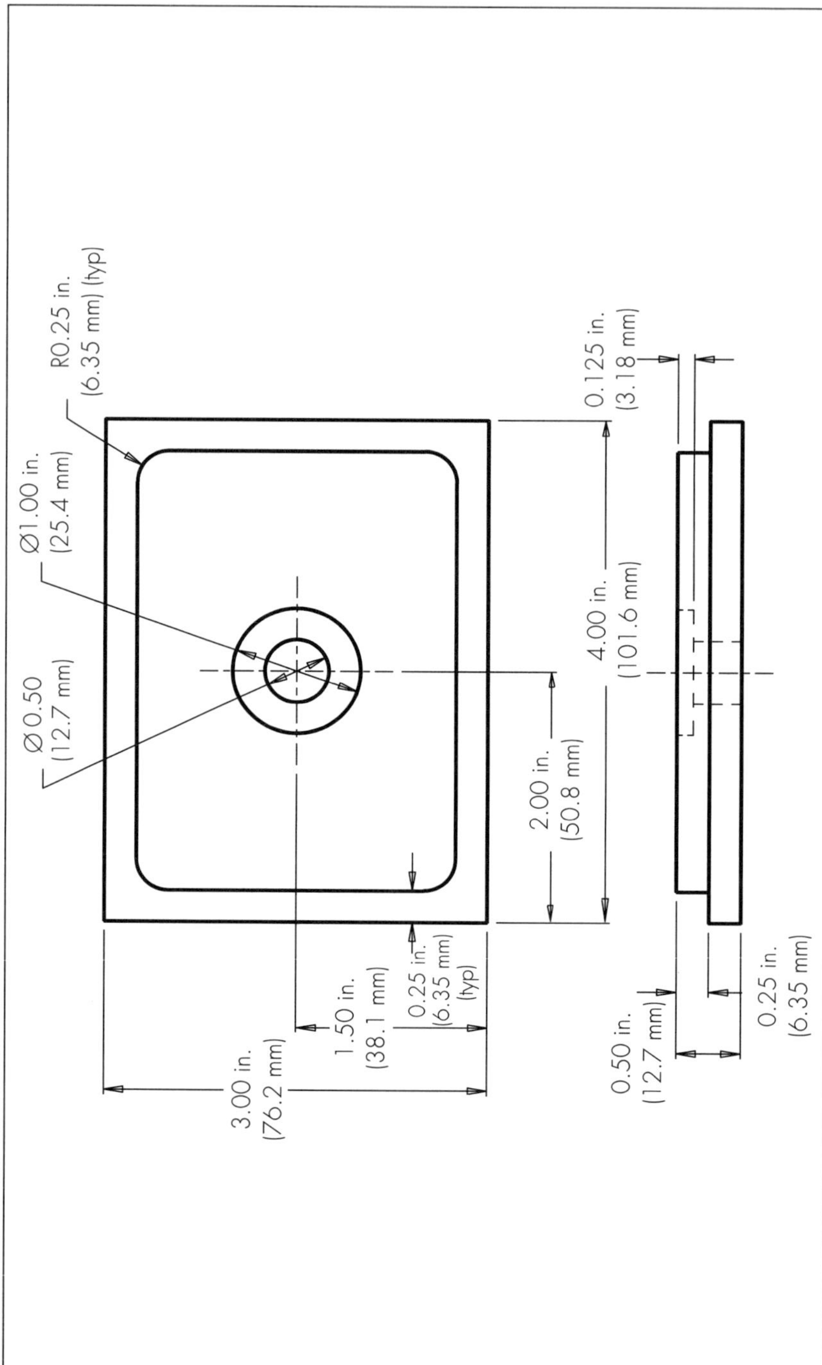

Figure 3-4. This is the drawing the programmer would reference for a program to mill the outside rectangular shape of the example workpiece.

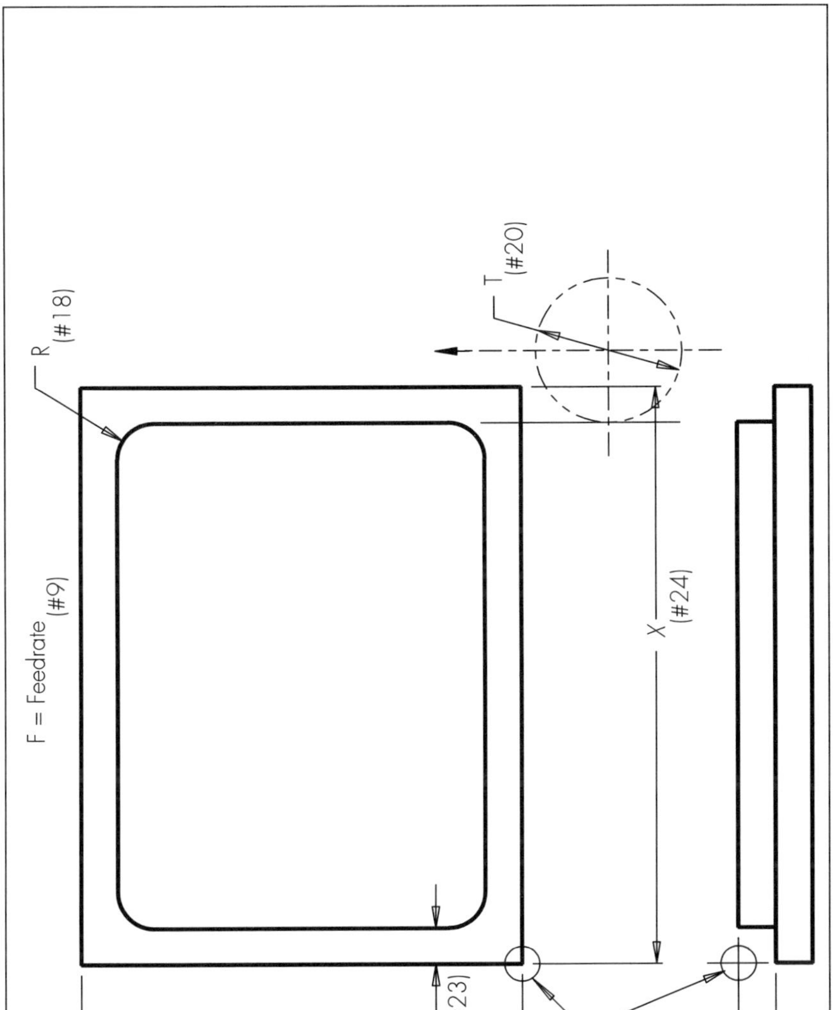

Figure 3-5. Shown here are the letter address arguments to be used with the custom macro B version of this application.

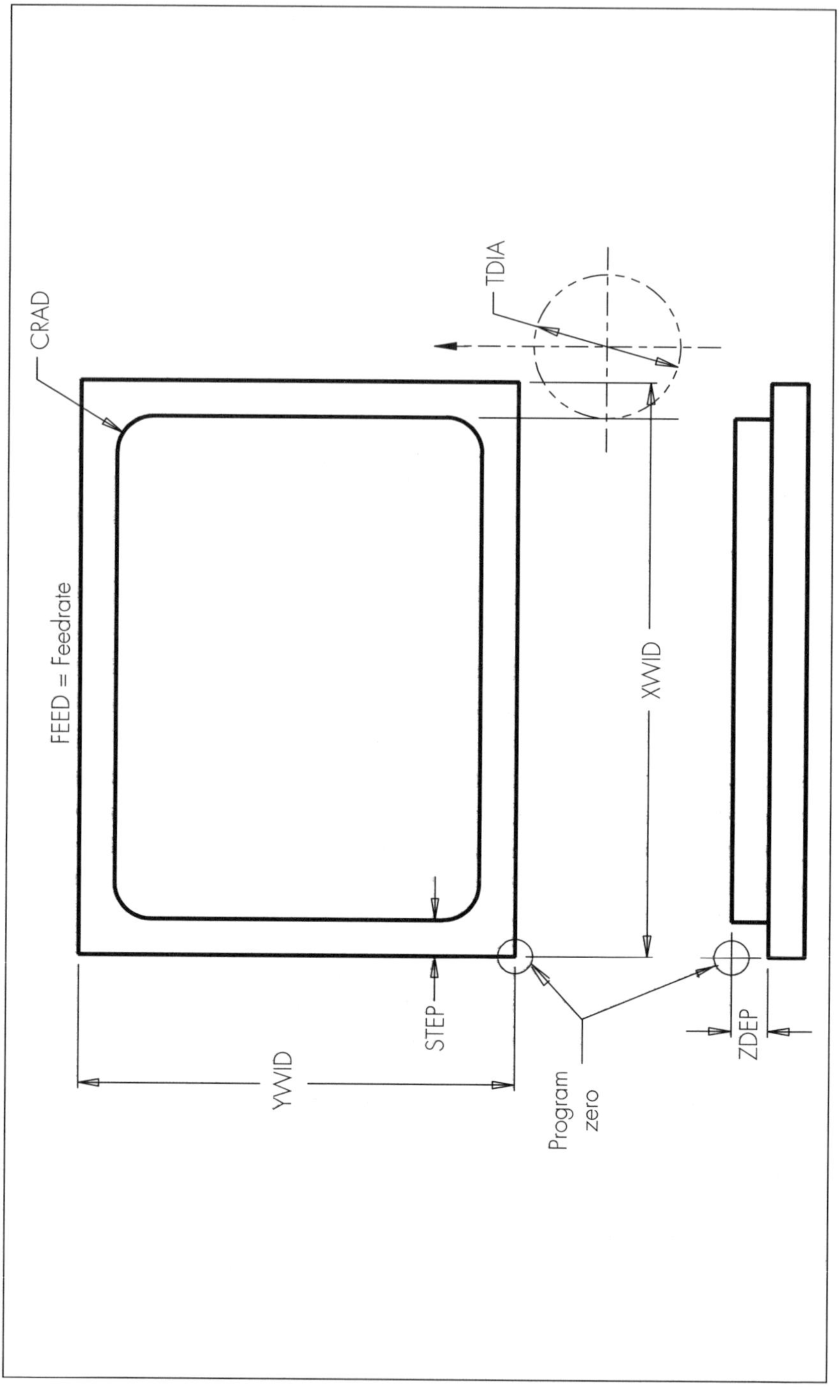

Figure 3-6. In user task 2, local variable arguments are used for this application.

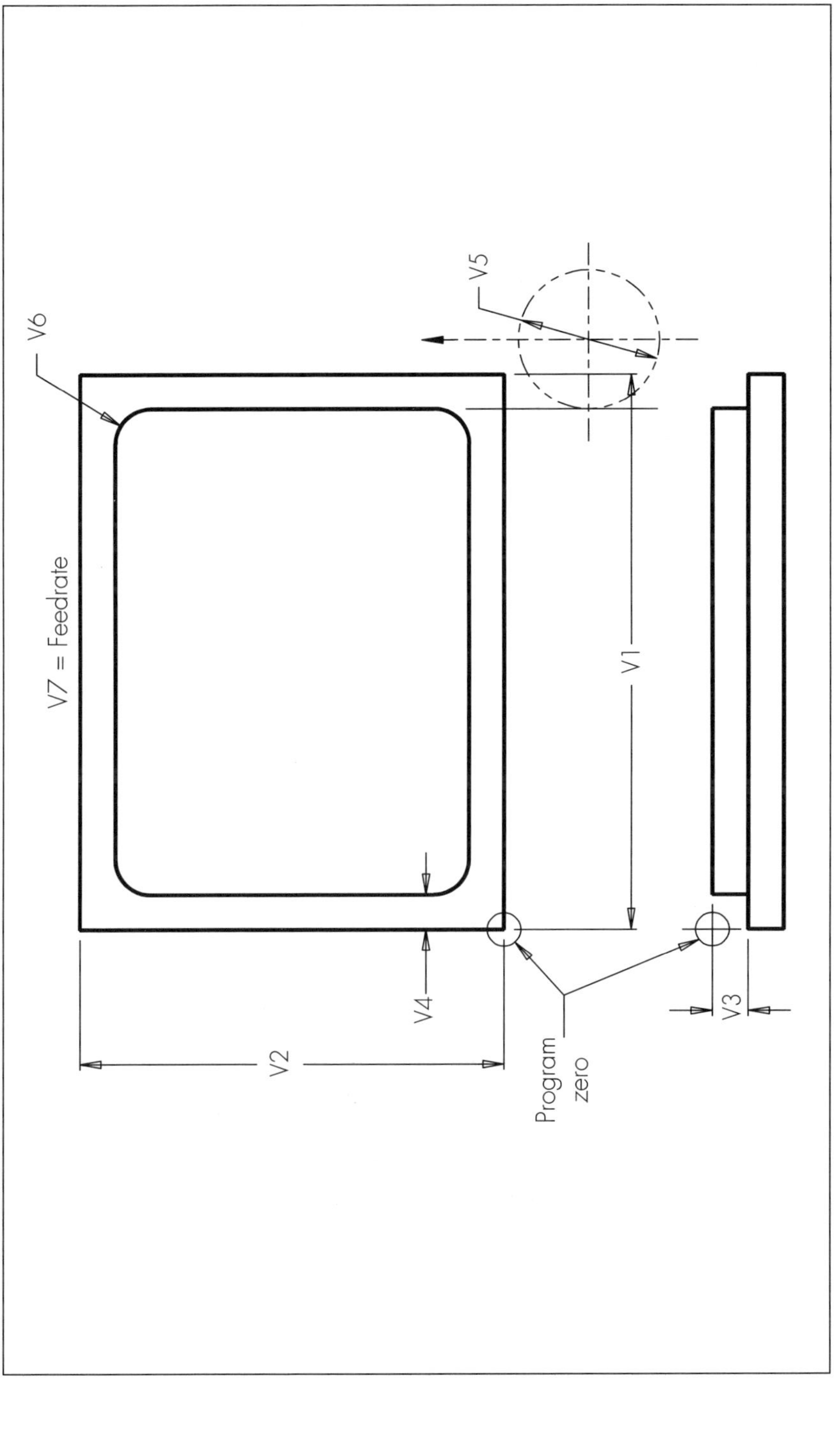

Figure 3-7. Users of Fadal's macro version of this application must specify V-series variable arguments such as shown here.

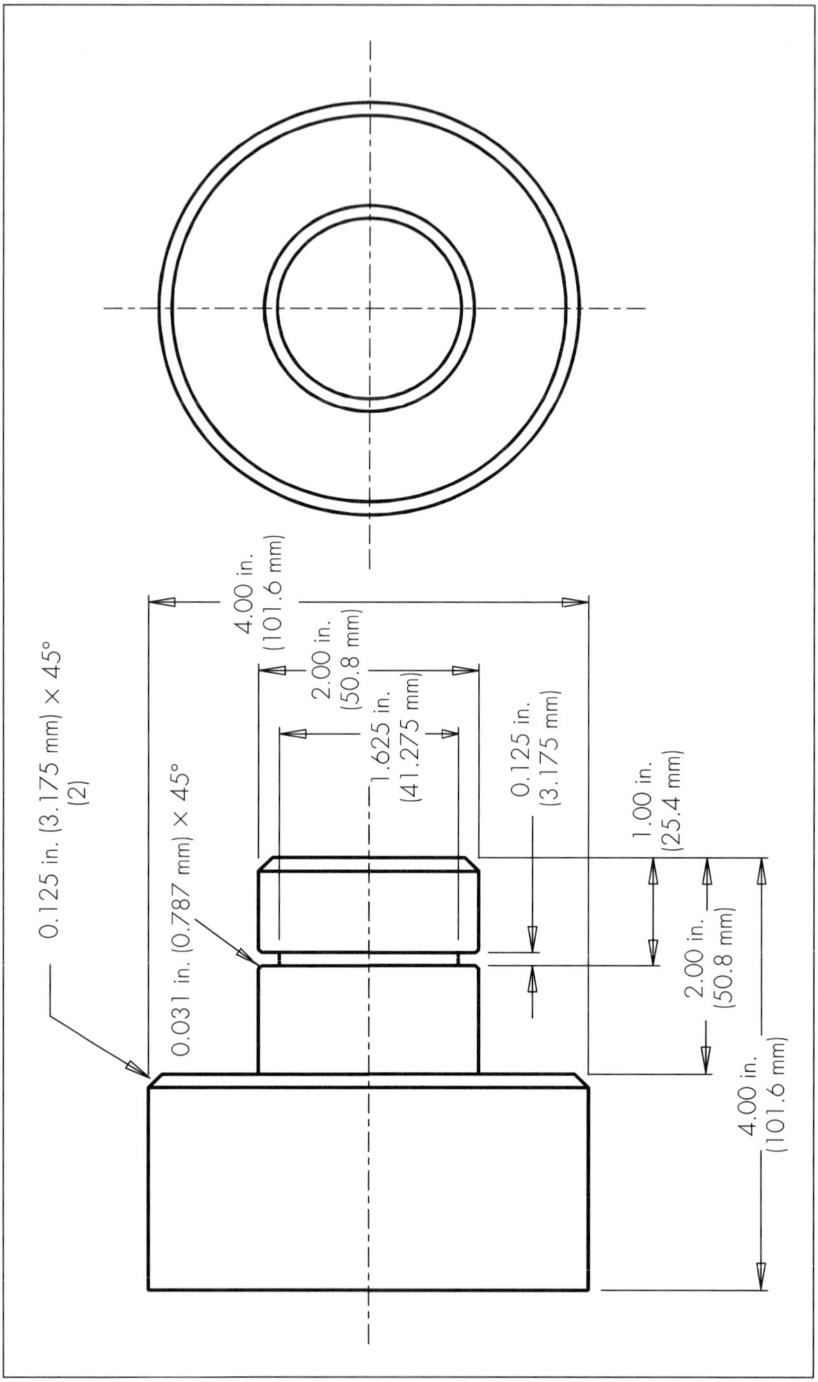

Figure 3-8. User-created canned-cycle parametric programming substantially simplifies programs calling for grooving operations on workpieces with varying dimensions.

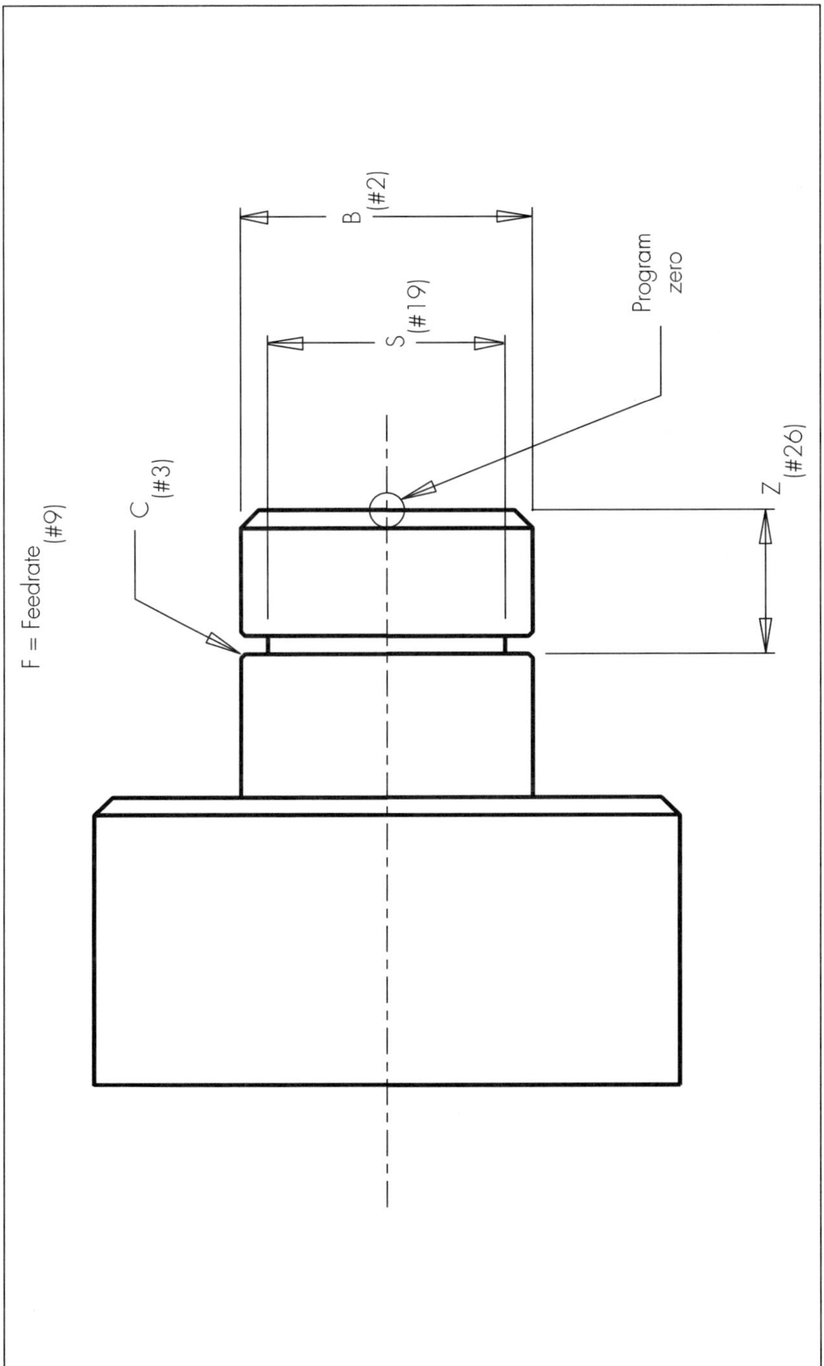

Figure 3-9. Custom macro B letter address arguments to be used with the grooving application.

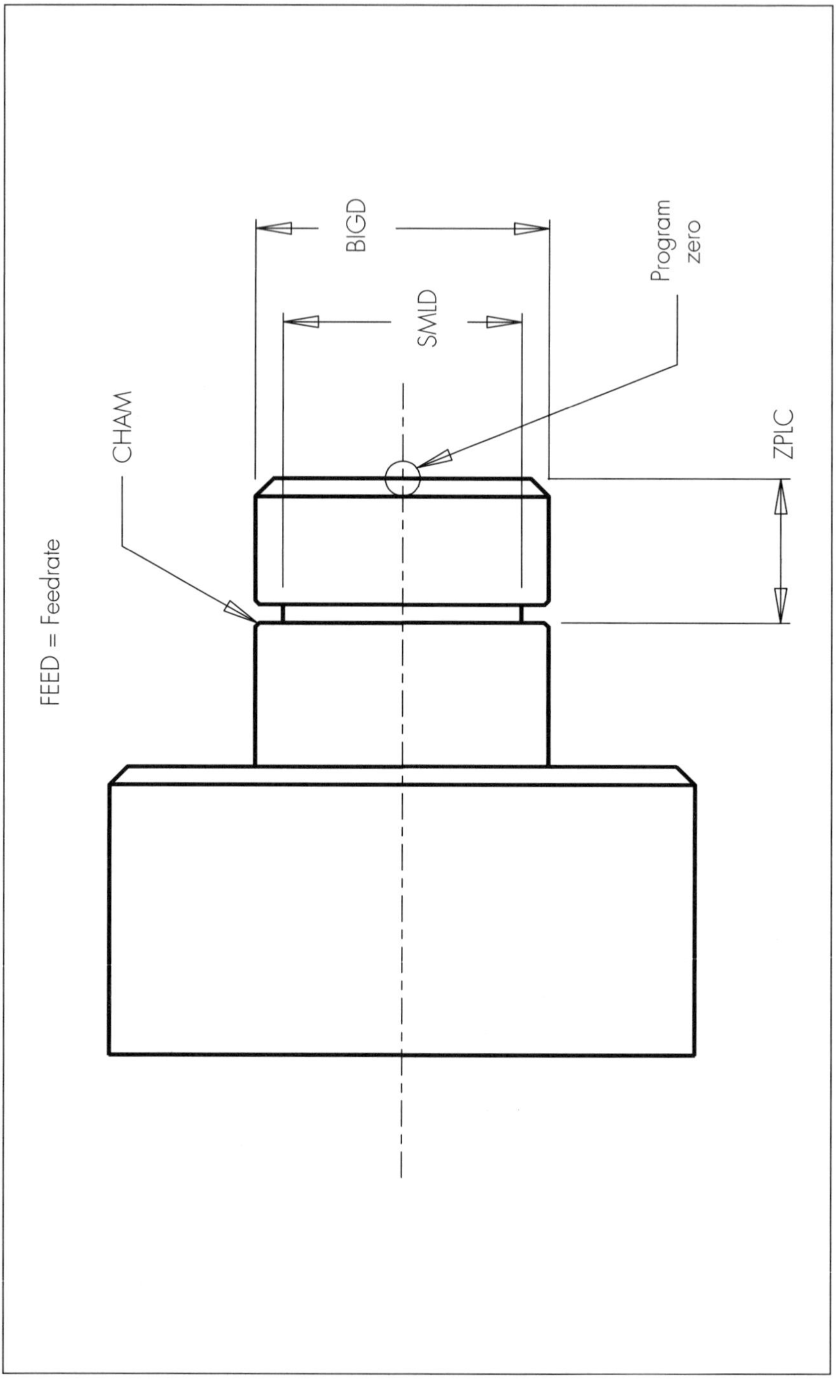

Figure 3-10. The local variable arguments of user task 2 version applied to the grooving operation.

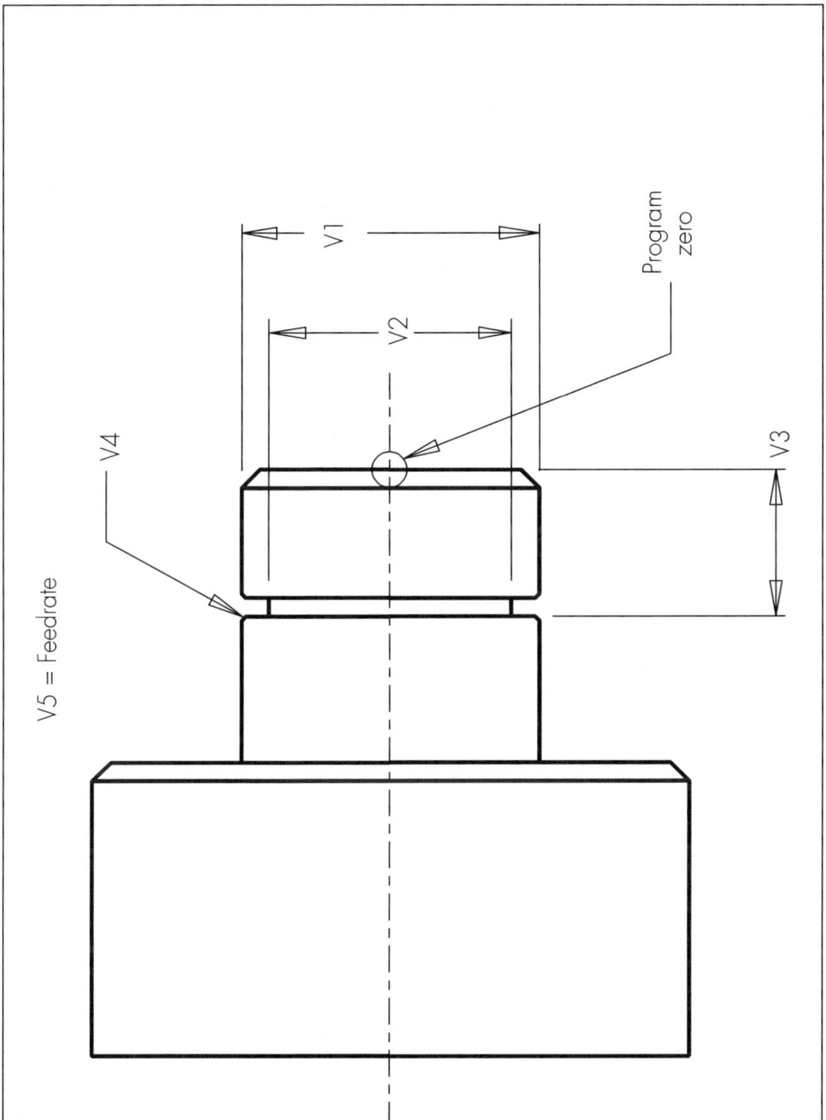

Figure 3-11. Shown are the V-series variable arguments used with Fadal's macro version of the grooving application.

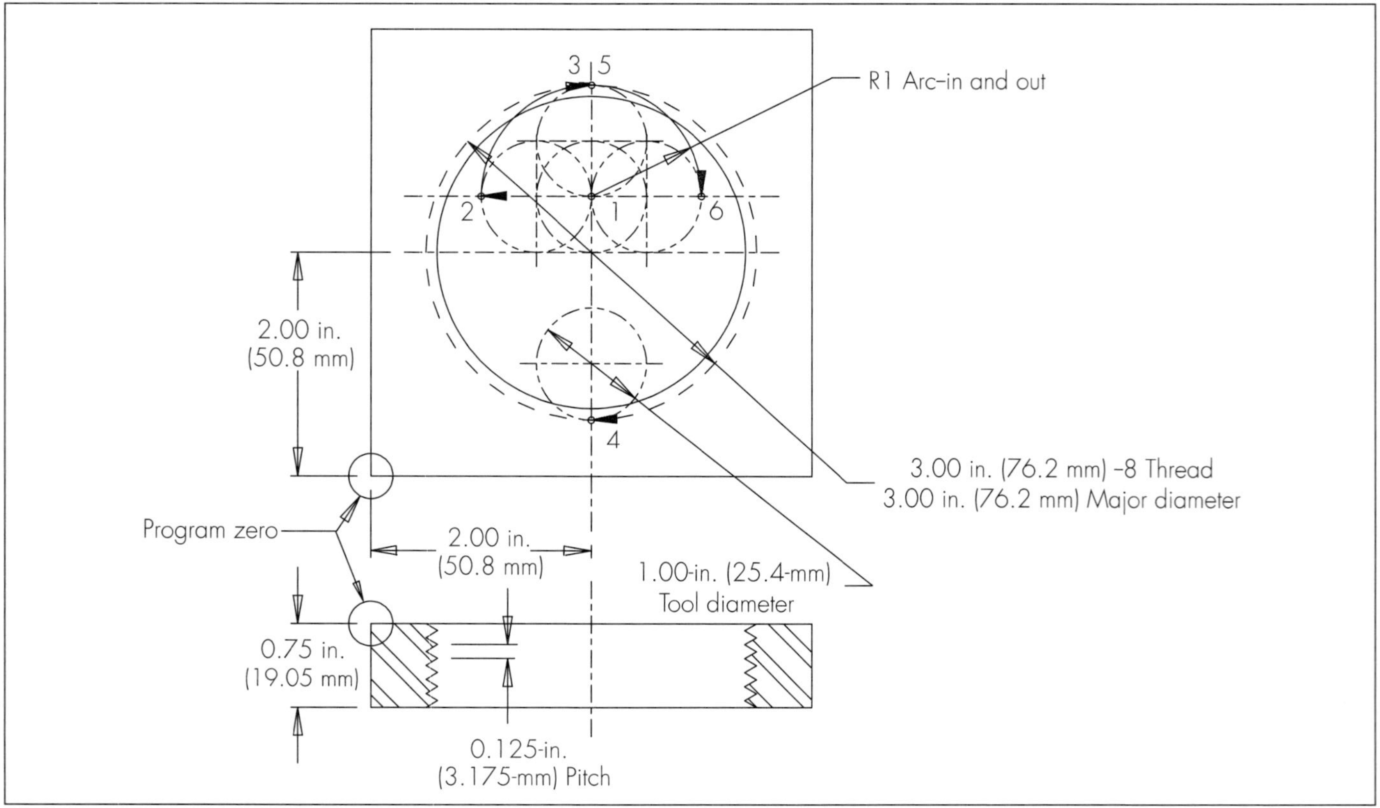

Figure 3-12. Another example of a user-created canned-cycle application showing the workpiece and tool movements used with an internal thread-milling cycle.

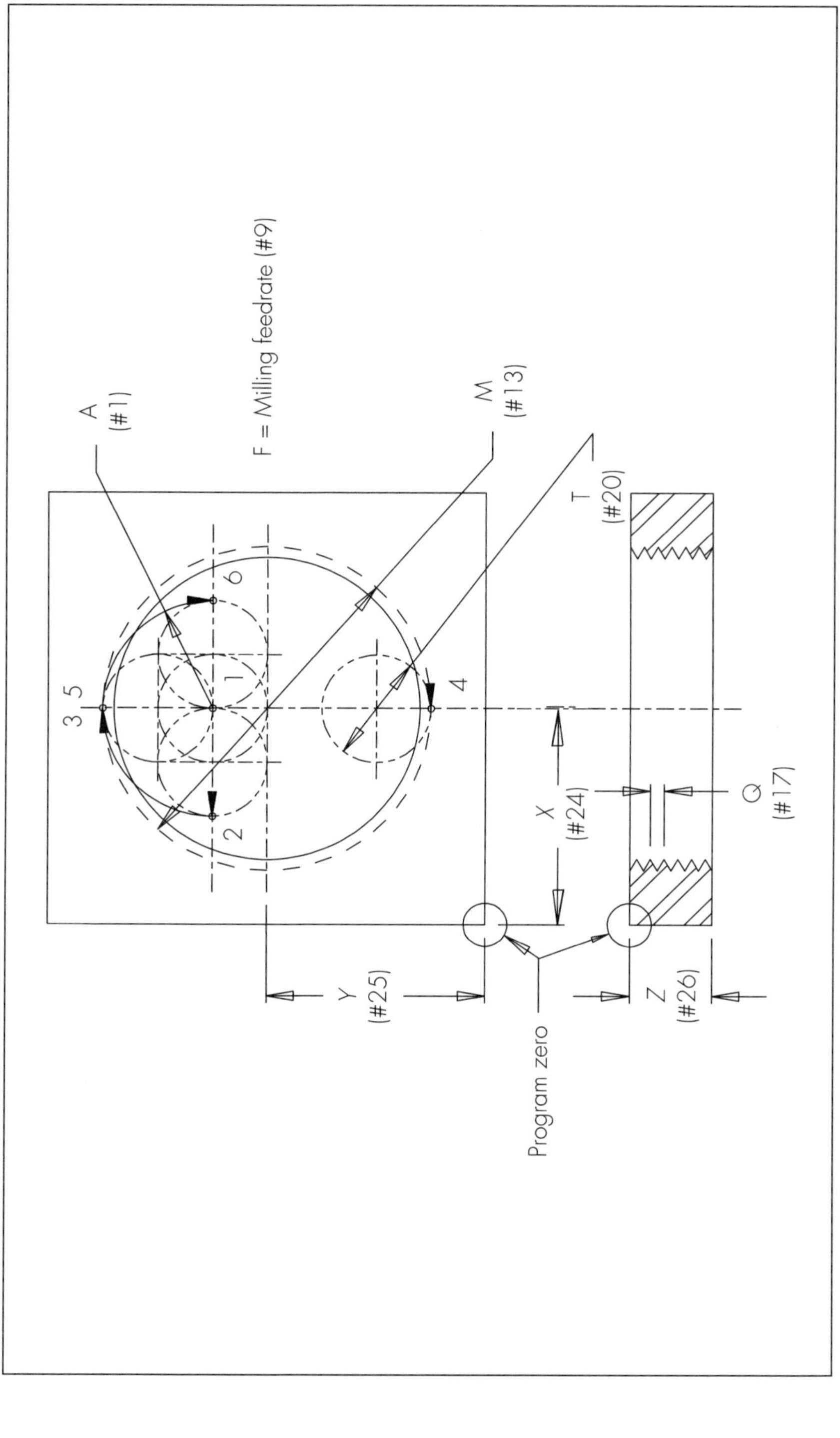

Figure 3-13. The letter address arguments to be used with the custom macro B version of the thread-milling application.

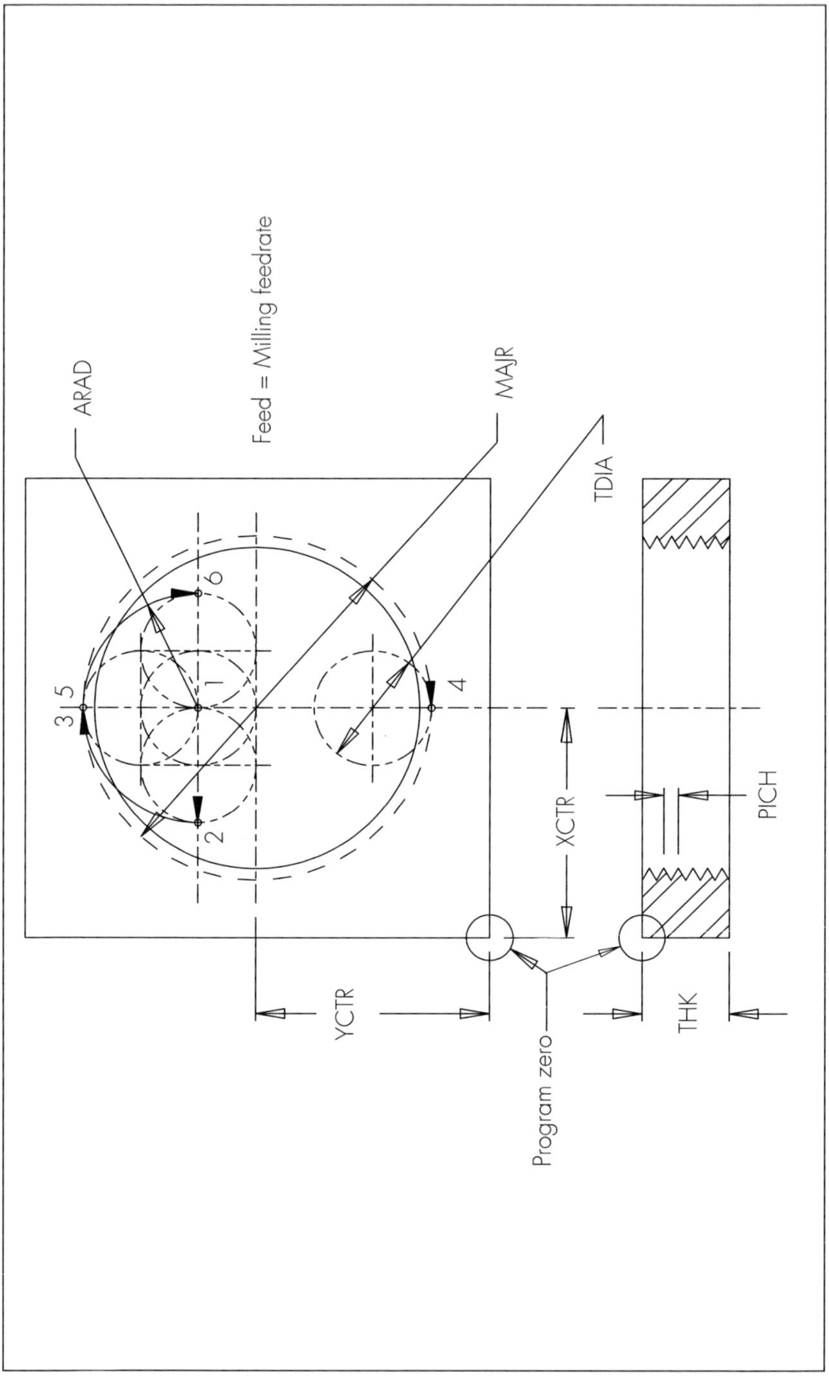

Figure 3-14. The local variable arguments to be used with the user task 2 version of the thread-milling application.

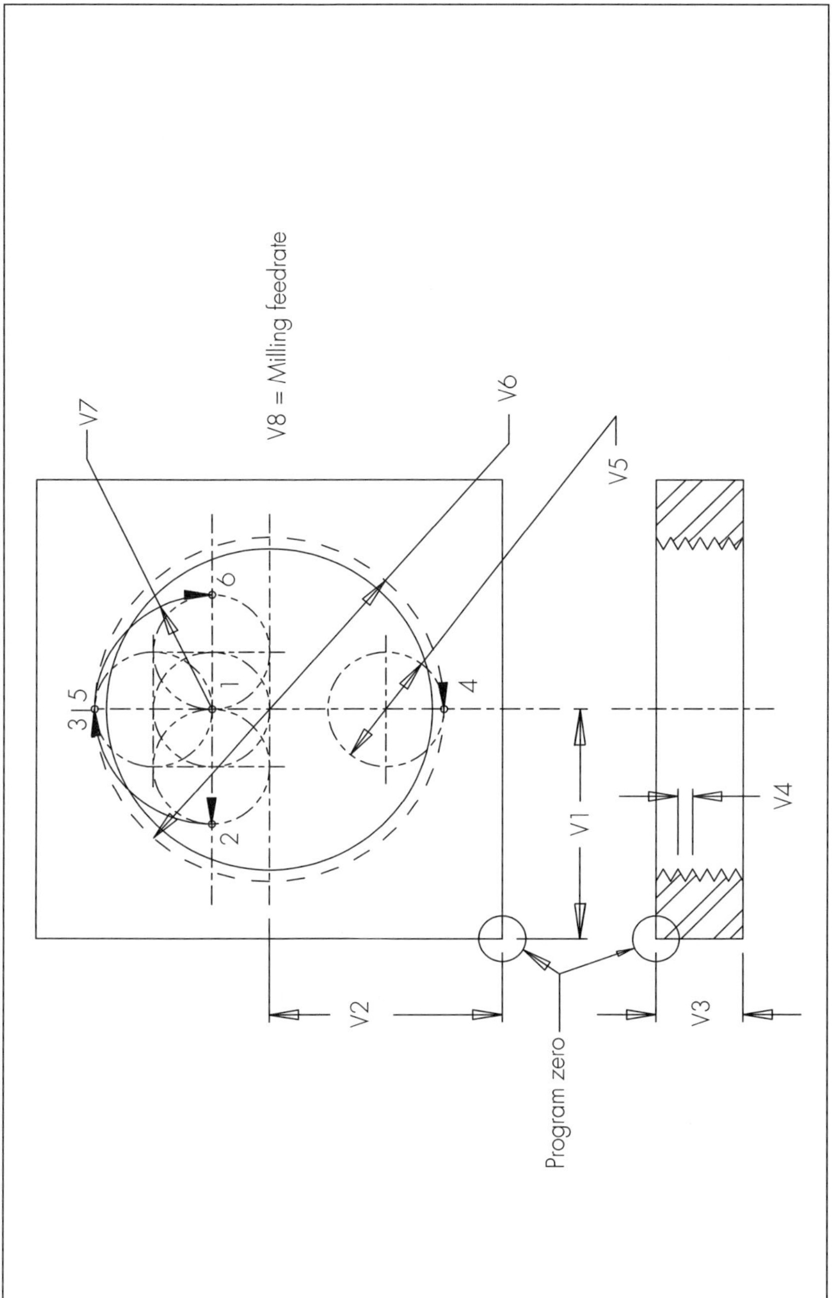

Figure 3-15. V-series variable arguments to be used with Fadal's macro version of the thread-milling application.

Chapter 4

Logic and Program Flow Control

As might be presumed, the general flow (order of execution) for a standard CNC program is from beginning to end. When a program is put into operation, the control will read, interpret, and execute the very first command in your program. Then it will proceed to the next command—read, interpret, and execute. Then on to the next. It is by this sequential order that CNC programs are normally executed.

With parametric programming, you have the ability to change the order of program execution. You can have your parametric programs make *tests* to determine which of two or more possible outcomes it will follow. You can even make the control repeat a series of commands until some criterion related to the series of commands is satisfied. We call this powerful ability to change the order of program execution *program flow control.*

Most parametric programming languages offer several commands related to program flow control. While the actual syntax for each command varies from one version to another, the need for each command type remains remarkably similar. In this chapter, we introduce program flow command types and show examples of when they are used.

STATEMENT LABELS

Most program flow commands require unique *pointers* or *markers* to designate positions in the program that are to be used to control program flow. We call these markers *statement labels.* Statement labels specify points in the program where the control is to continue executing the program. The use of statement labels will be clarified in greater detail as we discuss *unconditional branching* a bit later.

STATEMENT LABELS IN FANUC'S CUSTOM MACRO B

Fanuc uses sequence numbers (N words) for statement labels. Though this text shows a sequence number in almost every command of most programs for the purpose of documentation, to conserve memory space it is a good idea to use sequence numbers in *only* those commands that require statement labels. One technique that works well is to take advantage of the numerical value of the sequence number to help you find the statement label in the program. This is especially helpful in very long programs.

For example, use sequence numbers below 10 for statement labels close to the beginning of the custom macro program, use sequence numbers labels around 50 in the middle of the custom macro program, and use sequence numbers over 90 close to the end of the custom macro program. If you use this technique, you can easily find the statement label being referenced.

Keep in mind that while sequence numbers are quite flexible in normal CNC programming, there are two strict rules about their use as statement labels. First, if a command references a statement label, the statement label must *exist* in the program. If it doesn't an alarm will sound. Second, you must be careful not to use the same statement label twice in your program. While duplicate sequence numbers will not affect the flow of normal CNC programs, duplicate statement labels in parametric programs can result in unpredictable program flow or alarm generation.

STATEMENT LABELS IN OKUMA'S USER TASK 2

User task 2 also allows the use of sequence numbers for statement labels. And if you use sequence numbers as statement labels, follow the suggestions given in the previous discussion of custom macro B statement labels.

User task 2 provides the additional ability to use a more logical *label name* as a statement label. While the label name must also begin with the letter N (just like a sequence number), the next four characters allow much more flexibility in statement label naming.

If using statement label names (instead of numbers), the first character (following the N) must be a letter of the alphabet. The next three characters can be letters or numbers. Examples of allowable statement label names include NSTRT, NHERE, NPOS1, and NEXIT.

Note that statement labels cannot be duplicated in the program. If you use the statement label NSTRT at one point in the program, for example, you *cannot* use it in any other position in the same program. Additionally, if you reference a statement label that does not exist, an alarm will sound.

STATEMENT LABELS IN FADAL'S MACRO

Statement labels in macro must begin with a colon (:). Any series of letters and numbers can follow, meaning you have a great deal of flexibility in statement label naming. You'll want to take advantage of this flexibility, and choose statement label names that make sense. Examples of allowable statement label names include :HERE, :LOOPBEGIN, :LOOPEND, and :EXIT. In Fadal's macro, note that *the statement label is the only word that can appear in the macro command.*

Statement labels cannot be duplicated in the program. If you use the statement label :EXIT at one point in the program, for example, you *cannot* use it in any other position in the same program. And, again, if you reference a statement label that does not exist, an alarm will sound.

UNCONDITIONAL BRANCHING

As the name implies, the unconditional branching command will cause the control to *branch* or *jump* from one location in the program to another location in the same program. The location to which branching is done must be marked with a statement label. Most versions of parametric programming (including all addressed by this text) specify unconditional branching with a *GOTO* statement. The syntax for GOTO changes just slightly from one version to another. Here are three examples.

In Fanuc's custom macro B:

```
.
.
GOTO 50 (Branch to statement label N50. Note no N word)
.
.
.
N50 G00 X . . . (Statement label N50)
```

In Okuma's user task 2:

```
.
.
GOTO NSTRT (Branch to statement label NSTRT)
.
.
.
NSTRT G00 X . . . (Statement label NSTRT)
```

In Fadal's macro:

```
.
.
#GOTO :EXIT 'Branch to statement label :EXIT
.
.
.
```

```
#:EXIT 'Only the statement label is allowed on this line!
G00 X . . .
.
.
```

APPLICATIONS FOR UNCONDITIONAL BRANCHING

While most applications for unconditional branching are related to parametric programming applications, there are some times when it can be helpful to know about unconditional branching even with standard CNC programs.

Changing machining order

For example, have you ever made a processing mistake with the sequence of machining operations when developing a complicated program? Consider this *incorrect* turning center process:

1. Rough face and turn
2. Finish face and turn
3. Drill a 2-in. (50.8-mm) diameter hole
4. Rough bore
5. Finish bore

As you develop the CNC program, say it does not occur to you that you are breaking an important rule of basic machining practice that states: *rough everything before you finish anything*. So you write the program (shown here in Fanuc format) with the incorrect process.

Incorrect program shown in custom macro B format:

```
O0001 (Program number)
N005 T0101 (Index to rough turn and face tool)
N010 G96 S400 M03 (Select constant surface speed, start spindle)
N015 G00 X3.2 Z0.005 (Rapid to approach position)
N020 G01 X – 0.062 F0.010 (Rough face)
N025 G00 X2.79 Z0.1 (Rapid to rough turn diameter)
N030 G01 Z – 0.995 (Rough turn)
N035 X3.1 (Feed up face)
N040 G00 X6.0 Z5.0 (Rapid to tool-change position)
N045 M01 (Optional stop)
N050 T0202 (Index to finish face and turn tool)
N055 G96 S500 M03 (Start spindle)
N060 G00 X3.2 Z0 (Rapid to approach position)
N065 G01 X2.15 F0.006 (Face just past diameter to be bored)
N070 G00 X2.75 Z0.1 (Rapid to finish turn diameter)
N075 G01 Z – 1.0 (Finish turn)
```

```
N080 X3.1 (Feed up face)
N085 G00 X6.0 Z5.0 (Rapid to tool-change position)
N090 M01 (Optional stop)
N095 T0303 (Index to 2-in. [50.8-mm] drill)
N100 G97 S400 M03 (Start spindle)
N105 G00 X0 Z0.1 (Rapid to approach position)
N110 G01 Z – 1.35 F0.008 (Drill hole)
N115 G00 Z0.1 (Rapid out of hole)
N120 G00 X6.0 Z5.0 (Rapid to tool-change position)
N125 M01 (Optional stop)
N130 T0404 (Index to rough boring bar)
N135 G96 S400 M03 (Start spindle)
N140 G00 X1.21 Z0.1 (Rapid to approach position)
N145 G01 Z – 0.995 F0.010 (Rough bore hole)
N150 X1.9 (Feed down face)
N155 G00 Z0.1 (Rapid out of hole)
N160 X6.0 Z5.0 (Rapid to tool-change position)
N165 M01 (Optional stop)
N170 T0505 (Index to finish boring bar)
N175 G96 S500 M03 (Start spindle)
N180 G00 X1.25 Z0.1 (Rapid to finish turn diameter)
N185 G01 Z – 1.0 F0.006 (Finish bore)
N190 X1.9 (Feed down face)
N195 G00 Z0.1 (Rapid out of hole)
N200 X6.0 Z5.0 (Rapid to tool-change position)
N205 M30 (End of program)
```

It is not until after you take this program out to the machine and run it that you discover the mistake. Now your very expensive CNC turning center will be sitting idle while you change the process of this program. While some CNC controls have very nice cut, copy, and paste functions for CNC program segments, many do not. Most programmers will have to go back to their personal computer-based CNC text editor or CAM system to fix the program.

The unconditional branching (GOTO) command will make it possible to easily change this program's execution order right at the CNC control by adding three simple lines. The following sequence shows the corrected process. The resulting modified program is then shown.

1. Rough face and turn
2. Drill 2-in. (50.8-mm)-diameter hole
3. Rough bore
4. Finish bore
5. Finish face and turn

Modified program shown in custom macro B format:

O0001 (Program number)
N005 T0101 (Index to rough turn and face tool)
N010 G96 S400 M03 (Select constant surface speed, start spindle)
N015 G00 X3.2 Z0.005 (Rapid to approach position)
N020 G01 X – 0.062 F0.010 (Rough face)
N025 G00 X2.79 Z0.1 (Rapid to rough turn diameter)
N030 G01 Z – 0.995 (Rough turn)
N035 X3.1 (Feed up face)
N040 G00 X6.0 Z5.0 (Rapid to tool-change position)
N045 M01 (Optional stop)
N048 GOTO 095 (Branch to beginning of drilling operation)
N050 T0202 (Index to finish face and turn tool)
N055 G96 S500 M03 (Start spindle)
N060 G00 X3.2 Z0 (Rapid to approach position)
N065 G01 X2.15 F0.006 (Face just past diameter to be bored)
N070 G00 X2.75 Z0.1 (Rapid to finish turn diameter)
N075 G01 Z – 1.0 (Finish turn)
N080 X3.1 (Feed up face)
N085 G00 X6.0 Z5.0 (Rapid to tool-change position)
N090 M01 (Optional stop)
N093 GOTO 205 (Branch to end of program)
N095 T0303 (Index to 2-in. [50.8-mm] drill)
N100 G97 S400 M03 (Start spindle)
N105 G00 X0 Z0.1 (Rapid to approach position)
N110 G01 Z – 1.35 F0.008 (Drill hole)
N115 G00 Z0.1 (Rapid out of hole)
N120 G00 X6.0 Z5.0 (Rapid to tool-change position)
N125 M01 (Optional stop)
N130 T0404 (Index to rough boring bar)
N135 G96 S400 M03 (Start spindle)
N140 G00 X1.21 Z0.1 (Rapid to approach position)
N145 G01 Z – 0.995 F0.010 (Rough bore hole)
N150 X1.9 (Feed down face)
N155 G00 Z0.1 (Rapid out of hole)
N160 X6.0 Z5.0 (Rapid to tool-change position)
N165 M01 (Optional stop)
N170 T0505 (Index to finish boring bar)
N175 G96 S500 M03 (Start spindle)
N180 G00 X1.25 Z0.1 (Rapid to finish turn diameter)
N185 G01 Z – 1.0 F0.006 (Finish bore)
N190 X1.9 (Feed down face)

N195 G00 Z0.1 (Rapid out of hole)
N200 X6.0 Z5.0 (Rapid to tool-change position)
N203 GOTO 050 (Branch to beginning of finish face and turn)
N205 M30 (End of program)

Note the addition of the three unconditional branching statements in lines N048, N093, and N203. These commands cause the control to machine the workpiece in the desired order.

Using optional block skip to exit a program

As you know, optional block skip (also called block delete) is specified by the slash code (/). When the control reads the slash code, it then looks to the position of the optional block skip switch on the control panel. If the switch is on, the control skips the command that begins with the slash code. When it is off, the control executes the command.

There are times when it can be helpful to repeat a series of commands. Of course, subprogramming easily allows this, and as long as the number of times the subprogram must be executed is known, programming is very simple. However, if the programmer is unsure about *how many* times the subprogram must be executed, programming becomes more difficult.

For illustration, consider the operation of trial boring on a machining center. With trial boring, the programmer gives the operator a series of special commands (in a subprogram) to perform the trial boring operation. However, the hole may or may not be perfectly to size after the first try. It may take several tries to get the hole right to size. In this case, the programmer does not know precisely *how many* times the subprogram must be executed. Consider these commands in custom macro B format:

.
.
/N065 M95 P1000 (Jump to trial boring subprogram)
/N070 **GOTO 065** (Go back to line N065)
.
.

As long as the optional block skip switch is turned off, the subprogram will be executed. When the operator eventually gets the hole on size, he or she will turn on the optional block skip switch and the control ignores line N070 and proceeds with the balance of the program.

With conditional branching

Though *unconditional* branching can sometimes be helpful even with standard CNC programs, we concentrate in here on true parametric programming applications. By far the most popular application for unconditional branching is in conjunction with *conditional* branching, the focus of the following discussion.

CONDITIONAL BRANCHING

Like conditional branching in any computer programming language, conditional branching in parametric programming allows a decision-making capability. The parametric programmer can have commands in the program perform tests, and the program will be executed differently based on the results of these tests.

The purposes of testing are extremely varied and we cannot overstress the importance of conditional branching for testing purposes. You can test just about *anything* from within your parametric program. And based upon the results of the test, you can make the program execute differently. This feature has fantastic implications.

SYNTAX FOR CONDITIONAL BRANCHING STATEMENTS

Before you can fully comprehend specific applications for conditional branching, you must understand its general syntax. All versions of parametric programming addressed in this text use an *IF* statement for conditional branching. However, some versions are more flexible than others in the use of this statement. All applications discussed in this text, however, can be handled with conditional branching, regardless of how flexible the IF statement.

The conditional expression

All IF statements require a *conditional expression*. The result of the conditional expression will be either true or false. If the result is true, the command to the right of the IF statement is executed; if false, the command (on the next line) following the IF statement is executed.

Conditional expressions require *conditional operators*, six of which are generally included in most versions of parametric programming. Conditional operators include less than (LT), less than or equal to (LE), equal to (EQ), greater than or equal to (GE), greater than (GT), and not equal to (NE). Note that in Fadal's macro, you can also use the same conditional operator symbols used in BASIC (< for less than, <= for less than or equal to, = for equal to, >= for greater than or equal to, > for greater than, and <> for not equal to).

The six conditional operators provide great flexibility in the method by which you develop conditional branching commands. In most cases, the same test can be written in several ways. While this flexibility is good, it can sometimes be confusing, especially when you are testing with the *not equal to* (NE or <>) conditional operator. Always remember that the test within the conditional branching statement is to determine whether the *conditional expression* is true or false. When testing against a negative condition, it is easy to make mistakes.

Conditional branching in Fanuc's custom macro B

While newer Fanuc controls do allow a little more flexibility with conditional branching (and you will want to consult your Fanuc manual for more information), we focus our presentation here on only those functions of conditional branching that will work with those Fanuc controls equipped with custom macro B.

As mentioned earlier, if the conditional branching statement is true, the command to the right of the statement is executed. With custom macro B, you are limited to placing a GOTO statement to the right of the IF statement (newer controls additionally allow other custom macro B statements).

To illustrate, suppose you wish to write a grooving custom macro for your turning center that can take advantage of right-hand or left-hand tooling. You prefer to include a special argument in the call statement (H for *hand*) to specify the tooling style. Your criteria states that if H is set to one, right-hand tooling is being used. If H is set to zero, left-hand tooling is being used.

If, of course, right-hand tooling is being used, the spindle must be running in a forward (M03) direction. If left-hand tooling is being used, the spindle must be running in reverse (M04). Following is a portion of the custom macro program that performs the test and starts the spindle in the appropriate direction. (Note that letter address H must be referenced by local variable #11 in the custom macro.)

```
.
.
.
IF [#11 EQ 1.] GOTO 5
M04
GOTO 6
N5 M03
N6 . . .
.
.
```

Note that brackets ([]) must be used to enclose the conditional expression. This conditional branching statement uses the *equal to* (EQ) logical operator, which makes it quite easy to evaluate the results of the test (true or false). If the programmer using this custom macro includes an H1.0 in the call statement, the control will evaluate the conditional branching statement as true. In this case, the GOTO statement next to the IF statement will be executed. In N5, notice that the M03 will start the spindle in the forward direction (as required for right-hand tools).

On the other hand, if the person using this parametric program sets the value of H to zero (or anything else), the IF statement's conditional expression will be evaluated as false. In this case, the command following the IF statement will be executed, starting the spindle in the reverse direction (which is required for left-hand tools).

Finally, note the need for the *GOTO 6* statement. If the conditional expression is evaluated as false and the spindle starts in the reverse direction, we want the control to skip the other possibility. The GOTO statement is needed in this case to keep the control from immediately starting the spindle in the forward direction. While this use of the GOTO statement does not advance the notion of very *structured* programming, it does show that just about any situation requiring conditional branching can be handled with the IF statement in custom macro B.

The flexibility provided by the six conditional operators is clearly demonstrated by the following right-hand versus left-hand tooling test, which can be handled in several ways. Consider these different commands that render the same results:

```
.
.
IF [#11 NE 1.] GOTO 5
M03
GOTO 6
N5 M04
N6 . . .
.
.
```

Or these:

```
.
.
IF [#11 GT 0] GOTO 5
M04
GOTO 6
N5 M03
N6 . . .
.
.
```

Which is *best?* In some cases, it truly does not matter which logical operator you choose to test with. This means that you can set up the test in any way that works and is comfortable to you. However, there are many instances when you will really want to think through your test to search for the hidden benefit(s) of one logical operator over another. In the left-hand versus right-hand tooling case, for example, let's say for the sake of illustration that you almost always run right-hand tooling. Less than 10 percent of your jobs require a left-hand tool. In this case, you might elect to set up the test in such a way that the person using your custom macro must only specify the argument for tooling hand (H in our case) if they are using left-hand tooling. If right-hand tooling is used, the argument H can be omitted. Consider these commands:

```
.
.
IF [#11 EQ 0] GOTO 5
M03
GOTO 6
N5 M04
N6 . . .
.
.
```

Remember that if H is omitted from the call statement, the value of #11 will be vacant (which is different from zero). In this case, the logical expression of the IF statement will be evaluated as false and the spindle will be started in the forward direction.

If this is your first exposure to conditional branching, do not be overly concerned with developing the best possible logical expression for every IF statement you write. With time and experience, setting up more appropriate conditional branching statements will get easier. Also, if you get and read a beginner's book on BASIC programming, you will find much information about how to write effective conditional branching statements.

Conditional branching in Okuma's user task 2

We mentioned earlier that if the conditional expression in the IF statement is evaluated as true, the command to the right of the statement is executed. With user task 2, you are limited to placing a GOTO statement to the right of the IF statement. The following is a specific example that shows the syntax.

Let's say you wish to write a custom macro to execute a grooving operation for your turning center that can use right-hand *or* left-hand tooling. You wish to include a special local variable argument in the call statement, TLHD for *tool hand*, for example, to specify the tooling style. If TLHD is set to one, right-hand tooling is being used. If TLHD is set to zero, left-hand tooling is being used.

If, of course, right-hand tooling is being used, the spindle must be running in a forward (M03) direction. If left-hand tooling is being used, the spindle must be running in reverse (M04). A portion of the user task program that makes the test and starts the spindle in the appropriate direction reads like this:

```
.
.
.
IF [TLHD EQ 1.] GOTO NFWD
M04
```

```
GOTO NCONT
NFWD M03
NCONT . . .
.
.
```

Note that brackets ([]) must be used to enclose the conditional expression. This conditional branching statement uses the *equal to* (EQ) logical operator, which makes it quite easy to evaluate the results of the test (true or false). If the programmer using this user task program includes TLHD=1.0 in the CALL command, the control will evaluate the conditional branching statement as true. In this case, the GOTO statement next to the IF statement will be executed. In the line beginning with statement label NFWD, notice that the M03 will start the spindle in the forward direction (as required for right-hand tools).

On the other hand, if the person using this parametric program sets the value of TLHD to zero (or anything else), the IF statement's conditional expression will be evaluated as false. In this case, the command (on the next line) following the IF statement will be executed, starting the spindle in the reverse direction (as required for left-hand tools).

Finally, note the need for the *GOTO NCONT* statement (we're using NCONT to represent *continue*). If the conditional expression is evaluated as false, and the spindle starts in the reverse direction, we must command the control to skip the other possibility. This GOTO statement is needed to keep the control from immediately starting the spindle in the forward direction. Again while the use of this GOTO statement does not promote a very *structured* programming approach, it does show that just about any situation requiring conditional branching can be handled with the IF statement in user task 2.

As with custom macro B, the six conditional operators in user task 2 also provide much flexibility. The following right-hand versus left-hand tooling test can be handled in several ways. These different commands render the same results:

```
.
.
IF [TLHD NE 1.] GOTO NREV
M03
GOTO NCONT
NREV M04
NCONT . . .
.
.
```

Or these:

```
.
.
IF [TLHD GT 0] GOTO NFWD
M04
GOTO NCONT
NFWD M03
NCONT . . .
.
.
```

Which should you select? In some cases, it truly does not matter which logical operator you choose to test with. You can set up the test in any way that works and in a way that is comfortable to you. However, there may be many instances when you will really want to think through your test to search for the hidden benefit(s) of one logical operator over another.

For example, user task 2 provides a slight variation to the IF statement for testing whether a given local variable argument is currently active (whether it has a value). The command

```
IF TLHD GOTO NCONT
```

for example, simply tests to see if local variable argument TLHD is currently active (note that brackets are not required in this command). If it is, the control will execute the GOTO NCONT command. If it is not active (not specified in the CALL command), the control will proceed to the very next command in the user task program.

We'll illustrate using the same example as we did for custom macro B (left-hand versus right-hand tooling, almost always running right-hand tooling). Again, only 10 percent of your jobs require a left-hand tool. In this case, you might elect to set up your test in such a way that the person using your user task program must only specify the argument for tooling hand (TLHD) if they are using left-hand tooling. If right-hand tooling is used, the argument TLHD can be omitted. Consider these commands in the user task program:

```
.
.
IF TLHD GOTO NTST1
M03
GOTO NCONT
NTST1 IF [TLHD EQ 1.] GOTO NFWD
M04
GOTO CONT
```

```
NFWD M03
M04
NCONT . . .
 .
 .
```

In the first IF statement, we determine whether local variable TLHD is active. If it is, we still have to determine how it is set, so we branch to the second test. If it is not active, however, we can assume that the person using this user task program is using right-hand tooling (TLHD is left out of the CALL command), and start the spindle in the forward direction. Then we skip the unneeded test.

Don't be overly concerned with developing the best possible logical expression for every IF statement you write. With time and experience, setting up more appropriate conditional branching statements will get easier. Also, with a fundamental understanding of BASIC programming, you will find writing effective conditional branching statements much easier.

Conditional branching in Fadal's macro

As we stated at the outset of this section, if the conditional branching statement is true, the command to the right of the statement is executed. Of the three versions of parametric programming addressed in this text, Fadal's macro provides the most flexibility regarding how the conditional expressions within the IF statement can be developed. With macro, you can even include multiple conditional expressions in one IF statement. Additionally, if the expression is evaluated as true, you are not limited to specifying a GOTO statement to the right of the IF statement. Any macro command can be given. In this introductory discussion, we focus our presentation on only the most important functions of Fadal's IF statement.

Let's begin with the simplest form of the IF statement, again using the grooving macro for illustration. You want to write a macro to perform a grooving operation for your turning center that can take advantage of right-hand or left-hand tooling. In the program you wish to include a special argument in the call statement, let's say V1, to represent the tooling style. If V1 is set to one, right-hand tooling is being used. If V1 is set to zero, left-hand tooling is being used.

If right-hand tooling is being used, the spindle, of course, has to be running in a forward (M03) direction. Conversely, if left-hand tooling is being used, the spindle must be running in reverse (M04). Here is a portion of the macro that makes the test and starts the spindle running in the appropriate direction:

```
.
.
.
#IF V1 = 1. THEN GOTO :FWD
M04
#GOTO :CONT
```

```
#:FWD
M03
#:CONT
.
.
```

This conditional branching statement uses the *equal to* (= or EQ) logical operator, which makes it quite simple to evaluate the results of the test (true or false). If the programmer using this macro incorporates V1.0 in the call statement, the control will evaluate the conditional branching statement as true. In this case the GOTO statement next to the IF statement will be executed. In the command after the statement label :FWD, notice that the M03 will start the spindle in the forward direction (as required for right-hand tools).

On the other hand, if the person using this parametric program sets the value of V1 to zero (or anything else), the IF statement's conditional expression will be evaluated as false. In this case, the command (on the next line) following the IF statement will be executed, starting the spindle in the reverse direction (as required for left-hand tools).

Finally, note the need for the *GOTO :CONT* statement. If the conditional expression is evaluated as false, and the spindle starts in the reverse direction, we need the control to skip the other possibility. This GOTO statement is needed to keep the control from immediately starting the spindle in the forward direction. Again, this GOTO statement does not advance the cause of structured programming. With the flexibility allowed by Fadal's macro programming, there are other, more structured ways to develop this test.

The flexibility provided by the six conditional operators is apparent also in macro. The following right-hand versus left-hand tooling test can be handled in several ways. Consider these different commands that render the same results:

```
.
.
#IF V1<>1. THEN GOTO :REV
M03
#GOTO :CONT
#:REV
M04
#:CONT
.
.
```

Or these:

```
.
.
```

```
#IF V1>0 THEN GOTO :FWD
M04
#GOTO :CONT
#:FWD
M03
#:CONT . . .
.
.
```

Is one better than any other? In some cases, it truly does not matter which logical operator you choose to test with, meaning you can set up the test in any way that works, and in any way that is comfortable to you. However, there are again times when you will really want to think through your test to search for the hidden benefit(s) of one logical operator over another. And with all the power provided by Fadal's macro, the choices are almost endless, owing to a couple of special macro features.

First, the macro IF statement allows you to include macro commands to the right of the IF statement (not just the GOTO statement). Consider, for instance, these commands:

```
.
.
#IF V1 = 1 THEN R1 = 3
#IF V1<>1 THEN R1 = 4
M + R1
.
.
```

Notice that if the conditional expression is true, the equality statement will be read. (One of these statements must be true.) In either case, the value of R1 is set to the appropriate spindle in M-code number.

Second, you are allowed to set up multiple conditional expressions with macro using AND, OR, and NOT logical operators. AND requires that all of the conditional expressions be true before the statement to the right of the IF statement will be executed. OR requires just one of the conditional expressions to be true before the statement to the right of the IF statement will be executed. NOT simply reverses the conditional expression being performed. Though they have nothing to do with the right-hand versus left-hand tooling example, these logical operators serve to illustrate what is possible with this very powerful command:

```
#IF (V1>V2) AND (V3 = 0) THEN GOTO :LOOP
```

Notice that both conditional expressions must be enclosed in parentheses. With AND, only if both conditional expressions are true will the GOTO statement be executed.

#IF (V1>V2) OR (V3 = 0) THEN GOTO :LOOP

If either of the conditional expressions is true, the GOTO statement will be executed.

#IF NOT(V1 = 0) THEN V2 = 33.25

Only if V1 is not equal to zero will the equality statement be executed.

#IF (V1>V2) AND (V3 = 0) AND (V4>5) THEN GOTO :LOOP

Only in the event that all three conditional expressions are true will the control execute the GOTO statement.

It bears repeating that if this is your first exposure to conditional branching, do not be overly concerned with developing the best possible logical expression for every IF statement you write. With time and experience, setting up more appropriate conditional branching statements will get easier. Also, if you follow the suggestion given earlier and read a beginner's book on BASIC, you will find information about how to write effective conditional branching statements.

APPLICATIONS FOR CONDITIONAL BRANCHING

Conditional branching is used for many purposes. In this section we expose you to as many applications as space permits. Generally speaking, all applications fit into one of these six categories.

Argument flags

There will be times when you will need to set up criteria within your parametric program for two or more possibilities. The user of your parametric program will then use a *flag* to tell the program which of the possibilities he or she wants at the present time. In essence, the programmer is *flagging down* the parametric program, specifying how it should behave with a flag.

In the section on conditional branching syntax, our example cited a tooling style flag. The flag argument just discussed specifies which hand of tooling is being used. The conditional branching statement(s) in the parametric program example sets up criteria for both possibilities, and the flag specifies which possibility is required at any given time.

Though it is not mandatory, it is good practice to set up your flags in such a way that a setting of 1.0 represents a yes, on, or in other way positive answer to the condition. Use a setting of zero to represent a no, off, or in other way negative answer to the condition. This is simply one of a host of instances in which argument flags are helpful. The following paragraphs cite some others.

To control coolant. For the sake of illustration, let's say you are developing a user-created canned cycle and you know that at times the material being machined will be steel (which requires coolant) and at other times the material will be cast iron (which does not require coolant). You could easily set a flag argument in the main program's call statement to specify whether or not coolant is required. Consider these commands given in the three different versions of parametric programming we focus on in this text.

In custom macro B. This example uses letter address argument C (represented by local variable #3) as the coolant flag argument. In the call statement (G65), if C is equal to 1.0, coolant is required. If C is equal to zero, coolant should be left off. Here is one way to handle this application with custom macro commands:

```
.
.
IF [#3 EQ 1.] GOTO 5
M09
GOTO 6
N5 M08
N6 . . .
.
.
```

In user task 2. Here we use local variable argument CTFL (for coolant flow) as the coolant flag argument. In the CALL command, if CTFL is equal to 1.0, coolant is required. If CTFL is equal to zero, coolant should be left off. One way to handle this application with user task commands is:

```
.
.
IF [CTFL EQ 1.] GOTO NCYES
M09
GOTO NCONT
NCYES M08
NCONT . . .
.
.
```

In Fadal's macro. The coolant flag argument in this example is the V-series variable argument V1. In the main program, if V1 is set equal to 1.0, coolant is required. If V1 is equal to zero, coolant should be left off. Using these macro commands, the argument part of the program for this application looks like this:

```
.
.
#IF V1 = 1 THEN GOTO :COOLANTON
M09
#GOTO :CONT
#:COOLANTON
M08
#:CONT
.
.
```

To climb or conventional mill. You may want to set up criteria in your parametric program to allow both climb and conventional milling. If the flag argument is set to one, the parametric program should climb mill. If set to zero, the program should conventional mill.

In custom macro B. This example uses letter address argument M (represented by local variable #13) as the milling-type flag argument. In the call statement (G65), if M is equal to 1.0, climb milling is required. If M is equal to zero, conventional milling is required. Here is one way to handle this application with custom macro parametric programming commands:

```
.
.
IF [#13 EQ 1.] GOTO 5
(Set up calculations and/or motions for conventional milling)
.
.
GOTO 6
N5 (Set up calculations and/or motions for climb milling
.
.
N6 . . .
.
.
```

In user task 2. This example uses local variable argument MILL (for milling type) as the climb-milling flag argument. In the CALL command, if MILL is equal to 1.0, climb milling is required. If MILL is equal to zero, conventional milling is required. One way to handle this application with user task commands is shown here:

```
.
.
IF [MILL EQ 1.] GOTO NCLIM
(Set up calculations and/or motions for conventional milling)
.
.
GOTO NCONT
NCLIM (Set up calculations and/or motions for climb milling)
.
.
NCONT . . .
.
.
```

In Fadal's macro. The V-series variable argument V2 is used as the milling-type flag argument in this example. In the main program, if V2 is set equal to 1.0, climb milling is required. If V2 is equal to zero, conventional milling is to be done. The macro commands look like this:

```
.
.
#IF V2 = 1 THEN GOTO :CLIMMILL
(Set up calculations and/or motions for conventional milling)
.
.
#GOTO :CONT
#:CLIMMILL
(Set up calculations and/or motions for climb milling)
.
.
#:CONT
.
.
```

There are, of course, many applications for argument flags. Here we describe just a few more.

If developing a user-created canned cycle for circle milling, you may wish to use a flag to specify whether there is a hole currently in the workpiece. If there is, the milling cutter can rapid to depth. If there is not, it must feed to depth.

If developing a peck drilling cycle, you may wish to include a flag argument to specify whether deep hole or chip-breaking peck drilling is required. Truly, any application calling for two or more possibilities is probably a good candidate for an argument flag.

If developing a threading parametric program for a turning center, you may wish to use a flag to specify whether or not the thread ends in a thread relief. If it does, you will have the threading tool pull straight out of the workpiece.

If not, you will have the threading tool feed out at a 45 degree angle, minimizing stress on the threading tool.

Global flags

As you know, most arguments (with the exception of Fadal's V-series variables) are *local*. They are forgotten as soon as the control finishes executing the parametric program. However, there are times when you may want a given flag to be *global*, available to any program at any time.

If, for example, you have a machine tool (machining center *or* turning center) that does not allow complete manual control of coolant during a program's execution, and if coolant is commanded from within a CNC program (with M08), coolant will come on. This can be messy during verification of the program if the machine is not fully guarded, especially if the operator has no convenient way to get the coolant to stop flowing.

Even though you may have CNC machines that allow adequate manual control of coolant, this example is a great application for a global flag. A special nonvolatile (global) variable (a #500 series in custom macro, or a V-series variable in user task 2 or macro) can be set by the setup person when a dry run is to be done. This can be done right at the machine through the display screen and keyboard. Whenever coolant is activated in the CNC program, a special test can be performed which will determine the value of the global variable. If the variable is set to allow coolant to come on, the coolant command will be given. If not, the coolant command will be skipped.

In Custom macro B. We show just one example of this application. You should be able to easily convert it to any version of parametric programming. We'll use permanent common variable #500 as the global coolant flag variable. The setup person will set #500 to 1.0 prior to doing a dry run to keep coolant from dispensing. After the program is verified and production is to be run, the setup person will set #500 back to zero. Consider these commands.

```
.
.
IF [#500 EQ 1.0] GOTO 5
M08
N5 . . .
.
.
```

Upon closer examination, you'll find that this is very similar to using the optional block skip function. If, for example, the command

```
/M08
```

is programmed, the coolant will only come on if the optional block skip switch is turned off. However, most controls allow only one optional block skip function. If you need more, you're out of luck. Global flags have no such limitation. And anything you can use the optional block skip to control, you can use a global flag to control as well—and you can have as many global flags as you need.

Call statement error checking

Even a perfectly written parametric program will fail if the person using it does not set arguments properly. If, for example, the user leaves out a very important argument, it can be difficult to predict just how the parametric program will behave. In similar fashion, if a basic mistake in an argument setting is made, it can result in disaster. While a beginning parametric programmer will be most concerned with simply getting his or her first few programs to work, a well-written parametric program will use tests to determine if certain mistakes have been made or if other problems exist.

Testing mandatory arguments. Any argument that must be supplied to the parametric program is called a *mandatory argument*. These are arguments that commonly change from one use of the parametric program to the next and are extremely important to the success of the program. We recommend that you test all mandatory arguments for the purpose to confirm that they are included in the call statement. If they are not, most versions of parametric programming give you the option of actually generating an alarm or printing a message that halts execution of the program. Here are some examples.

In Fanuc's custom macro B. To confirm that mandatory arguments are assigned with custom macro B, you must understand the meaning of vacancy. Remember from Chapter 2, if a letter address argument is not assigned in the call statement, its corresponding local variable is set to vacant (having no value). The custom macro B representation for vacancy is #0. You can use #0 to test for vacancy in the body of the custom macro.

To generate the actual alarm, a special *system variable* is used. While we devote an entire chapter to CNC features (including all kinds of system variables), you must be familiar with this particular type of system variable to understand our example.

The alarm-generating system variable is #3000. If the control reads a #3000 command, an alarm will be generated. Like any program storage (PS) alarm, the alarm generated by #3000 will put the machine in a true alarm state. The operator must clear the alarm (by pressing reset) to clear the custom macro generated alarm. The syntax for the #3000 system variable is:

#3000 = 100 (ARGUMENT A MISSING IN CALL)

Again, if the control reads this command, it will generate an alarm. The value (100) is an *alarm number* of your choosing and can range from 100 to 255 (some control models allow other alarm numbers as well). The message in parentheses (all capitals) is the alarm documenting message. It must be 26 characters or fewer. If this command is read, the control will switch to the alarm page and the following alarm message will appear on the control screen.

MC-100 ARGUMENT A MISSING IN CALL

With an understanding of vacant variables and alarm generation, you should be able to also understand this example of how to test whether mandatory arguments have been left out of the call statement. Here is the right-side milling custom macro call statement shown in Chapter 2. For this example, we'll make every argument in the call statement a mandatory argument.

N025 G65 P1000 X0. Y0. Z0. W5.0 H3.0 T0.5 D1.0 F5.0 (Mill right side)

Remember *X* is represented by #24, *Y* by #25, *Z* by #26, W by #23, H by #11, T by #20, D by #7, and F by #9. A very explicit way of testing mandatory arguments is to set up a specific alarm generating command for each argument. At the beginning of custom macro O1000, you can have the following series of commands test that each local variable representing an argument has a value.

```
O1000 (Program number)
IF [#24 NE #0] GOTO 1
#3000 = 100 (X MISSING IN CALL)
N1 IF [#25 NE #0] GOTO 2
#3000 = 101 (Y MISSING IN CALL)
N2 IF [#26 NE #0] GOTO 3
#3000 = 102 (Z MISSING IN CALL)
N3 IF [#23 NE #0] GOTO 4
#3000 = 103 (W MISSING IN CALL)
N4 IF [#11 NE #0] GOTO 5
#3000 = 104 (H MISSING IN CALL)
N5 IF [#20 NE #0] GOTO 6
#3000 = 105 (T MISSING IN CALL)
```

```
N6 IF [#7 NE #0] GOTO 7
#3000 = 106 (D MISSING IN CALL)
N7 IF [#9 NE #0] GOTO 8
#3000 = 107 (F MISSING IN CALL)
N8 . . .
.
.
.
```

If any of these IF statements is evaluated as false (a local variable is vacant), an alarm will be generated. In each case, a specific alarm is generated for each missing argument. Some programmers feel this may be a little *too* specific, so to minimize commands, you may elect to generate one generic alarm to signal if *any* of the arguments is missing. This should suffice to let the operator know what is wrong. The following program shows how.

```
O1000 (Program number)
IF [#24 EQ #0] GOTO 98
IF [#25 EQ #0] GOTO 98
IF [#26 EQ #0] GOTO 98
IF [#23 EQ #0] GOTO 98
IF [#11 EQ #0] GOTO 98
IF [#20 EQ #0] GOTO 98
IF [#7 EQ #0] GOTO 98
IF [#9 EQ #0] GOTO 98
.
.
.
.
GOTO 99
N98 #3000 = 107 (DATA MISSING IN CALL)
N99 M99 (End of custom macro)
```

In this example, if any of the IF statements is true, the generic alarm DATA MISSING IN CALL is generated.

In Okuma's user task 2. Keep in mind that if you try to reference an inactive local variable with user task 2 (one that has not been assigned), an alarm will sound. This in itself may provide enough protection against forgetting to assign mandatory arguments. However, the machine may be right in the middle of a machining operation when it reads a command including an inactive local variable and sounds the alarm. It is wiser to test whether mandatory arguments are assigned *before* you allow machining to begin.

To confirm that mandatory arguments are assigned with user task 2, you must understand the special conditional branching command that is used to determine whether or not a local variable is active. The syntax for this command is:

IF DIA1 GOTO N1

In this command, local variable DIA1 is being tested. If it is active, the command GOTO N1 is executed. If it is not, the next command in the user task program is executed.

To generate the actual alarm, two special *system variables* are used. While we devote an entire chapter to CNC features (including all kinds of system variables), you must be familiar with these particular system variables to understand the example we give.

The alarm generating system variable is VDOUT. Actually, VDOUT is used for several purposes, including alarm generation. If the value set to VDOUT is 991, alarm type C is generated (which is appropriate for program-related alarms). In addition, you are allowed to specify an alarm code number within the VDOUT command.

The actual message printed on the display screen is commanded by another system variable called VUACM (VUACM stands for **V**ariable for **U**ser **A**larm Co**M**ment). You are allowed up to 16 characters in the message.

If the control reads these system variables, an alarm will be generated. Like any program storage alarm, the alarm generated will put the machine in a true alarm state. The operator must clear the alarm (by pressing reset) to clear the user task generated alarm. The syntax for VUACM and VDOUT is:

VUACM[1] = 'DIA1 IS MISSING'
VDOUT[991] = 100

The number 1 in the brackets of the VUACM system variable command specifies the character number of the first character of the message (from 1 to 16). Note that the message must be enclosed in single quotation marks. Again, the 991 in the VDOUT system variable setting specifies alarm type C and the alarm number included in the message will be 100. Here is precisely how the message will appear on the display screen:

3202 Alarm C User reserve code 100 DIA1 IS MISSING

The 3202 in this message is Okuma's alarm number for type C user reserved alarm codes.

With an understanding of the special conditional branching command for local variables and alarm generation, you should understand this example of how to test whether mandatory arguments have been left out of the call statement. Following is the right-side milling custom macro call statement shown in Chapter 2. Again, for this example, we'll make every argument in this call statement a mandatory argument.

```
N025 CALL O1000 XZER = 0 YZER = 0 ZZER = 0 WIDT = 5.0 HGHT = 3.0
   THK = 0.5 TDIA = 1.0 FEED = 5.0 (Mill right side)
```

You also can establish very explicit ways of testing mandatory arguments by creating specific alarm generating commands for each argument. For instance, at the beginning of user task program O1000, you can have this series of commands test that each local variable has a value.

```
O1000 (Program number)
IF XZER THEN GOTO N1
VUACM[1] = 'XZER IS MISSING'
VDOUT[991] = 100
N1 IF YZER THEN GOTO N2
VUACM[1] = 'YZER IS MISSING'
VDOUT[991] = 101
N2 IF ZZER THEN GOTO N3
VUACM[1] = 'ZZER IS MISSING'
VDOUT[991]= 102
N3 IF WIDT THEN GOTO N4
VUACM[1] = 'WIDT IS MISSING'
VDOUT[991]= 103
N4 IF HGHT THEN GOTO N5
VUACM[1] = 'HGHT IS MISSING'
VDOUT[991] = 104
N5 IF THK THEN GOTO N6
VUACM[1] = 'THK IS MISSING'
VDOUT[991] = 105
N6 IF TDIA THEN GOTO N7
VUACM[1] = 'TDIA IS MISSING'
VDOUT[991] = 106
N7 IF TDIA THEN GOTO 8
VUACM[1] = 'TDIA IS MISSING'
VDOUT[991] = 103
N8 . . .
.
.
.
```

Notice in each case that a specific alarm is generated for each missing argument. This, to some programmers is a little *too* specific. To minimize commands, you may elect to generate one generic alarm if any of the arguments is missing. Here is another example that shows how.

```
O1000 (Program number)
IF XZER THEN GOTO N1
GOTO NALM
N1 IF YZER THEN GOTO N2
GOTO NALM
N2 IF ZZER THEN GOTO N3
GOTO N98
N3 IF WIDT THEN GOTO N4
GOTO NALM
N4 IF HGHT THEN GOTO N5
GOTO NALM
N5 IF THK THEN GOTO N6
GOTO NALM
N6 IF TDIA THEN GOTO N7
GOTO NALM
N7 IF TDIA THEN GOTO N8
GOTO NALM
N8 . . .
.
.
.
GOTO NEXIT
NALM VUACM[1] = 'DATA IS MISSING'
VDOUT[991] = 100
NEXIT RTS (End of user task program)
```

In this example, if any of the IF statements is false, the generic alarm, DATA IS MISSING, is generated.

In Fadal's macro. While macro does not have a specific command for the purpose of generating alarms, it does have a very flexible PRINT command that allows messages of all kinds to be printed on the display screen. This, combined with the ability to halt the machine, should suffice to let the operator know that something is wrong.

To actually halt the program, you could use a simple M00 program stop. However, the operator could then simply press the cycle start button and the control would continue with the (incorrect) program. For this reason, we recommend using the INPUT command with some special logic that keeps the machine stopped if the operator tries to activate the cycle after getting an alarm message.

One other problem that presents itself with macro when considering the testing of whether mandatory arguments have been assigned is that the V-series variables are *global*. They *never* lose their values as local variables do. There are two ways to resolve this problem.

First, you can use the CLEAR function prior to setting V-series variable arguments (usually in the main program). CLEAR lets you set series of specified variables to zero. Here are some examples:

```
#CLEAR 'Set all V-series variables to zero
#CLEAR V1 'Set V1 to zero
#CLEAR V1-V8 'Set V-series variables V1 through V8 to zero
```

Once all V-series variable arguments have been CLEARed in this manner, and after you assign their values, it can be assumed that its value will still be zero if the person using your macro program forgets to assign an argument. In the macro program itself, you can simply test each argument to find out if it is zero. If it is, you can generate the alarm message to halt the machine.

Unfortunately, there may be those arguments that must sometimes be set to zero. In the right-side milling example from Chapter 2, for example, we used V-series variable V3 to represent the top surface location of the workpiece in *Z*. This value is commonly set to zero, meaning that if we apply the test just discussed, we'll get an incorrect alarm message.

For this reason, you may elect to use a second method of testing each V-series variable argument. It involves more work, but is much more fail-safe. It involves determining a *range of allowable setting* for each argument. If any argument is out of its range, the alarm message will be given. You can actually combine the two to come up with the most appropriate test(s) for each argument.

Figure 4-1 (page 190) shows the V-series variable arguments from the right-side milling application of Chapter 2. (In the interest of reader continuity, we have not interrupted the written programs by placing figures as they are called out; the figures follow the sample programs.)

The trick is to establish an appropriate range for each variable, based on your own machining criteria. Here is the range for each V-series variable argument used in our example.

V1, V2, and V3—minus 10.0 through plus 10.0 (Since these are absolute positions in *X*, *Y*, and *Z*, they have a broad range.)
V4 and V5—plus 1.5 though plus 15.0 (The workpieces this company machines are never smaller than 1.5 in. [38.1 mm] or bigger than 15.0 in. [381 mm].)
V6—plus 0.25 through plus 1.5
V7—plus 0.375 through plus 1.5
V8—plus 0.5 through 15.0

While coming up with appropriate ranges takes a little time, this method is quite fail-safe and well worth the time it takes. Here is the beginning of the macro program that performs the mandatory argument testing. Note that an explicit alarm message is generated for each mandatory argument.

```
O1000 (Program number)
#IF (V1>–10.0) AND (V1<10.0) THEN GOTO :NXT1
#PRINT "V1 IS OUT OF RANGE"
#INPUT V99 'This command stops program
#GOTO :NOCONT 'Confirm that program cannot continue
#:NXT1
#IF (V2>–10.0) AND (V2<10.0) THEN GOTO :NXT2
#PRINT "V2 IS OUT OF RANGE"
#INPUT V99 'This command stops program
#GOTO :NOCONT 'Confirm that program cannot continue
#:NXT2
#IF (V3>–10.0) AND (V3<10.0) THEN GOTO :NXT3
#PRINT "V3 IS OUT OF RANGE"
#INPUT V99
#GOTO :NOCONT
#:NXT3
#IF (V4>1.5) AND (V4<15.0) THEN GOTO :NXT4
#PRINT "V4 IS OUT OF RANGE"
#INPUT V99
#GOTO :NOCONT
#:NXT4
#IF (V5>1.5) AND (V5<15.00) THEN GOTO :NXT5
#PRINT "V5 IS OUT OF RANGE"
#INPUT V99
#GOTO :NOCONT
#:NXT5
#IF (V6>0.25) AND (V6<1.5) THEN GOTO :NXT6
#PRINT "V6 IS OUT OF RANGE"
#INPUT V99
#GOTO :NOCONT
#:NXT6
#IF (V7>0.375) AND (V7<1.5) THEN GOTO :NXT7
#PRINT "V7 IS OUT OF RANGE"
#INPUT V99
#GOTO :NOCONT
#:NXT7
#IF (V8>0.5) AND (V8<15.00) THEN GOTO :NXT8
#PRINT "V8 IS OUT OF RANGE"
```

```
#INPUT V99
#GOTO :NOCONT
#:NXT8
.
.
.
.
.
#GOTO :CONT 'Skip no continue segment
#:NOCONT 'Eliminate possibility of continuing
#PRINT "YOU MUST RESET CONTROL!"
#INPUT V99
#GOTO :NOCONT
#:CONT
M99 (End of macro program)
```

Again, notice that in each case a specific alarm is generated for each misassigned argument. Some programmers feel this may be a little *too* specific. To minimize commands, you may elect to generate one generic alarm if any of the arguments is out of range. In the Fadal macro version of parametric programming, here is an example that shows how.

```
O1000 (Program number)
#IF (V1<–10.0) OR (V1>10.0) THEN GOTO :ALARM
#IF (V2<–10.0) OR (V2>10.0) THEN GOTO :ALARM
#IF (V3<–10.0) OR (V3>10.0) THEN GOTO :ALARM
#IF (V4<1.5) OR (V4>15.0) THEN GOTO :ALARM
#IF (V5<1.5) OR (V5>15.00) THEN GOTO :ALARM
#IF (V6<0.25 OR (V6>1.5) THEN GOTO :ALARM
#IF (V7<0.375) OR (V7>1.5) THEN GOTO :ALARM
#IF (V8<0.5) OR (V8>15.00) THEN GOTO :ALARM
.
.
.
#GOTO :CONT 'Skip alarm
#:ALARM 'Alarm generation commands
#PRINT "AN ARGUMENT IS OUT OF RANGE"
#INPUT V99 'This command will halt the program
#GOTO :ALARM
#:CONT
M99 (End of macro program)
```

In this example, if any of the IF statements is true, the generic alarm message AN ARGUMENT IS OUT OF RANGE is printed and the program will halt.

While it may be difficult for the operator to determine which of the arguments is incorrectly set, this technique minimizes the number of required commands.

Testing for erroneous argument settings. Leaving out a mandatory argument is just one of many possible mistakes a person using your parametric program could make when specifying arguments. Though you can get carried away with error-trapping routines, the lower the experience level of the person using your parametric program, the more important it is to make your program as fail-safe as possible. At the very least, you should guard against the most common mistakes. As your parametric program is put into production, be on the lookout for mistakes your people may be making. You can guard against almost any potential mistake with conditional branching. Here is a simple illustration.

Suppose you are developing a user-created canned cycle for multiple-pass grooving on a turning center. As you evaluate the error-trapping functions, it occurs to you that the grooving *tool width* must be equal to or smaller than the *groove width*. If it is not, you cannot predict how your parametric program will behave and you wish to generate an error trap for this potential mistake. For the three versions of parametric programming, here are three possible solutions.

In custom macro B. For this example, T (#20) is the tool width and W (#23) is the groove width.

```
.
.
.
IF [#20 LE #23] GOTO 1
#3000 = 100 (TOOL TOO WIDE)
N1 . . .
.
.
.
```

In user task 2. For this example, TWID is the tool width and GWID is the groove width.

```
.
.
.
IF [TWID LE GWID] GOTO N1
VUACM[1] = 'TOOL TOO WIDE'
VDOUT[991] = 100
N1 . . .
.
.
.
```

In Fadal's macro. For this example, V1 is the tool width and V2 is the groove width.

```
.
.
.
#IF V1<=V2 THEN GOTO :CONT
#:REPEAT
#PRINT “TOOL IS TOO WIDE”
#INPUT V99 ’Halts program activation
#GOTO :REPEAT ’Do not allow continuation
#:CONT . . .
.
.
.
```

What about *execution* time? As you have been reading this section on error trapping, you may have been questioning what effect error trapping will have on cycle time. Even with newer 32-bit controls, you will likely notice pauses in the program's execution while these logic commands are executed. With very low production quantities, the added cycle time may not be a problem. However, as these production quantities grow, the effects of logic and arithmetic commands on cycle time can be devastating, and you will want to do something about it.

One way is to use the optional block-skip function to skip any calculation and logic commands after the first workpiece is machined. The operator will leave the optional block-skip switch turned off for the first workpiece and all calculations and error checking will be done once. After the first workpiece, the block-skip switch will be turned on for the balance of the production run.

A better, more automatic way to minimize wasted execution time is to devise a special test (we'll call it *a check sum test*) to find out if any pertinent data has changed since the last execution of the program. If it has, all calculations and error checking will be done. If not, these commands are skipped. We elaborate on both methods later.

Setting default values

If you have ever worked with any kind of computer program application software (word processing, desktop publishing, drawing programs, etc.), you have likely seen *pull down menus*. Menus in software applications are commonly used to get data from the computer user. You have likely appreciated the fact that many of the actual entries on certain menus have default settings, meaning you must only enter data for those entries that are not currently set correctly. Default settings dramatically reduce the amount of data you must enter.

By incorporating conditional branching techniques, you can assign default values for arguments in most versions of parametric programming. This can dramatically reduce the number of arguments a programmer must assign. To be able to perform this function, however, you must have a firm method with which to determine whether an argument has been assigned or not, as is possible in custom macro B and user task 2.

Suppose, for example, you are developing a user-created canned cycle for pocket milling. As you look at your series of prints to evaluate the application, you notice that almost all workpieces require the pocket to be milled into the top (*Z* zero) surface of the workpiece. Just a very small percentage have the pocket at a different *Z* surface. Instead of forcing the person using your parametric program to specify the *Z* value for *all* programs, you want your program to assume the *Z* surface location is at zero unless the programmer specifies the *Z* surface argument. Here's how you do it.

In custom macro B. The letter address argument *Z* (#26) is used to pass the *Z* surface value to the custom macro.

```
.
.
.
IF [#26 NE #0] GOTO 1 (If Z is not vacant, skip default)
#26 = 0 (Set default value for Z to zero)
N1 . . . (Continue with program)
.
.
.
```

In user task 2. Local variable argument ZSUR is used to pass the *Z* surface to the user task program.

```
.
.
.
IF ZSUR GOTO N1 (If ZSUR is active, skip default)
ZSUR = 0 (Set default value for Z to zero)
N1 . . . (Continue with program)
.
.
.
```

In both cases, if the programmer leaves out the argument assigning the *Z* surface position, the control will assume it to be zero.

You may have noticed that setting default values for arguments is just the opposite of generating an alarm based on a missing mandatory argument. Instead of testing whether or not an argument has been assigned for the purpose of generating an alarm, you are testing whether it has been assigned for the purpose of setting its default value. We will call arguments that have default values set in the parametric program *defaulted arguments*. In well-written parametric programs, *all arguments supplied to the parametric program are mandatory arguments or defaulted arguments.*

Other kinds of error checking

Using conditional branching to test for missing or erroneous arguments is just one kind of error checking. With most versions of parametric programming, you can do much more. In fact, most allow you to set up an *error trap* for just about any kind of problem. One way to determine whether you have need for this kind of error checking is to track the specific problems your operators have had while running machines. What mistakes are they most prone to making?

Do your operators sometimes miss a decimal point when entering offset values (typing 1.0 instead of 0.1, for example)? You can test the values of all offsets to confirm they are in an acceptable range (more on how when we discuss system variables). If they are not, you can generate an alarm to halt the program.

Do your operators sometimes start the machine at an improper location? Possibly you have a large fixture on the table, over which the automatic tool changer (ATC) cannot be activated without a collision. You can test the machine's position (more on how when we discuss system variables) to determine if the ATC is above the fixture. If it is, you can generate an alarm.

On your multi-pallet machining centers, do your operators sometimes activate a program with the wrong pallet in position? Using a simple global flag that specifies the active pallet, you can test to ensure that the correct pallet is in position at the beginning of each program. If it is not, you can generate an alarm.

On your CNC turning centers, do your operators sometimes forget to size workpieces and update offsets at regular intervals? You could set up a counter that will cause the machine to go into an alarm state after a specified number of workpieces have been machined, telling the operator to perform sizing and tool maintenance for the appropriate tools.

These are just some of the countless tests you can perform in a parametric program to make your machines easier and safer to run. Not every detail of machine utilization can be tested, but do not be too quick to give up on an important test (especially if it is related to safety).

Loops

The last application for conditional branching is looping. Looping is the repeating of commands in a program a specified number of times. While some versions of parametric programming have a special command word for looping, most use conditional branching for this purpose. With custom macro B for example, there is a special WHILE statement for looping. However, it provides no special advantages over creating loops with conditional branching. For this reason, in this text we stress—exclusively—the use of conditional branching to generate loops, discussed in detail beginning on page 166.

CONDITIONAL BRANCHING EXAMPLE

Before moving on, we wish to show you an example of parametric programming that includes many of the conditional branching capabilities just discussed. It is an extension of the user-created canned cycle for thread milling shown at the end of Chapter 3. We give this thread-milling canned-cycle application much more power, make it safer to use, and, in general, easier to work with. All thanks to conditional branching. Figure 4-2 (page 191) shows the same drawing used for the example in Chapter 3. Here is how the criteria will change.

1. Set default for workpiece thickness to 0.75 in. (19.05 mm). We'll establish that well over half the workpieces to be machined are of this thickness. Note that setting default values is only possible in parametric programming languages like custom macro B and user task 2 that allow firm testing of whether an argument has been specified in the call statement or not. This is not possible in Fadal's macro.
2. Set default for tool diameter to 1.0 in. (25.4 mm). We'll establish that almost all workpieces require a 1.0-in.-diameter cutter. Again, this is only possible with parametric programming languages like custom macro B and user task 2 that allow firm testing of whether an argument has been specified in the call statement or not. This feature is not part of Fadal's macro.
3. All other arguments are mandatory arguments. Test them to confirm that they have been set.
4. Add logic for top-to-bottom or bottom-to-top milling. This will be done with a special flag (see documentation for each program for the flag name). This kind of logic can be helpful if milling blind and through holes. For *through* holes it is acceptable to mill from top to bottom. For *blind* holes, it is better to mill from bottom to top. Note that the actual direction of milling (cw versus ccw) also must change. If the flag is set to one (1.0) we have a blind hole, and milling must occur from bottom to top. If this flag is set to zero (0), milling must occur from top to bottom.

In custom macro B. Figure 4-3 (page 192) shows the letter address arguments of this application. Note the addition of argument B, which controls the milling direction.

A portion of the main program that calls the thread milling custom macro follows. Note in line N170, that the T and *Z* arguments are left out, assuming the custom macro will use their preset values. Also notice that argument B is set to 1.0, meaning we have a blind hole to thread mill.

```
O0006 (Program number)
.
.
(Machine hole to get ready for thread milling)
.
.
N150 T04 M06 (Place thread milling cutter in spindle)
N155 G90 G54 S400 M03 T05 (Select absolute mode, coordinate system, start
   spindle, get next tool ready)
```

N160 G00 X2.0 Y2.0 (Rapid over to center of hole)
N165 G43 H04 Z.1 (Rapid up to workpiece in *Z*)
N170 G65 P1006 X2.0 Y2.0 Q0.125 M3.0 A1.0 F5.0 B1.0 (Call custom macro and mill thread)
N175 G91 G28 Z0 M19 (Return to tool-change position)
N180 M01 (Optional stop)
.
.
.

To mill the thread, this modified custom macro B program is used.

O1006 (Program number)
(PRESET *Z* AND T)
IF [#26 NE #0] GOTO 5 (Test if *Z* is vacant)
#26 = 0.75 (Preset *Z* to 0.75)
N5 IF [#20 NE #0] GOTO 6 (Test if T is vacant)
#20 = 1.0 (Preset T to 1.0)
(TEST MANDATORY VARIABLES and GENERATE ALARM IF MISSING)
N6 IF [#24 EQ #0] GOTO 98 (If *X* is missing, generate alarm)
IF [#25 EQ #0] GOTO 98 (If *Y* is missing, generate alarm)
IF [#17 EQ #0] GOTO 98 (If Q is missing, generate alarm)
IF [#13 EQ #0] GOTO 98 (If M is missing, generate alarm)
IF [#1 EQ #0] GOTO 98 (If A is missing, generate alarm)
IF [#9 EQ #0] GOTO 98 (If F is missing, generate alarm)
IF [#2 EQ #0] GOTO 98 (If B is missing, generate alarm)
#20 = #20/2 (Set #20 to radius of tool instead of diameter)
#13 = #13/2 (Set #13 to radius of major diameter)
IF [#2 EQ 1.0] GOTO 20 (Test if blind hole)
(CALCULATIONS FOR TOP TO BOTTOM)
#101 = #24 – #1 + #20 (*X* position of tool center at approach position)
#102 = #24 + #1 – #20 (*X* position of tool center at retract position)
#103 = #25 + #13 – #20 (*Y* position of tool center at points 3 and 5)
#104 = #25 + #13 – #1 (*Y* position of tool center at points 2 and 6)
#105 = – [#26 + #17/4] (*Z* position of point 3)
#106 = #105 – #17/2 (*Z* position of point 4)
#107 = #106 – #17/2 (*Z* position of point 5)
#108 = #107 – #17/4 (*Z* position of point 6)
#109 = #1 – #20 (Radius of approach tool path)
#110 = #13 – #20 (Radius of tool path for milling thread)
#111 = #25 – #13 + #20 (*Y* position of tool center at point 4)
#112 = 2 (Set G code number for circular motion [G03])

```
GOTO 60
(CALCULATIONS FOR BOTTOM TO TOP)
N20 #101 = #24 + #1 – #20 (X position of tool center at approach position)
#102 = #24 – #1 + #20 (X position of tool center at retract position)
#103 = #25 + #13 – #20 (Y position of tool center at points 3 and 5)
#104 = #25 + #13 – #1 (Y position of tool center at points 2 and 6)
#105 = – [#26 – #17/4] (Z position of point 3)
#106 = #105 + #17/2 (Z position of point 4)
#107 = #106 + #17/2 (Z position of point 5)
#108 = #107 + #17/4 (Z position of point 2)
#109 = #1 – #20 (Radius of approach tool path)
#110 = #13 – #20 (Radius of tool path for milling thread)
#111 = #25 – #13 + #20 (Y position of tool center at point 4)
#112 = 3 (Set G code number for circular motion [G02])
(MOTIONS)
N60 G00 X#101 Y#104 (Move in XY to approach point)
Z – #26 (Rapid to start of thread milling in Z)
G#112 X#24 Y#103 Z#105 R#109 F#9 (Approach to point 3)
Y#111 Z#106 R#110 (Mill half way around circle to point 4)
Y#103 Z#107 R#110 (Mill the rest of the way to point 5)
X#102 Y#104 Z#108 R#109 (Retract from thread)
G00 X#24 Y#25 (Move to hole center)
Z0.1 (Rapid out of hole)
GOTO 99
(ALARM IF DATA IS MISSING)
N98 #3000 = 100 (DATA MISSING IN CALL)
N99 M99 (End of custom macro)
```

Note the impact these relatively small changes have on this program. About 15 commands are required for testing and setting default values. Additionally, the flag for blind versus through hole has a dramatic impact on this program. Notice, based on the findings of the flag test, how calculations are made and common variables are assigned for motions from bottom to top or from top to bottom. Even the G code number (2 or 3) is manipulated in this manner. Depending on how these variables are assigned, the same series of motion commands can be used to handle both possibilities.

In user task 2. Figure 4-4 (page 193) shows the local variable arguments related to this application. Note the addition of argument BHLE, which controls the milling direction.

The following is a portion of the main program that calls the thread milling user task program. Note in line N170, the TDIA and THK arguments are left out, assuming the custom macro will use their preset values. Notice also that argument BHLE is set to 1.0, meaning we have a blind hole to thread mill.

```
O0006 (Program number)
.
.
(Machine hole to get ready for thread milling)
.
.
N145 G15 H01 (Select coordinate system)
N150 T04 M06 (Place thread milling cutter in spindle)
N155 G90 S400 M03 T05 (Select absolute mode, start spindle, get next tool ready)
N160 G00 X2.0 Y2.0 (Rapid over to center of hole)
N165 G56 H04 Z.1 (Rapid up to workpiece in Z)
N170 CALL O1006 XCTR = 2.0 YCTR = 2.0 PICH = 0.125 MDIA = 3.0
   ARAD = 1.0 FEED = 5.0 BHLE = 1.0 (Call user task program and mill thread)
.
.
.
```

Here is the modified user task 2 program that mills the thread:

```
O1006 (Program number)
(PRESET THK AND TDIA)
IF THK GOTO N5 (Test if THK is active)
THK = 0.75 (Preset THK to 0.75)
N5 IF TDIA GOTO N6 (Test if TDIA is vacant)
TDIA = 1.0 (Preset TDIA to 1.0)
(TEST MANDATORY VARIABLES and GENERATE ALARM IF MISSING)
N6 IF XCTR GOTO N7 (If XCTR is active, GOTO N7)
GOTO N98 (If not, generate alarm)
N7 IF YCTR GOTO N8 (If YCTR is active, GOTO N8)
GOTO N98 (If not, generate alarm)
N8 IF PICH GOTO N9 (If PICH is active, GOTO N9)
GOTO N98 (If not, generate alarm)
N9 IF MAJR GOTO N10 (If MAJR is active, GOTO N10)
GOTO N98 (If not, generate alarm)
N10 IF ARAD GOTO N11 (If ARAD is active, GOTO N11)
GOTO N98 (If not, generate alarm)
N11 IF FEED GOTO N12 (If FEED is active, GOTO N12)
GOTO N98 (If not, generate alarm)
N12 IF BHLE GOTO N13 (If BHLE is active, GOTO N13)
GOTO N98 (If not, generate alarm)
N13 TRAD = TDIA/2 (Set TRAD to radius of tool)
MRAD = MDIA/2 (Set MRAD to radius of major diameter)
IF [BHLE EQ 1.0] GOTO N20 (Test if blind hole)
```

```
(CALCULATIONS FOR TOP TO BOTTOM)
XAPP = XCTR – ARAD + TRAD (X position of tool center at the approach
   position)
XRET = XCTR + ARAD – TRAD (X position of tool center at retract position)
YP35 = YCTR + MRAD – TRAD (Y position of tool center at points 3 and 5)
YP26 = YCTR + MRAD – ARAD (Y position of tool center at points 2 and 6)
ZPT3 = – [THK + PICH/4] (Z position of point 3)
ZPT4 = ZPT3 – PICH/2 (Z position of point 4)
ZPT5 = ZPT4 – PICH/2 (Z position of point 5)
ZPT6 = ZPT5 – PICH/4 (Z position of point 6)
TAAR = ARAD – TRAD (Radius of approach tool path)
TMPR = MRAD – TRAD (Radius of tool path for milling thread)
YPT4 = YCTR – MRAD + TRAD (Y position of tool center at point 4)
GCD = 2 (Set G code number for circular motion [G02])
GOTO N60
(CALCULATIONS FOR BOTTOM TO TOP)
N20 XAPP = XCTR + ARAD  TRAD (X position of tool center at approach
   position)
XRET = XCTR – ARAD + TRAD (X position of tool center at retract position)
YP35 = YCTR + MRAD – TRAD (Y position of tool center at points 3 and 5)
YP26 = YCTR + MRAD – ARAD (Y position of tool center at points 2 and 6)
ZPT3 = –[THK – PICH/4] (Z position of point 3)
ZPT4 = ZPT3 + PICH/2 (Z position of point 4)
ZPT5 = ZPT4 + PICH/2 (Z position of point 5)
ZPT6 = ZPT5 + PICH/4 (Z position of point 6)
TAAR = ARAD – TRAD (Radius of approach tool path)
TMPR = MRAD – TRAD (Radius of tool path for milling thread)
YPT4 = YCTR – MRAD + TRAD (Y position of tool center at point 4)
GCD = 3 (Set G code number for circular motion [G03])
(MOTIONS)
N60 G00 X = XAPP Y = YP26 (Move in XY to approach point)
Z = – THK (Rapid to start of thread milling in Z)
G = GCD X = XCTR Y = YP35 Z = ZPT3 R = TAAR F = FEED (Approach to
   point 3)
Y = YPT4 Z = ZPT4 R= TMPR (Mill half way around circle to point 4)
Y = YP35 Z = ZPT5 R = TMPR (Mill the rest of the way to point 5)
X = XRET Y = YP26 Z = ZPT6 R = TAAR (Retract from thread)
G00 X = XCTR Y = YCTR (Move to hole center)
Z0.1 (Rapid out of hole)
GOTO N99
(ALARM IF DATA IS MISSING)
N98 VUACM[1] = 'DATA MISSING'
VDOUT[991] = 100
N99 RTS (End of user task program)
```

Again, these relatively small changes have a significant impact on this program. About 25 commands are required for testing and setting default values. Moreover, the flag for blind versus through hole has a dramatic impact on the program. Notice, based on the findings of the flag test, how calculations are made and local variables are assigned for motions from bottom to top or from top to bottom. Even the G-code number (2 or 3) is manipulated in this manner. Depending on how these variables are assigned, the same series of motion commands can be used to handle both possibilities.

In Fadal's macro. Figure 4-5 (page 194) shows the V-series variable arguments related to this application. Note the addition of argument V9, which controls the milling direction.

Since there is no way to firmly identify whether a variable has or has not been assigned (all variables are global in macro; they *never* lose their values), there is no way to set default values in a fail-safe manner. For this reason, we bypass the setting of default values. For the same reason, we cannot test for generating alarms if V-series variables have not been assigned. However, we can (and will) test to ensure that all arguments fall in an appropriate range.

Here is a portion of the main program that calls the same thread-milling macro. In line N168, argument V9 is set to 1.0, defining a blind hole to thread mill.

O0006 (Program number)

.

.

(Machine hole to get ready for thread milling)

.

.

N150 T04 M06 (Place thread milling cutter in spindle)
N155 G90 G54 S400 M03 T05 (Select absolute mode, coordinate system, start spindle, get next tool ready)
N160 G00 X2.0 Y2.0 (Rapid over to center of hole)
N165 G43 H04 Z.1 (Rapid up to workpiece in *Z*)
N170 #V1 = 2.0 'Assign argument V1
N175 #V2 = 2.0 'Assign argument V2
N180 #V3 = 0.75 'Assign argument V3
N185 #V4 = 0.125 'Assign argument V4
N190 #V5 = 1.0 'Assign argument V5
N195 #V6 = 3.0 'Assign argument V6
N200 #V7 = 1.0 'Assign argument V7
N205 #V8 = 5.0 'Assign argument V8
N210 #V9 = 1.0 'Assign argument V9
N215 M98 P1006 (Call macro program and mill thread)
N220 G91 G28 Z0 M19 (Return to tool-change position)

```
N225 M01 (Optional stop)
.
.
.
```

Here is the modified macro program to mill the threads:

```
O1006 (Program number)
(TEST ARGUMENTS)
#IF (V1>–10.0) AND (V1<10.0) THEN GOTO :NXT1 'Test range of V1
#PRINT "V1 IS OUT OF RANGE" 'Print message
#INPUT V99 'This command halts program
#GOTO :NOCONT 'Confirm that program cannot continue
#:NXT1 'Continuation point
#IF (V2>–10.0) AND (V2<10.0) THEN GOTO :NXT2 'Test range of V2
#PRINT "V2 IS OUT OF RANGE"
#INPUT V99 'This command stops program
#GOTO :NOCONT 'Confirm that program cannot continue
#:NXT2
#IF (V3>–0) AND (V3<2.5) THEN GOTO :NXT3 'Test range of V3
#PRINT "V3 IS OUT OF RANGE"
#INPUT V99
#GOTO :NOCONT
#:NXT3
#IF (V4>0) AND (V4<0.25) THEN GOTO :NXT4 'Test range of V4
#PRINT "V4 IS OUT OF RANGE"
#INPUT V99
#GOTO :NOCONT
#:NXT4
#IF (V5>0.5) AND (V5<2.0) THEN GOTO :NXT5 'Test range of V5
#PRINT "V5 IS OUT OF RANGE"
#INPUT V99
#GOTO :NOCONT
#:NXT5
#IF (V6>1.0) AND (V6<5.0) THEN GOTO :NXT6 'Test range of V6
#PRINT "V6 IS OUT OF RANGE"
#INPUT V99
#GOTO :NOCONT
#:NXT6
#IF (V7>0.375) AND (V7<4.0) THEN GOTO :NXT7 'Test range of V7
#PRINT "V7 IS OUT OF RANGE"
#INPUT V99
#GOTO :NOCONT
```

```
#:NXT7
#IF (V8>0.5) AND (V8<15.00) THEN GOTO :NXT8 'Test range of V8
#PRINT "V8 IS OUT OF RANGE"
#INPUT V99
#GOTO :NOCONT
#:NXT8
#IF (V9>=0) AND (V9<=1.0) THEN GOTO :NXT9 'Test range of V9
#PRINT "V9 IS OUT OF RANGE"
#INPUT V99
#GOTO :NOCONT
#:NXT9
#IF V9 = 1.0 THEN GOTO :BLINDHOLE 'Test if blind hole
(MOVEMENTS FOR THROUGH HOLE)
#V5 = #V5/2 'Set V5 to the radius of the cutter instead of its diameter
#R9 = V3 'Initial Z position
#R1 = V1 'Set R1 to X center of hole
#R2 = V2 + V6/2 – V7 'Set R2 to Y center of approach radius
G00 X + R1 Y + R2 (Rapid to center of approach radius, point 1)
Z – R9 (Rapid to beginning Z position)
#R9 = R9 + V4/4 (Calculate Z position at end of approach radius)
#R1 = V1 – V7 + V5 'Set R1 to X position approach radius
#R2 = V8 * 5 'Set R2 to fast moving feedrate
G01 X + R1 F + R2 (Fast feed to beginning of approach radius, point 2)
#R1 = V1 'Set R1 to X center of hole
#R2 = V2 + V6/2 – V5 'Set R2 to Y end point of approach radius
#R3 = V7 – V5 'Set R3 to tool path radius for circular motion
#R4 = V8 'Set R4 to milling feedrate
G02 X + R1 Y + R2 Z – R9 R + R3 F + R4 (Circular approach to beginning of
   thread, point 3)
#R9 = R9 + V4/2 'Calculate Z position half way around thread
#R1 = V2 – V6/2 + V5 'Set R1 to Y position half way around thread
#R2 = V6/2 – V5 'Calculate radius for circular move
G02 Y + R1 Z – R9 R + R2 (Mill half way around thread, point 4)
#R9 = R9 + V4/2 'Calculate Z position at end of thread
#R1 = V2 + V6/2 – V5 'Set R1 to Y end of circular move
G02 Y + R1 Z – R9 R + R2 (Mill balance of thread, point 5)
#R9 = R9 + V4/4 'Calculate Z position at end of retract radius
#R1 = V1 + V7 – V5 'Calculate X end point of retract radius
#R2 = V2 + V6/2 – V7 'Calculate Y end point of retract radius
#R3 = V7 – V5 'Calculate radius of retract
G02 X + R1 Y + R2 Z – R9 R + R3 (Retract to point 6)
#R1 = V1 'Store X center position in R1
G00 X + R1 (Return to center of approach radius)
```

```
Z0.1 (Rapid to clearance position above workpiece)
GOTO :CONT 'Skip blind hole possibility
#:BLINDHOLE
(MOVEMENTS FOR BLIND HOLE)
#V5 = #V5/2 'Set V5 to the radius of the cutter instead of its diameter
#R9 = V3 'Initial Z position
#R1 = V1 'Set R1 to X center of hole
#R2 = V2 + V6/2 – V7 'Set R2 to Y center of approach radius
G00 X + R1 Y + R2 (Rapid to center of approach radius, point 1)
Z – R9 (Rapid to beginning Z position)
#R9 = R9 – V4/4 (Calculate Z position at end of approach radius)
#R1 = V1 + V7 – V5 'Set R1 to X position approach radius
#R2 = V8 * 5 'Set R2 to fast moving feedrate
G01 X + R1 F + R2 (Fast feed to beginning of approach radius, point 2)
#R1 = V1 'Set R1 to X center of hole
#R2 = V2 + V6/2 – V5 'Set R2 to Y end point of approach radius
#R3 = V7 – V5 'Set R3 to tool path radius for circular motion
#R4 = V8 'Set R4 to milling feedrate
G03 X + R1 Y + R2 Z – R9 R + R3 F + R4 (Circular approach to beginning of
   thread, point 3)
#R9 = R9 – V4/2 'Calculate Z position half way around thread
#R1 = V2 – V6/2 + V5 'Set R1 to Y position half way around thread
#R2 = V6/2 – V5 'Calculate radius for circular move
G03 Y + R1 Z – R9 R + R2 (Mill half way around thread, point 4)
#R9 = R9 – V4/2 'Calculate Z position at end of thread
#R1 = V2 + V6/2 – V5 'Set R1 to Y end of circular move
G03 Y + R1 Z – R9 R + R2 (Mill balance of thread, point 5)
#R9 = R9 – V4/4 'Calculate Z position at end of retract radius
#R1 = V1 – V7 + V5 'Calculate X end point of retract radius
#R2 = V2 + V6/2 – V7 'Calculate Y end point of retract radius
#R3 = V7 – V5 'Calculate radius of retract
G03 X + R1 Y + R2 Z – R9 R +  R3 (Retract to point 6)
#R1 = V1 'Store X center position in R1
G00 X + R1 (Return to center of approach radius)
Z0.1 (Rapid to clearance position above workpiece)
#GOTO :CONT 'Skip no continue segment
#:NOCONT 'Eliminate possibility of continuing
#PRINT "YOU MUST RESET CONTROL!"
#INPUT V99 'Halt program
#GOTO :NOCONT 'Go back to message
#:CONT 'Continuation point
M99 (End of macro program)
```

GENERATING LOOPS

To help you understand looping, first consider subprogramming commands. Most CNC controls allow you to repeat subprograms a specified number of times. With Fanuc format, for example, an L word specifies the number of times the subprogram must be executed. The command

```
N050 M98 P1000 L5
```

tells the control to execute the program O1000 five times.

Keep in mind that repeating subprograms is rather limited. While you may be able to play some tricks with incremental techniques, if *anything* changes from one execution of the subprogram to the next, you cannot use subprogramming techniques to handle the application. With looping, there are no such limitations.

Like subprograms, loops allow you to repeat a series of commands a specified number of times. However, with looping, virtually *anything* can change each time through the loop. Before we show helpful applications for looping, you must understand the basic steps to develop a loop.

There are many, many ways to develop loops. And it is hard to argue with any technique that works. However, we recommend using a highly structured and consistent method of loop creation. With this method, all loops will have a *counter*. This counter is a variable, and will be an integer (whole) number (1, 2, 3, etc.). Another variable (or constant number) will represent the number of times the loop must be executed. This variable or constant also will be an integer. After each execution of the loop, the counter will be *stepped*. The conditional branching statement will be used to test the counter and determine when the loop has been executed the required number of times. Here is how you develop a loop in step-by-step manner.

1. Initialize counter and anything that changes each time through the loop.
2. Test to see if the loop is completed.
3. Perform calculations (if necessary).
4. Generate motions (if necessary).
5. Step counter and everything that changes each time through the loop.
6. Go back to the test.

We recommend testing early on to determine if the loop is finished (though some programmers elect to test at the end of the loop). Testing early might help you avoid machine motion problems caused by bad looping criteria. If you test first, at least there is the chance that the test will render a condition that will not allow machining to occur.

While it can be complicated enough to develop the logic for the loop itself, there are almost always other functions that must be addressed in a loop. For now, we will concentrate on the loop structure itself. Here is an example loop that simply counts to 10. We show this loop written in the three parametric programming versions addressed by this text. In all cases, notice that the counter variable is an integer, as is the constant number of times the loop is to be executed (10).

In custom macro B:

```
O1001 (Program number)
#100 = 1 (Initialize loop counter)
(Initialize anything else that changes)
N1 IF [#100 GT 10.0] GOTO 99 (Test if finished)
(Calculate, if necessary)
(Perform motions for this time through the loop, if necessary)
.
.
.
#100 = #100 + 1 (Step counter by one)
(Step anything else that changes)
GOTO 1 (Go back to test)
N99 M99 (End of custom macro)
```

In user task 2:

```
O1001 (Program number)
CTR1 = 1 (Initialize loop counter)
(Initialize anything else that changes)
N1 IF [CTR1 GT 10.0] GOTO N99 (Test if finished)
(Calculate, if necessary)
(Perform motions for this time through the loop, if necessary)
.
.
.
CTR1 = CTR1 + 1 (Step counter by one)
(Step anything else that changes)
GOTO N1 (Go back to test)
N99 RTS (End of user task program)
```

In Fadal's macro:

```
O1001 (Program number)
#V1 = 1 'Initialize loop counter
(Initialize anything else that changes)
```

```
#:LOOPBEGIN 'Statement label for loop beginning
#IF V1>10.0 THEN GOTO :LOOPEND 'Test if finished
(Calculate, if necessary)
(Perform motions for this time through the loop, if necessary)
.
.
.
#V1 = V1 + 1 'Step counter by one
(Step anything else that changes)
#GOTO :LOOPBEGIN 'Go back to test
#:LOOPEND 'End of loop
M99 (End of macro)
```

In all cases, the counter is initialized to one. The first time the conditional branching command is executed, the counter (one) is not greater than 10 and the conditional expression is evaluated as false. The loop is executed the first time. At the completion of this execution, the counter is stepped by one. Its new value will be two. When the test is executed the second time, it will still be false (two is not greater than 10). The loop is executed a second time. This continues until the loop has been executed 10 times. *After* the tenth execution, the counter's value will be eleven. When the test is made, it's conditional expression will be true and the loop will be exited.

SOLVING PROBLEMS RELATED TO LOOPS

If you have had experience with computer programming languages, you know that there is quite a bit more to creating loops than just setting and stepping a counter and testing to see when the loop is finished. Unfortunately, no two loops will be written in precisely the same manner, meaning you must treat every loop application as special. But if you understand and follow the guidelines we show, you should be able to develop even complex looping applications with structured and consistent methods. We recommend that you have firm answers to these questions *before* you begin writing a given loop.

When is a loop required?

There are two distinct times when loops are used. With one, looping is a convenience. It simply minimizes redundant commands. With the other, looping is mandatory. There is no other feasible way to handle the application.

When you wish to minimize commands. In the first instance, loops can be used (though they may not be mandatory) to dramatically shorten the number of redundant commands in a program. A few commands executed over and over again can sometimes save hundreds or thousands of commands. Earlier in this chapter, for example, we discussed how you can test arguments to see if they have been assigned.

In our discussion of Fadal's macro argument testing (V1 through V8), we had to program many commands just to test eight variables. Imagine having to test 50 variables. A loop can minimize the number of commands you would need. To keep this example simple, let's say all variables to be tested must be in the range of from minus 10 to plus 10 (–10.0 through +10.0). If any are not, you will want to generate the alarm message and halt the program. Consider these commands in Fadal's macro format:

```
O0001 (Program number)
N1 #V60 = 1 'Counter
N2 #:LOOPBEGIN 'Begin loop for argument testing
N3 #IF V60>50 THEN GOTO :EXIT 'Test if finished with loop
N4 #:REPEAT 'Beginning of repeated alarm
N5 #IF (V(V60)>–10.0) AND (V(V60)<10.0) THEN GOTO :NEXT 'Test cur-
    rent argument
N6 #PRINT "ARGUMENT V", V60, "IS OUT OF RANGE" 'Print alarm mes-
    sage
N7 #INPUT V99 'Halt program
N8 #GOTO :REPEAT 'Be sure program does not continue
N9 #:NEXT
N10 #V60 = V60 + 1
N11 #GOTO :LOOPBEGIN
N12 #:EXIT
N13 . . .
.
.
.
```

This series of commands actually tests 50 variables, from V1 through V50. If any are not in the desired range, a specific alarm message is printed and the program will halt.

In line N1, notice that V-series variable V60 is being used as the counter. In this application, it just so happens that the counter value (1, 2, 3, through 50) corresponds to the variable numbers being tested. So in line N5, V60 is referenced to specify the variable number being tested. The first time N5 is executed V1 will be the variable tested. The next time, V2 will be tested. This will be repeated until the loop is satisfied (after 50 loop executions). If any one of the variables is found to be out of range, a specific alarm message (from line N60) will be generated. Notice that V60 also is referenced in line N6 to specify which V-series variable is not correctly set.

While this application for looping may or may not excite you, the main point is that looping can dramatically reduce the need for redundant commands. While the commands needed to test 50 variables could be written in long-hand fashion without looping, a great number of error-prone commands would be required.

When the number of loop repetitions is a variable. Looping also is used when the number of loop executions varies from one looping series to the next. This is the more important reason for looping, since there is no other (feasible) way to handle the varying number of repeated commands.

For example, suppose you are developing a user-created canned cycle for multi-pass grooving on a turning center. As you look over your set of prints showing workpieces that will use your user-created canned cycle, you notice that some grooves will require just one plunging pass, while others need two, three, or even four passes. It would be time ill-spent to develop a separate parametric program for each number of plunging passes. Instead, you can develop a loop that will allow the number of passes to change from one loop execution to the next.

How many times must the loop be executed?

In some applications, this question is very easy to answer. If, for example, you are developing a user-created canned cycle for machining bolt-hole circles, and if the number of holes on the bolt-hole circle varies from one execution of your canned cycle to the next, you will need to create a loop to machine the holes. In this case, the number of times the loop must be executed is equal to the number of holes on the bolt-hole pattern (one execution of the loop for each hole).

On the other hand, it can sometimes be quite difficult to determine *up front* the number of times a loop must be executed, based on all variations the loop must handle. Consider, for example, a deep-hole peck drilling cycle. Figure 4-6 (page 195) shows the application.

This is a perfect application for a loop. Based on the hole depth and peck depth specifications, any number of loop executions will be required. If, for example, the hole depth is 3.0 in. (76.2 mm) and the peck depth is 1.0 in. (25.4 mm), how many loop executions (pecks) will be required? It should be easy to see that if you divide the hole depth by the peck depth (3.0/1.0), you come up with the number of pecks. In this case, three. If the hole depth is 4.0 in. (101.6 mm) and the peck depth is 0.5 in. (12.7 mm), eight loop executions (pecks) will be required. If the hole depth is 5.0 in. (127 mm) and the peck depth is 0.5 in., 10 loop executions will be required.

It would appear that determining the number of loop executions for the peck drilling loop is quite easy to calculate: simply divide the hole depth by the peck depth. In our example cases, it happens that the depth of the hole is evenly divisible by the peck depth. This, of course, will not always be the case. What if the hole depth is 3.05 in. (77.47 mm) and the peck depth is 1.0 in. (25.4 mm)?

One (rather poor) way to handle this problem is to force the user of this parametric program to ensure that the hole depth is evenly divisible by the peck depth. This will force anyone using the parametric program to perform calculations prior to using your program, but it makes your loop easier to write. This is the reverse of good programming practice. Parametric programs should always be written in a way that makes it easiest for the person *using* your parametric program, not the way that makes it easiest for you to write.

As you consider this application further, it may occur to you that if you round the result of the hole depth divided by the peck depth, you'll come up with an integer equal to the number of executions (pecks) required for the loop. However, you also will have to recalculate the peck depth by dividing the hole depth by the newly calculated number of pecks. Here are the calculations in all three versions of parametric programming addressed in this text. We also include the counter variable as well as the test that determines when the loop is completed.

In custom macro B. We use Z (#26) to represent the hole depth and Q (#17) to represent the peck depth.

```
O1001 (Deep hole peck drilling cycle)
#100 = ROUND[#26/#17] (Number of pecks)
#17 = #26/#100 (Recalculated peck depth)
#101 = 1 (Initialize counter for loop)
.
.
N1 IF [#101 GT #100] GOTO 99 (If finished, exit loop)
.
.
.
N99 M99 (End of custom macro)
```

In user task 2. We use HDEP to represent the hole depth and PDEP to represent the peck depth.

```
O1001 (Deep hole drilling cycle)
PECK = ROUND[HDEP/PDEP] (Number of pecks)
PDEP = HDEP/PECK (Recalculated peck depth)
PASS = 1 (Initialize counter for loop)
.
.
N1 IF [PASS GT PECK] GOTO NEND
.
.
.
NEND RTS (End of user task program)
```

In Fadal's macro. We use V1 to represent the hole depth and V2 to represent the peck depth.

```
O1001 (Deep hole drilling cycle)
#SET RNDO
#V3 = RND (V1/V2) 'Number of pecks
#V2 = V1/V3 'Recalculated peck depth
#V4 = 1 'Initialize counter for loop
.
.
#:LOOPBEGIN
#IF V4>V3 THEN GOTO :LOOPEND
.
.
.
#:LOOPEND
M99 (End of macro)
```

This is exactly the kind of challenge you will face as you begin developing your own loops. Though each new loop will present its own special challenges, *one of the first challenges to overcome is determining how many times the loop will be executed.* We urge you to conquer this challenge before moving on with the loop.

Warning! You may be tempted to set up this peck drilling loop in such a way that it will test against the *Z*-axis position instead of a nice round number of passes. This same temptation will come any time your loop performs multiple machining operations. However, this solution to the number-of-passes problem almost always renders a rather poorly formatted ending to your loop.

For example, suppose the hole depth is 3.05 in. (77.47 mm) and the peck depth is 1 in. (25.4 mm). Your loop may plunge the tool the first time and then test to see if the hole is finished. Since 1.0 is not greater than 3.05, the depth value is stepped and the loop is executed a second time. After the second pass, the test shows that 2.0 is still not greater than 3.05. The depth value is again stepped and the loop is executed again. After the third pass, the hole will be 3.0 in. (76.2 mm) deep; not quite to depth yet, but the loop cannot make a full pass without going too deep.

This is what we mean by a poor loop ending. The loop must be exited, yet there is still machining left to do, meaning more motion commands must be programmed even after the loop. It is *always* better to completely solve the problem of the number of loop executions and by how much counters must be stepped *before* you begin writing the loop. Letting the loop test against a nice round number of executions will render a structured, consistent result.

What must be initialized and stepped?

Anything that changes each time through the loop must be initialized prior to the IF statement and stepped at the end of the loop. Beginners tend to have problems visualizing things that change from one loop execution to the next. You may find yourself in the middle of writing your loop before it occurs to you that some loop function must change during each loop execution. If you find yourself with this type of problem, it will help to write the first two or three passes you want the loop to perform *long-hand* with hard and fixed CNC commands. This will allow you to analyze what you need your loop to do and will make it easy to see what changes during each pass.

With the deep-hole peck drilling cycle, for example, two *Z* surfaces will change after each loop execution. The current peck position will change, as will the rapid approach position prior to each pass. (The rapid approach position outside the hole will not change. It will remain 0.1 in. [2.54 mm] above the surface to be machined.)

Also you must determine *by how much* each of these variables will change after each execution of the loop. We will call the change amount for each changing variable the *increment value*. In both cases pertaining to the deep-hole peck drilling cycle, the increment value will be equal to the peck depth. That is, after each peck the new rapid approach position and peck depth position will advance the peck depth deeper into the hole.

Here is the continuation of the deep-hole peck drilling cycle, including the variables that must be initialized and stepped.

In custom macro B. We use *Z* (26) to represent the hole depth and Q (#17) to represent the peck depth.

O1001 (Deep-hole peck drilling cycle)
#100 = ROUND[#26/#17] (Number of pecks)
#17 = #26/#100 (Recalculated peck depth)
#101 = 1 (Initialize counter for loop)
#102 = 0.1 (Initialize current rapid approach position)
#103 = – [#17] (Initialize current hole bottom position)
.
N1 IF [#101 GT #100] GOTO 99 (If finished, exit loop)
.
.
.
N99 M99 (End of custom macro)

In user task 2. We use HDEP to represent the hole depth and PDEP to represent the peck depth.

O1001 (Deep-hole drilling cycle)
PECK = ROUND[HDEP/PDEP] (Number of pecks)
PDEP = HDEP/PECK (Recalculated peck depth)
PASS = 1 (Initialize counter for loop)
CURR = 0.1 (Initialize current rapid approach position)

```
CURZ = – [PDEP] (Initialize current hole bottom position)
.
N1 IF [PASS GT PECK] GOTO NEND
.
.
.
NEND RTS (End of user task program)
```

In Fadal's macro. We use V1 to represent the hole depth and V2 to represent the peck depth.

```
O1001 (Deep-hole drilling cycle)
#SET RNDO
#V3 = RND (V1/V2] 'Number of pecks
#V2 = V1/V3 'Recalculated peck depth
#V4 = 1 'Initialize counter for loop
#V5 = 0.1 'Initialize current rapid approach position
#V6 = V2 'Initialize current hole bottom position
.
#:LOOPBEGIN
#IF V4>V3 THEN GOTO :LOOPEND
.
.
.
#:LOOPEND
M99 (End of macro)
```

Are any other constants required for the loop?

Though the deep-hole peck drilling cycle requires no other constant values, you will usually need to perform even more calculations outside the loop (prior to the IF statement) to come up with values needed within the loop. In an evenly-spaced bolt-hole circle parametric program, for example, the incremental angular distance between the holes will be the increment value for the current angle being machined. That is, after each hole is machined, the current angle must be stepped by the incremental angular distance between the holes. The incremental angular distance between the holes can be calculated as 360 divided by the number of holes on the bolt-hole pattern. Since this calculation renders a constant number and is only required once, it is best to place it outside the loop (prior to the IF statement). You will commonly find it possible to perform calculations outside the loop, minimizing the need to perform redundant calculations within the loop.

Are any approach motions required prior to the loop?

Most machining loops require that the tool be positioned in a certain way prior to starting the loop. This initial positioning movement is commonly done outside the loop, prior to the IF statement. In our deep-hole peck drilling cycle, for example, if this cycle is to be used on a turning center, it will be necessary to bring the tool up close to the workpiece to be machined (center of the spindle in *X* and 0.1 in. (2.54 mm) away from the workpiece in *Z*). Here is the additional command in each version of the deep hole drilling cycle.

In custom macro B. We use *Z* (#26) to represent the hole depth and Q (#17) to represent the peck depth.

```
O1001 (Deep-hole peck drilling cycle)
#100 = ROUND[#26/#17] (Number of pecks)
#17 = #26/#100 (Recalculated peck depth)
#101 = 1 (Initialize counter for loop)
#102 = 0.1 (Initialize current rapid approach position)
#103 = – [#17] (Initialize current hole bottom position)
G00 X0 Z0.1 (Rapid up into position)
N1 IF [#101 GT #100] GOTO 99 (If finished, exit loop)
.
.
.
N99 M99 (End of custom macro)
```

In user task 2. We use HDEP to represent the hole depth and PDEP to represent the peck depth.

```
O1001 (Deep-hole peck drilling cycle)
PECK = ROUND [HDEP/PDEP] (Number of pecks)
PDEP = HDEP/PECK (Recalculated peck depth)
PASS = 1 (Initialize counter for loop)
CURR = 0.1 (Initialize current rapid approach position)
CURZ = – [PDEP] (Initialize current hole bottom position)
G00 X0 Z0.1 (Rapid up into position)
N1 IF [PASS GT PECK] GOTO NEND
.
.
.
NEND RTS (End of user task program)
```

In Fadal's macro. We use V1 to represent the hole depth and V2 to represent the peck depth.

```
O1001 (Deep-hole peck drilling cycle)
#SET RND0
#V3 = RND (V1/V2] 'Number of pecks
```

```
#V2 = V1/V3 'Recalculated peck depth
#V4 = 1 'Initialize counter for loop
#V5 = 0.1 'Initialize current rapid approach position
#V6 = V2 'Initialize current hole bottom position
G00 X0 Z0.1 (Rapid up into position)
#:LOOPBEGIN
#IF V4>V3 THEN GOTO :LOOPEND
.
.
.
#:LOOPEND
M99 (End of macro)
```

What must be done during each loop execution?

This may sound like a simple question, but as you approach your first few loops, it will likely seem difficult to determine just what must be done each time the loop is executed. To better understand, we break actions in the loop into three separate types.

Calculations for movements are commonly required during loops that perform machining operations. In almost all cases, something changes from one execution (machining pass) of the loop to the next. If positioning movements change, calculations must be used to specify the locations of the new positioning movements. In the case of the deep-hole drilling cycle, since the stepping of each variable handles all the differences from one loop execution to the next, there is no need to perform special calculations prior to the motion commands.

Motion commands also are required for loops that perform motion tasks. As you write your first few loops, be careful not to do too much during your motion commands. Make sure you isolate only those commands required to make one machining pass. You will notice a tendency to proceed right into the second or third pass. In the case of the deep-hole drilling cycle, the loop must perform three motions. First, it must rapid the tool to its (new) approach position in Z. Second, it must feed the tool to its current bottom position. Finally, it must retract the tool out of the hole.

Counter and changing variable stepping must be done at the completion of each loop execution. Be sure you step everything that changes each time through the loop. In our loop for the deep-hole drilling example, along with the counter, we must step the current approach position and the current bottom position. After stepping everything that changes, make sure to send the control back to the IF statement to repeat the test that determines when the loop is finished (with a GOTO statement).

Here is the completed loop for deep-hole drilling in all three versions of parametric programming addressed in this text.

In custom macro B. We use Z (#26) to represent the hole depth and Q (#17) to represent the peck depth.

```
O1001 (Deep-hole peck drilling cycle)
#100 = ROUND [#26/#17] (Number of pecks)
#17 = #26/#100 (Recalculated peck depth)
#101 = 1 (Initialize counter for loop)
#102 = 0.1 (Initialize current rapid approach position)
#103 = – [#17] (Initialize current hole bottom position)
G00 X0 Z0.1 (Rapid up into position)
N1 IF [#101 GT #100] GOTO 99 (If finished, exit loop)
G00 Z#102 (Rapid to current approach position)
G01 Z#103 F0.007 (Plunge to current bottom position)
G00 Z0.1 (Rapid out of the hole)
#101 = #101 + 1 (Step loop counter)
#102 = #102 – #17 (Step current rapid approach position)
#103 = #103 – #17 (Step current hole bottom position)
GOTO 1 (Go back to test)
N99 M99 (End of custom macro)
```

In user task 2. We use HDEP to represent the hole depth and PDEP to represent the peck depth.

```
O1001 (Deep-hole peck drilling cycle)
PECK = ROUND [HDEP/PDEP] (Number of pecks)
PDEP = HDEP/PECK (Recalculated peck depth)
PASS = 1 (Initialize counter for loop)
CURR = 0.1 (Initialize current rapid approach position)
CURZ = – [PDEP] (Initialize current hole bottom position)
G00 X0 Z0.1 (Rapid up into position)
N1 IF [PASS GT PECK] GOTO NEND
G00 Z = CURR (Rapid to current approach position)
G01 Z = CURZ F0.007 (Plunge to current bottom position)
G00 Z0.1 (Rapid out of the hole)
PASS = PASS + 1 (Step loop counter)
CURR = CURR – PDEP (Step current rapid approach position)
CURZ = CURZ – PDEP (Step current hole bottom position)
GOTO N1 (Go back to test)
NEND RTS (End of user task program)
```

In Fadal's macro. We use V1 to represent the hole depth and V2 to represent the peck depth.

```
O1001 (Deep-hole peck drilling cycle)
#SET RNDO
#V3 = RND (V1/V2] 'Number of pecks
#V2 = V1/V3 'Recalculated peck depth
#V4 = 1 'Initialize counter for loop
#V5 = 0.1 'Initialize current rapid approach position
#V6 = – (V2) 'Initialize current hole bottom position
G00 X0 Z0.1 (Rapid up into position)
#:LOOPBEGIN
#IF V4>V3 THEN GOTO :LOOPEND
#R1 = V5 'Set R1 to the value of V5
G00 Z+R1 (Rapid to current approach position)
#R1 = V6 'Set R1 to value of V6
G01 Z+V6 F0.007 (Plunge to current bottom position)
G00 Z0.1 (Rapid out of the hole)
#V4 = V4 + 1 'Step loop counter
#V5 = V5 – V2 'Step current rapid approach position
#V6 = V6 – V2 'Step current hole bottom position
GOTO :LOOPBEGIN 'Go back to test
#:LOOPEND
M99 (End of macro)
```

LOOP EXAMPLES

We include two real-world examples of parametric programs that include loops, one for machining centers and one for turning centers. Both the bolt-hole circle and multiple-pass grooving examples that follow should reinforce your understanding of how loops function. For each example, we answer the questions posed earlier about loop criteria. It's important to remember, however, that no two looping applications will be the same, and each will challenge you with different problems. If you follow our guidelines for developing loops, and if you answer all important questions relative to the loop, you should be able to tackle even the most complex loops.

Bolt-hole circle example

Figure 4-7 (page 195) shows the drawing for this application.

When is a loop required? This application requires a loop because we wish our bolt-hole machining parametric program to machine any number of holes. It would not be feasible to try to write a separate parametric program for each number of holes.

How many times will the loop be executed? For this application, the answer to this question is simple. The loop will be executed once per hole.

What must be initialized and stepped? Aside from the loop counter, the current angle (initialized as the value of the starting angle), must be initialized and stepped. While the *XY* position will change for each hole, it will be a function (sine and cosine) of the current angle.

Are there any constants for the loop? Yes; the incremental angular distance (calculated as 360 divided by the number of holes) must be calculated before the current angle can be stepped. Also, the *Z* surfaces can be set as constants to avoid redundant calculations within the loop.

Are any approach movements required before the loop? In our application, we will assume that the CNC programmer has not positioned the tool in *XY* prior to calling this parametric program. However, we will assume the spindle is running, tool-length compensation is instated, and the tool is ready to begin machining.

What must be done during the loop itself? The loop must first calculate the *XY* position of the current hole based on the current angle. In *X*, this will be done with the formula *cosine of the angle times the radius plus the dimension in X to the center of the bolt-hole pattern.* In *Y* the sine function will be used. Next the hole will be machined using the canned cycle type specified in the call statement. Finally, the counter will be stepped (by one) as will the current angle (by the incremental angular distance between the holes).

In custom macro B. Figure 4-8 (page 196) shows the letter address arguments to be used in this example.

Notice that the cycle type is specified with argument C. The number corresponding to C represents the G code number for the desired canned cycle. For example, C81. requests a standard drilling cycle; C84. requests a tapping cycle; and so on. Here is a portion of a main program that uses the bolt-hole circle custom macro.

O0008 (Main program)
N005 G54 G90 S800 M03 T02 (Select coordinate system, absolute mode, start spindle, get next tool ready)
N010 G00 X3.0 Y2.5 (Rapid to center of bolt hole pattern)
N015 G43 H01 Z.1 (Instate tool length compensation, rapid up to workpiece)
N020 G65 P1008 X3.0 Y2.5 Z0 R1.75 D0.75 A45.0 H8.0 C81. F5.0 (Machine entire bolt-hole pattern with drilling cycle)
N025. . .
.
.
.

In custom macro B. X is represented by local variable #24, *Y* by #25, *Z* by #26, R by #18, D by #7, A by #1, H by #11, C by #3, and F by #9.

O1008 (Custom macro to machine bolt-hole circle)
#101 = 1 (Initialize counter)

```
#102 = #1 (Initialize current angle to A)
#103 = 360/#11 (Constant for incremental angular distance between holes)
#104 = #26 + 0.1 (Constant for rapid approach plane)
#105 = #26 – #7 (Constant for Z bottom position of hole)
N1 IF [#101 GT #11] GOTO 99 (Test if loop is finished)
#110 = #24 + COS [#102] * #18 (Calculate X position for current hole based on
   current angle)
#111 = #25 + SIN [#102] * #18 (Calculate Y position for current hole based on
   current angle)
G#3 X#110 Y#111 R#104 Z#105 F#9 (Machine current hole)
G80 (Cancel cycle)
#101 = #101 + 1 (Step counter)
#102 = #102 + #103 (Step current angle)
GOTO 1 (Go back to test at loop beginning)
N99 M99 (End of custom macro)
```

In user task 2. Figure 4-9 (page 197) shows the local variable arguments to be used in this example.

Notice that the cycle type is specified with argument CYCL. The number corresponding to CYCL represents the G-code number for the desired canned cycle. For example, CYCL = 81. requests a standard drilling cycle; CYCL = 84. requests a tapping cycle; and so on. Here is a portion of a main program that uses the bolt-hole circle user task program.

O0008 (Main program)
N005 G15 H01 (Select coordinate system)
N010 G90 S800 M03 T02 (Select absolute mode, start spindle, get next tool ready)
N015 G00 X3.0 Y2.5 (Rapid to center of bolt-hole pattern)
N020 G56 H01 Z.1 (Instate tool-length compensation, rapid up to workpiece)
N025 CALL O1008 XCTR = 3.0 YCTR = 2.5 ZSUR = 0 RAD = 1.75 HDEP = 0.75 AGL = 45.0 HLES = 8.0 CYCL = 81. FEED = 5.0 (Machine entire bolt-hole pattern with drilling cycle)
N030. . .
.
.
.

This is the user task 2 program:

```
O1008 (User task program to machine bolt-hole circle)
HLE = 1 (Initialize counter)
CURA = AGL (Initialize current angle to AGL)
```

```
IAGL = 360/HLES (Constant for incremental angular distance between holes)
RAPP = ZSUR + 0.1 (Constant for rapid approach plane)
ZBTM = ZSUR – HDEP (Constant for Z bottom position of hole)
N1 IF [HLE GT HLES] GOTO N99 (Test if loop is finished)
CURX = XCTR + COS [CURA] * RAD (Calculate X position for current hole
   based on current angle)
CURY = YCTR + SIN [CURA] * RAD (Calculate Y position for current hole
   based on current angle)
G = CYCL X = CURX Y = CURY R = RAPP Z = ZBTM F = FEED (Machine
   current hole)
G80 (Cancel cycle)
HLE = HLE + 1 (Step counter)
CURA = CURA + IAGL (Step current angle)
GOTO N1 (Go back to test at loop beginning)
N99 M99 (End of user task program)
```

In Fadal's macro. Figure 4-10 (page 198) shows the V-series variable arguments to be used in this example.

Notice that the cycle type is specified with argument V8. The number corresponding to V8 represents the G-code number for the desired canned cycle. For example, V8 = 81. requests a standard drilling cycle; V8 = 84. requests a tapping cycle; and so on. A portion of a main program using the bolt-hole macro follows.

```
O0008 (Main program)
N005 G54 G90 S800 M03 T02 (Select coordinate system, absolute mode, start
   spindle, get next tool ready)
N010 G00 X3.0 Y2.5 (Rapid to center of bolt-hole pattern)
N015 G43 H01 Z.1 (Instate tool-length compensation, rapid up to workpiece)
N020 #V1 = 3.0 'Assign argument V1
N025 #V2 = 2.5 'Assign argument V2
N030 #V3 = 0 'Assign argument V3
N035 #V4 = 1.75 'Assign argument V4
N040 #V5 = 45.0 'Assign argument V5
N045 #V6 = 8.0 'Assign argument V6
N050 #V7 = 5.0 'Assign argument V7
N055 #V8 = 81.0 'Assign argument V8
N060 #V9 = 0.75 'Assign argument V9
N065 M98 P1008 (Machine entire bolt-hole pattern with drilling cycle)
N070. . .
.
.
.
```

Here is the macro to machine the bolt-hole circle.

```
O1008 (Macro to machine bolt-hole circle)
#V21 = 1 'Initialize counter
#V22 = V5 'Initialize current angle to V5
#V23 = 360/V6 'Constant for incremental angular distance between holes
#R4 = V7 'Set R4 to feedrate
#R5 = V8 'Set R5 to cycle type
#R6 = V3 + 0.1 'Constant for rapid approach plane
#R7 = V3 – V9 'Constant for Z bottom position of hole
#:LOOPBEGIN 'Beginning point of loop
#IF [V21>V6] THEN GOTO :EXIT 'Test if loop is finished
#R8 = V1 + COS (V22) * V4 'Calculate X position for current hole based on
   current angle
#R9 = V2 + SIN (V22) * V4 'Calculate Y position for current hole based on cur-
   rent angle
G + R5 X + R8 Y + R9 R0 + R6 Z + R7 F + R4 (Machine current hole)
G80 (Cancel cycle)
#V21 = V21 + 1 'Step counter
#V22 = V22 + V23 'Step current angle
#GOTO :LOOPBEGIN 'Go back to test at loop beginning
#:EXIT 'End of loop
M99 (End of macro)
```

Multiple-pass grooving example

When is a loop required? Figure 4-11 (page 199) shows the drawing for a multiple-pass grooving application. This application may or may not require a loop. If the width of the grooving tool is equal to the width of the groove, no loop is required, and only one plunging pass will be made. However, if the groove width is wider than the tool width, multiple passes will be required. The loop will handle any number of required passes.

How many times will the loop be executed? This question is more difficult. Suppose, for example, the groove width is 0.25 in. (6.35 mm) and the tool width is 0.125 in. (3.175 mm). Most programmers would agree that *three* passes would be required, since there must be some overlap between passes. Figure 4-12 (page 200) shows the argument assignments used for stepping the grooving tool in custom macro B and one method of determining how many passes are required (with *# of passes*) when the groove is wider than the grooving tool. The 0.02-in. (0.51-mm) value in this formula is the minimum overlap between passes.

What must be initialized and stepped? Aside from the loop counter, the current *Z* position of the grooving tool must be stepped. Figure 4-12 also shows the formula to determine by how much the *Z* position must be stepped (with *PASS*).

Are there any constants for the loop? Yes; the *X* positions for each pass remain the same, as so the amount to move over between passes (PASS).

Are any approach movements required before the loop? In this application, we will assume that the CNC programmer has positioned the tool up close to the workpiece, but not precisely where it must be to make the first pass, meaning an approach movement must be made before the loop.

What must be done during the loop itself? The loop must position the tool to the current *Z* point, plunge the groove, pause, and get the tool out of the groove. Then the counter must be stepped by one and the current *Z* position must be stepped by the proper move over amount (PASS) shown in Figure 4-12.

This portion of the main program calls the custom macro to machine a groove:

```
O0009 (Program number)
.
.
(Turning operations to get workpiece ready to groove)
.
.
N155 T0505 (Index to grooving tool)
N160 G96 S500 M03 (Start spindle at 500 sfm)
N165 G00 X4.0 Z0.1 (Rapid up close to workpiece)
N170 G65 P1009 B1.5 S1.0 Z1.0W0.25T0.125 C0.03 F0.005 (Machine groove)
N175 G00 X6.0 Z5.0 (Rapid to tool-change position
N180 M01 (Optional stop)
N185 . . .
.
.
.
```

In the following custom macro B program for this application, letter address argument B will be represented in the custom macro by local variable #2, S by #19, T by #20, F by #9, C by #3, W by #23, and Z by #26.

```
O1009 (Custom macro to machine groove)
#101 = FUP [#23/[#20 – 0.02]] (Calculate number of passes)
#102 = #23 – #20 (Calculate stock remaining after first pass)
#103 = #102/[#101 – 1] (Calculate move over amount in Z)
#104 = 1 (Initialize pass counter)
#105 = #26 (Initialize current Z position to plunge)
#106 = #2 + 0.2 (Calculate clearance position in X)
G00 X#106 Z – [#105] (Rapid to first plunging position)
IF [#23 GT #20] GOTO 25 (Test if multiple passes are necessary)
```

```
(Only one pass is required)
G01 X#19 F#9 (Plunge groove)
G04 P500 (Pause to eliminate tool pressure)
G00 X#106 (Rapid out of groove)
GOTO 50 (Skip multiple passes)
N25 IF [#104 GT #101] GOTO 50 (Test if finished with multiple passes)
G00 Z – #105 (Rapid to current Z position for plunge)
G01 X#19 F#9 (Plunge groove in current position)
G04 P500 (Pause to eliminate tool pressure)
G00 X#106 (Rapid out of groove)
#105 = #105 – #103 (Step current plunging position in Z)
#104 = #104 + 1 (Step counter)
GOTO 25 (Go back to test if finished making plunging passes)
(Machine chamfers)
N50 G00 Z – [#26 + #3] (Rapid to left side of left chamfer)
G01 X#2 (Feed flush with big diameter)
X[#2 – [2 * #3]] Z – #26 (Plunge left chamfer)
G00 X#106 (Rapid out of groove)
Z – [#26 – #23 – #3 + #20] (Rapid to right side of right chamfer)
G01 X#2 (Feed flush with big diameter)
X[#2 – [2 * #3]] Z – [#26 -#23 + #20] (Plunge right chamfer)
G00 X#106 (Rapid out of groove)
M99 (End of custom macro)
```

In user task 2. Figure 4-13 (page 201) shows the local variable arguments to be used in the user task 2 version of this application.

Here is the portion of the main program that calls the user task program to machine a groove.

```
O0009 (Program number)
.
.
(Turning operations to get workpiece ready to groove)
.
.
N155 T0505 (Index to grooving tool)
N160 G96 S500 M03 (Start spindle at 500 sfm)
N165 G00 X4.0 Z0.1 (Rapid up close to workpiece)
N170 CALL O1009 BIGD = 1.5 SMLD = 1.0 ZPT = 1.0 GWID = 0.25 TWID
   = 0.125 CHAM = 0.03 FEED = 0.005 (Machine groove)
N175 G00 X6.0 Z5.0 (Rapid to tool-change position)
N180 M01 (Optional stop)
N185 . . .
```

.

.

.

The user task 2 program for this application looks like this:

```
O1009 (user task program to machine groove)
TPAS = FUP [GWID/[TWID – 0.02]] (Calculate number of passes)
STK = GWID – TWID (Calculate stock remaining after first pass)
ZMVE = STK/[TPAS – 1] (Calculate move over amount in Z)
CPAS = 1 (Initialize pass counter)
CURZ = ZPT (Initialize current Z position to plunge)
XCLR = BDIA + 0.2 (Calculate clearance position in X)
G00 X = XCLR Z = – [CURZ] (Rapid to first plunging position)
IF [GWID GT TWID] THEN GOTO N25 (Test if multiple passes are necessary)
(Only one pass is required)
G01 X = SMLD F = FEED (Plunge groove)
G04 P500 (Pause to eliminate tool pressure)
G00 X = XCLR (Rapid out of groove)
GOTO N50 (Skip multiple passes)
N25 IF [CPAS GT TPAS] GOTO N50 (Test if finished with multiple passes)
G00 Z = – CURZ (Rapid to current Z position for plunge)
G01 X = SMLD F = FEED (Plunge groove in current position)
G04 P500 (Pause to eliminate tool pressure)
G00 X = XCLR (Rapid out of groove)
CURZ = CURZ – ZMVE (Step current plunging position in Z)
CPAS = CPAS + 1 (Step counter)
GOTO N25 (Go back to test if finished making plunging passes)
(Machine chamfers)
N50 G00 Z = – [ZPT + CHAM] (Rapid to left side of left chamfer)
G01 X = BIGD (Feed flush with big diameter)
X = [BIGD – [2 * CHAM]] Z = – ZPT (Plunge left chamfer)
G00 X = XCLR (Rapid out of groove)
Z = – [ZPT – GWID – CHAM + TWID] (Rapid to right side of right chamfer)
G01 X = BIGD (Feed flush with big diameter)
X = [BIGD – [2 * CHAM]] Z = – [ZPT – GWID + TWID] (Plunge right chamfer)
G00 X = XCLR (Rapid out of groove)
RTS (End of user task program)
```

In Fadal's macro. Figure 4-14 (page 202) shows the V-series variable arguments to be used in this application.

The portion of the main program that calls the macro to machine a groove is:

```
O0009 (Program number)
.
.
(Turning operations to get workpiece ready to groove)
.
.
N155 T0505 (Index to grooving tool)
N160 G96 S500 M03 (Start spindle at 500 sfm)
N165 G00 X4.0 Z0.1 (Rapid up close to workpiece)
N170 #V1 = 1.5 'Assign argument V1
N175 #V2 = 1.0 'Assign argument V2
N180 #V3 = 1.0 'Assign argument V3
N185 #V4 = 0.24 'Assign argument V4
N190 #V5 = 0.125 'Assign argument V5
N195 #V6 = 0.03 'Assign argument V6
N200 #V7 = 0.005 'Assign argument V7
N175 M98 P1009 (Machine groove)
N180 G00 X6.0 Z5.0 (Rapid to tool-change position)
N185 M01 (Optional stop)
N190 . . .
.
.
.
```

The macro for this application looks like this:

```
O1009 (Macro to machine groove)
#V51 = INT [V4/[V5 – 0.02]] + 1 'Calculate number of passes
#V52 = V4 – V5 'Calculate stock remaining after first pass
#V53 = V52/[V51 – 1] 'Calculate move over amount in Z
#V54 = 1 'Initialize pass counter
#R4 = V3 'Initialize current Z position to plunge
#R5 = V1 + 0.2 'Calculate clearance position in X
#R6 = V7 'Set R6 to feedrate
G00 X + R5 Z – R4 (Rapid to first plunging position)
#R9 = V2 'Set R9 to plunge diameter
IF V4>V5 THEN GOTO :MULTPASS 'Test if multiple passes are necessary
(Only one pass is required)
G01 X + R9 F + R6 (Plunge groove)
G04 P500 (Pause to eliminate tool pressure)
G00 X + R5 (Rapid out of groove)
GOTO :CHAMFERS 'Skip multiple passes
#:MULTPASS 'Label to begin multiple pass loop
```

```
#IF V54>V51 THEN GOTO :CHAMFERS 'Test if finished with multiple passes
G00 Z – R4 (Rapid to current Z position for plunge)
G01 X + R9 F + R6 (Plunge groove in current position)
G04 P500 (Pause to eliminate tool pressure)
G00 X + R5 (Rapid out of groove)
#R4 = R4 – V53 'Step current plunging position in Z
#V54 = V54 + 1 'Step counter
#GOTO :MULTPASS (Go back to test if finished making plunging passes)
(Machine chamfers)
#:CHAMFERS 'Label to begin chamfers
#R1 = V3 + V6 'Calculate Z position at left side of left chamfer
G00 Z – R1 (Rapid to left side of left chamfer)
#R1 = V1 'Set R1 to big diameter
G01 X + R1 (Feed flush with big diameter)
#R2 = V1 – 2 * V6 'Calculate X bottom of chamfer
#R3 = V3 'Set R3 to left side of groove in Z
X + R2 Z – R3 (Plunge left chamfer)
G00 X + R5 (Rapid out of groove)
#R3 = V3 – V4 – V6 + V5 'Calculate Z of right side of right chamfer
Z-R3 (Rapid to right side of right chamfer)
G01 X + R1 (Feed flush with big diameter)
#R3 = V3 – V4 + V5 'Calculate Z end of right chamfer
X + R2 Z – R3 (Plunge right chamfer)
G00 X + R5 (Rapid out of groove)
M99 (End of macro)
```

NESTING LOOPS

As can be seen, loops add a great deal of power to your parametric programming capabilities. In the bolt-hole cycle, 10-12 commands can machine any number of holes. In the grooving cycle, about 10 commands can make any number of grooving passes. In some looping applications, just a few commands can cause thousands of machine movements.

Though we have focused much attention on looping, we wish to show one more powerful capability with looping, called loop nesting. Nesting a loop means incorporating one loop inside another; it is quite helpful in machining center applications.

A grid pattern example

Figure 4-15 (page 203) shows an instance when loop nesting can be helpful.

This workpiece requires the machining of several rows of evenly spaced holes. Let's say this is just one of many workpieces your company machines that require grid patterns of holes to be machined, and that the number of holes along *X* and *Y* vary from one workpiece to the next.

This application requires one loop inside another. The series of holes along the *X* axis, for example, can be machined in one loop that is inside the loop that machines the *Y* rows of holes. This, of course, could be reversed if preferred. Figure 4-16 (page 204) shows the custom macro B arguments we'll use for this example. (Only one version of this example will be shown.)

The object is to keep it simple. We'll keep certain things constant, like *Z* surface, hole depth, and feedrate, even though these values would likely need to be specified with arguments. Here is a portion of a main program that machines the grid pattern.

```
O0010 (Main program)
.
.
N050 T05 M06 (Place 1/8-in. (3.175-mm) drill in spindle)
N055 G54 G90 S400 M03 T06 (Select coordinate system, absolute mode, start
   spindle, get next tool ready)
N060 G00 X0 Y0 (Rapid over to workpiece in XY)
N065 G43 H05 Z.1 (Instate tool-length compensation, rapid up in Z)
N070 G65 P1010 X0.25 Y0.25 U0.25 V0.5 W10.0 H5.0 (Machine entire grid
   pattern)
.
.
.
```

Here is the custom macro that machines the entire grid pattern. Letter address *X* is represented by local variable #24, *Y* by #25, U by #21, V by #22, W by #23, and H by #11.

```
O1010 (Custom macro to machine grid pattern)
G00 X#24 Y#25 (Move to lower left hole location)
(Begin outside loop for rows of holes along Y)
#101 = 1 (Row counter for outside loop)
#102 = #25 (Current Y position for holes)
N1 IF [#101 GT #11] GOTO 99 (Test if Y row loop is finished)
(Begin inside loop for holes along X)
#111 = 1 (Hole counter for inside loop)
#112 = #24 (Current X position for hole)
N10 IF [#111 GT #23] GOTO 50 (Test if X row loop is finished)
G81 X#112 Y#102 R0.1 Z – 0.5 F4.5 (Drill current hole)
G80 (Cancel cycle)
#111 = #111 + 1 (Step hole counter for inside loop)
#112 = #112 + #21 (Step X position of hole)
```

```
GOTO 10 (Go back to test for inside loop)
N50 #101 = #101 + 1 (Step Y row counter for outside loop)
#102 = #102 + #22 (Step Y position for hole)
GOTO 1 (Go back to test for outside loop)
N99 M99 (End of custom macro)
```

The counter for the rows of holes along *Y* is #101. The counter for the holes along *X* is #111. Notice how the test in N1 begins the outside loop. The first time this command is read, the conditional expression will be false, which allows the entry to the inside loop. Holes along *X* will be machined until the test in line N10 is true (the first row of holes along *X* is finished). Then #101 is stepped and the second row of holes is machined. While this example may take some study, notice how powerful this short program truly is. Literally thousands of holes can be machined with one calling command to this very short program.

Again, we do not show this example in user task 2 or Fadal's macro. However, you should now be able to convert this program to any version of parametric programming with relative ease.

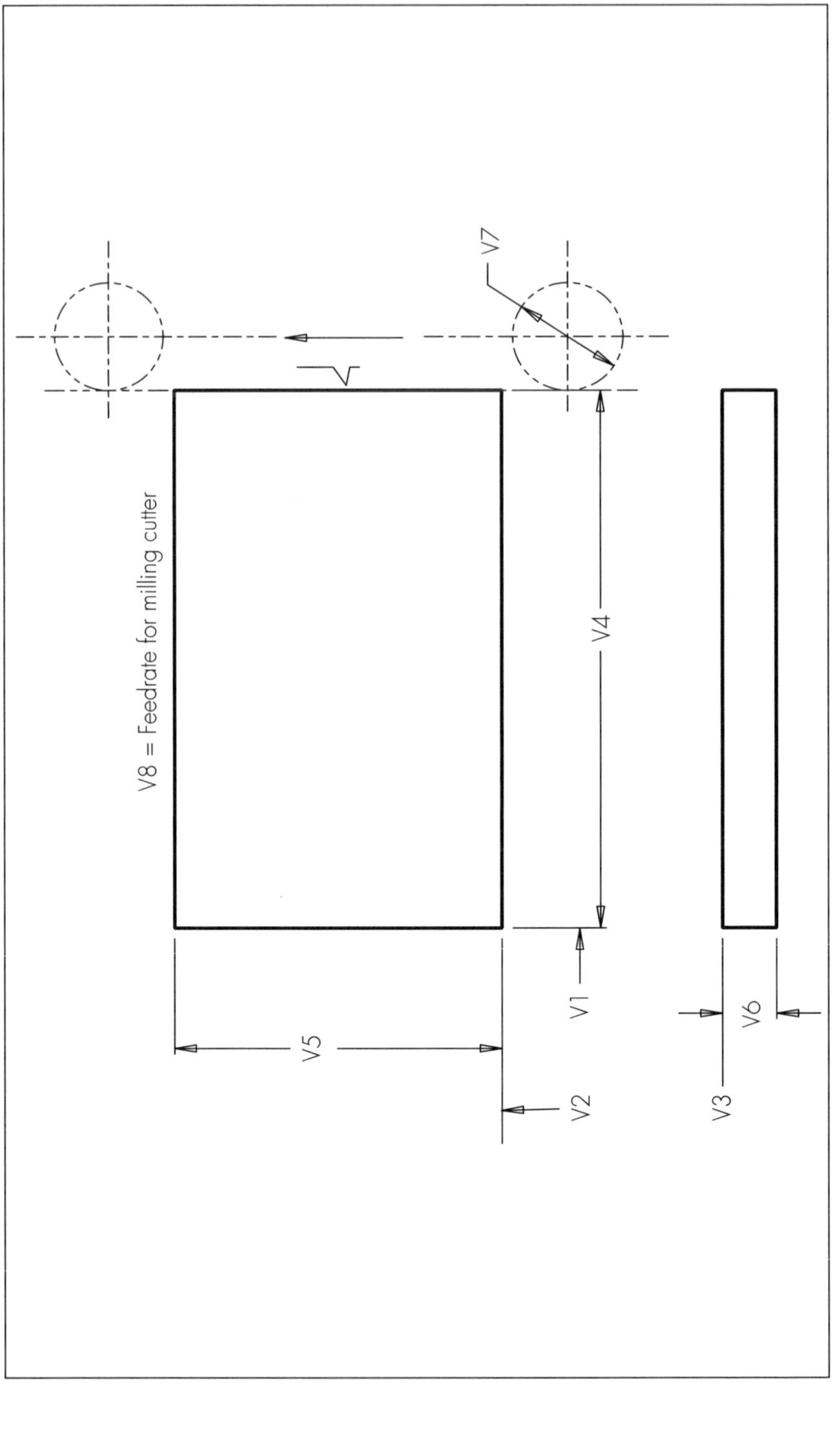

Figure 4-1. Testing V-series arguments in Fadal's macro involves establishing a range of allowable settings for each V argument.

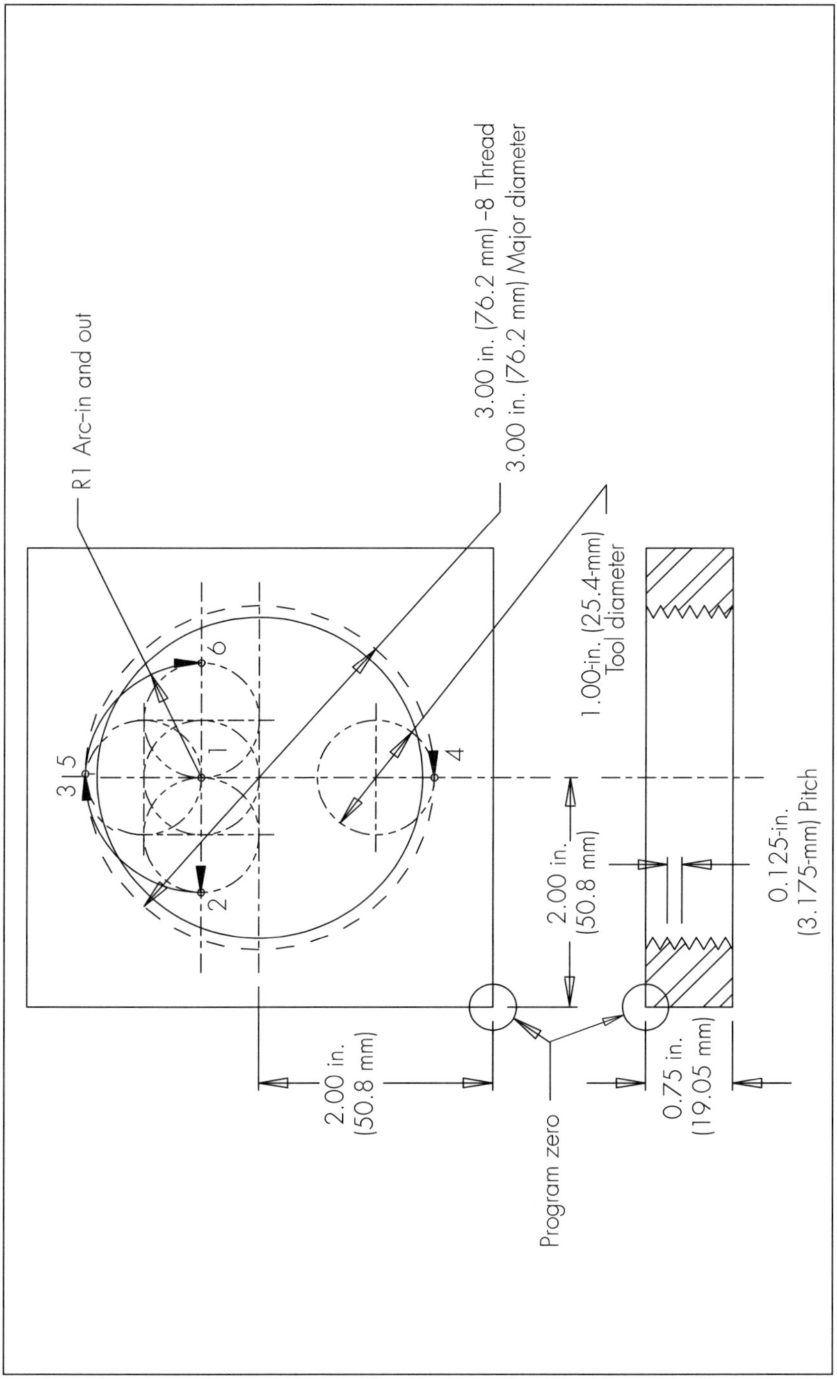

Figure 4-2. Engineering drawing of enhanced thread-milling application for which the user-created canned-cycle program will be written.

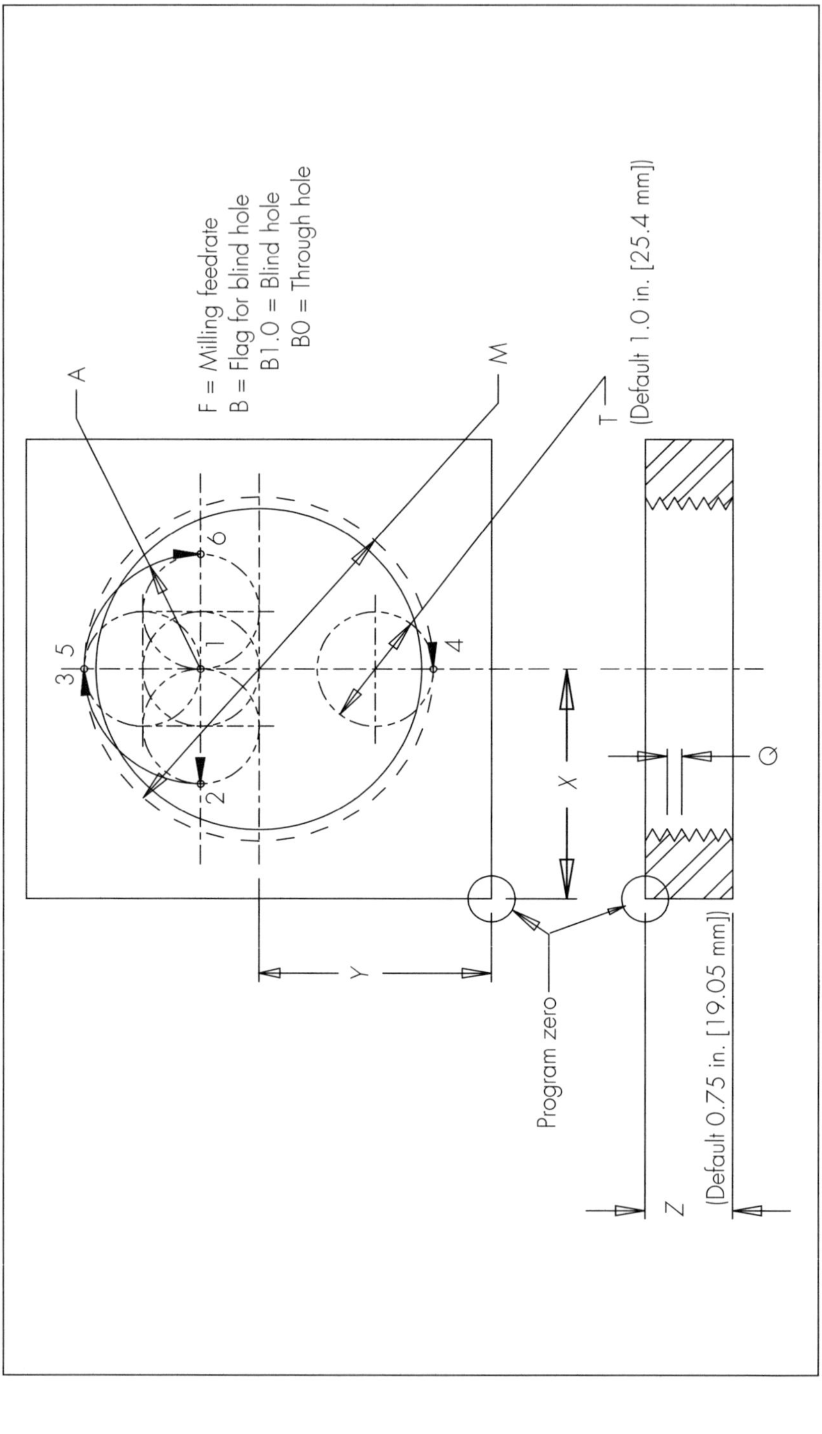

Figure 4-3. Arguments for the custom macro B version of the thread-milling application.

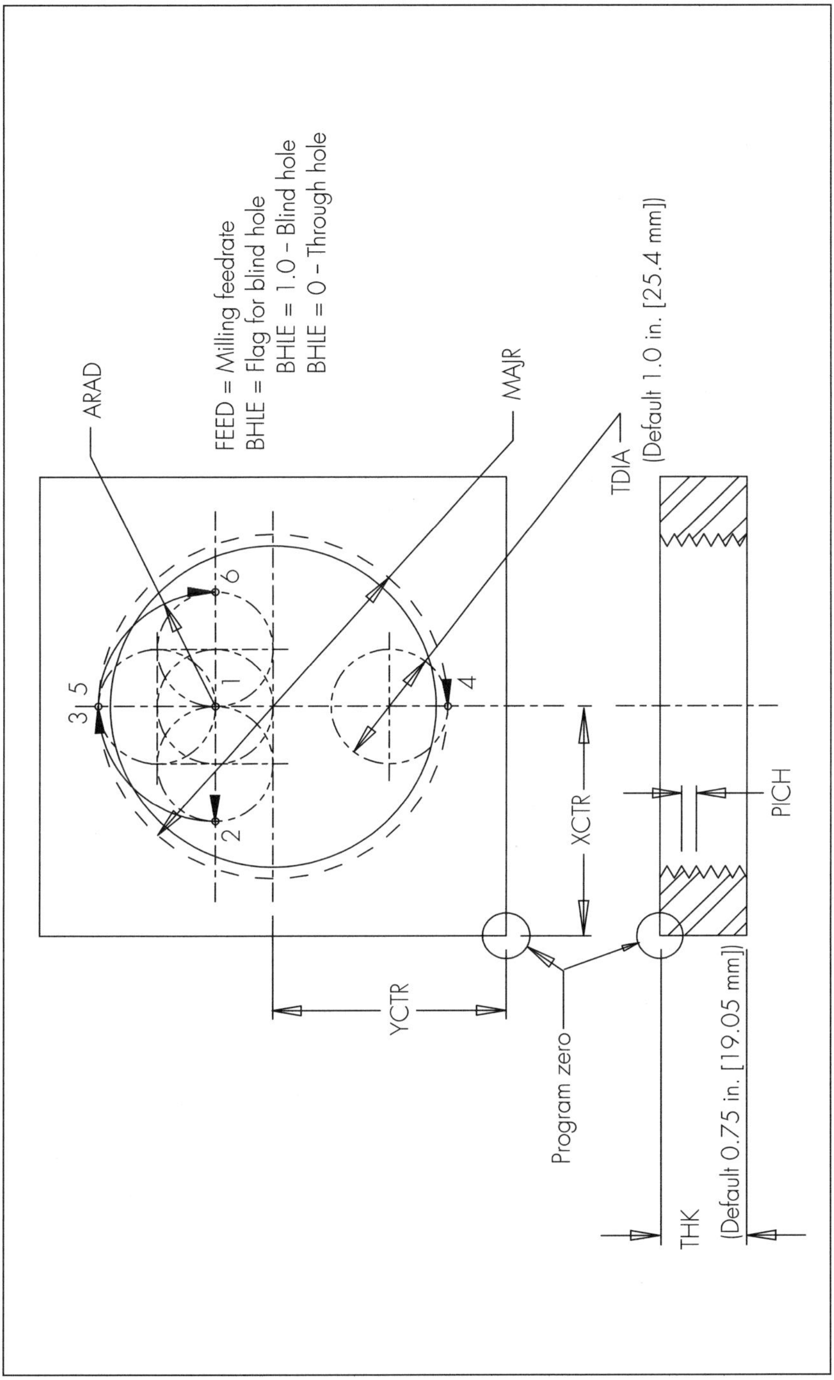

Figure 4-4. Arguments for the user task 2 version of the thread-milling application.

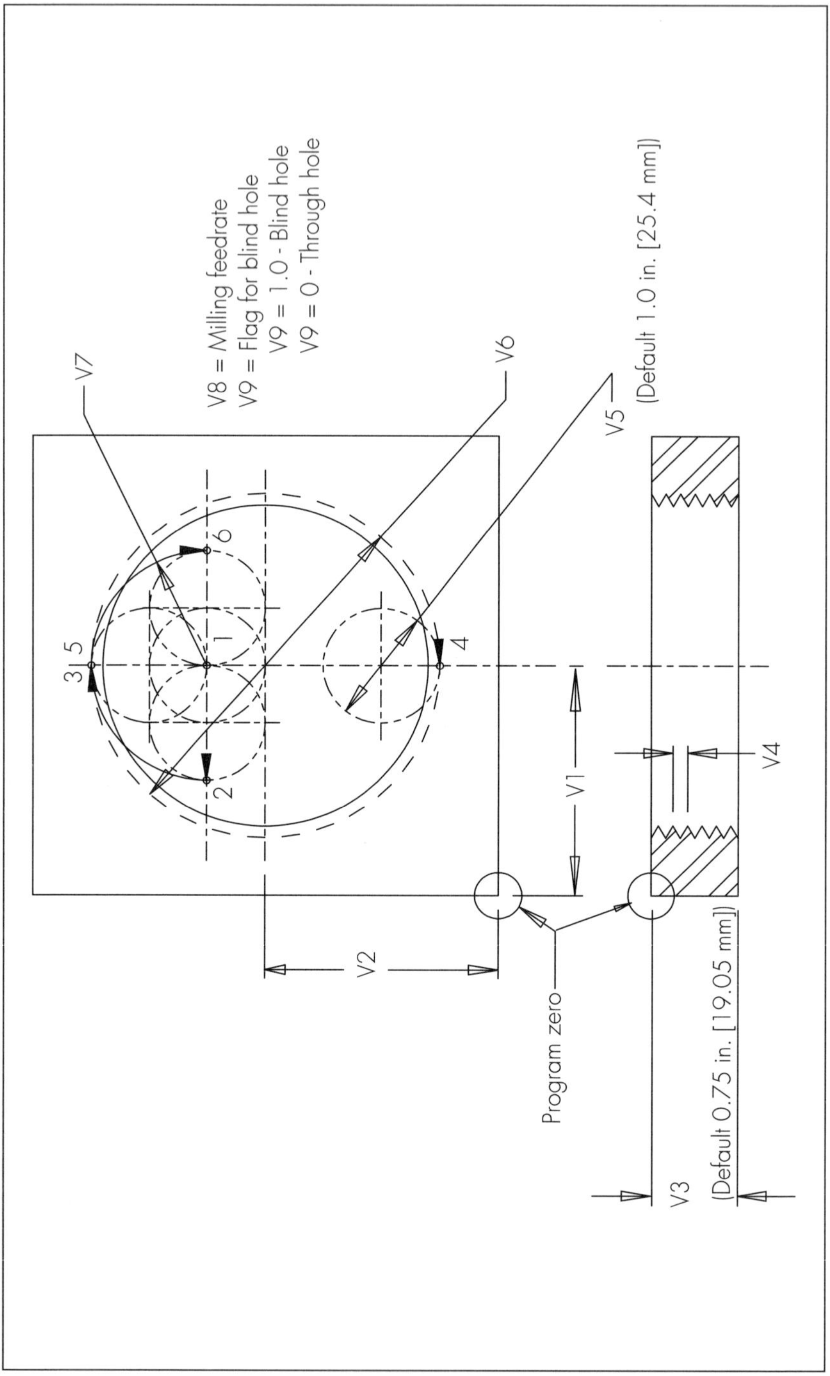

Figure 4-5. Arguments for the Fadal macro version of the thread-milling application.

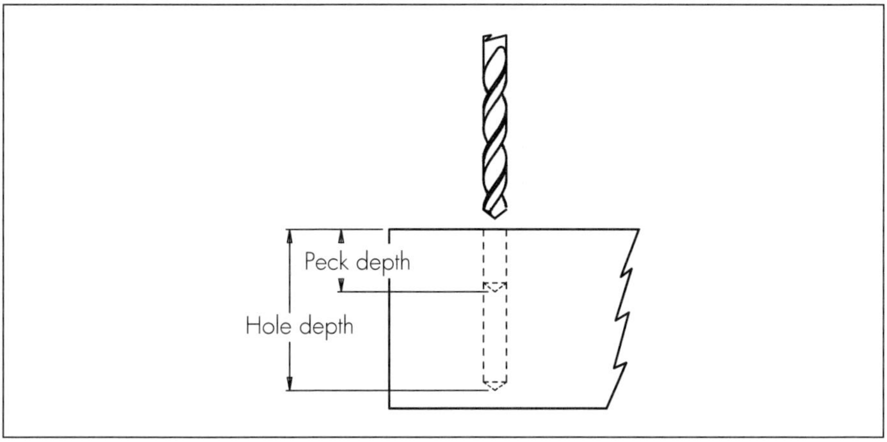

Figure 4-6. In a peck drilling loop, the number of loop executions must be calculated.

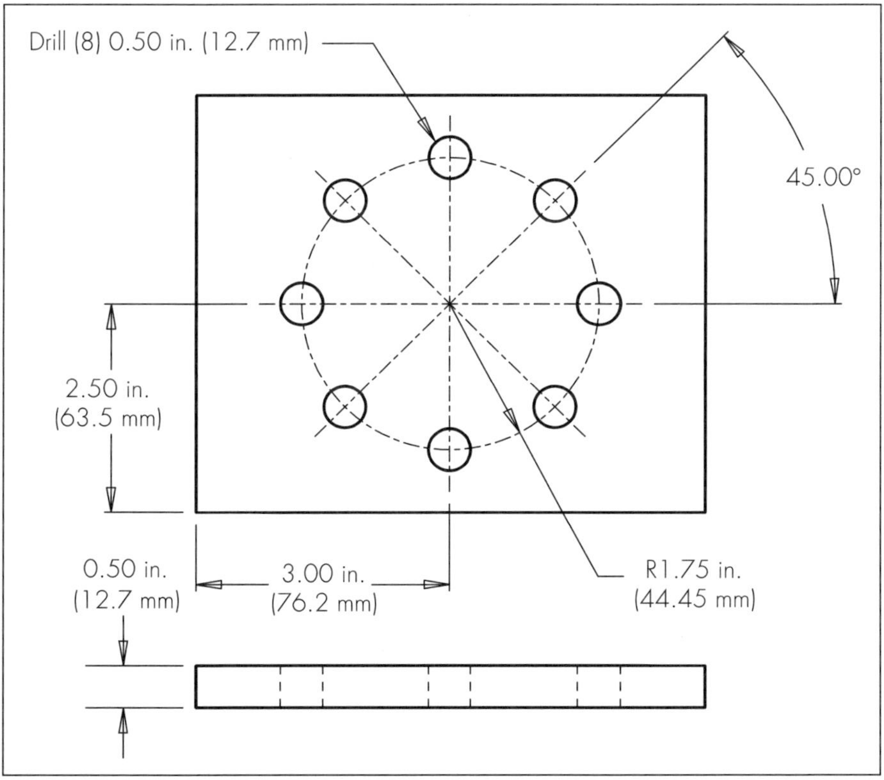

Figure 4-7. Engineer's drawing from which bolt-hole circle parametric program will be written.

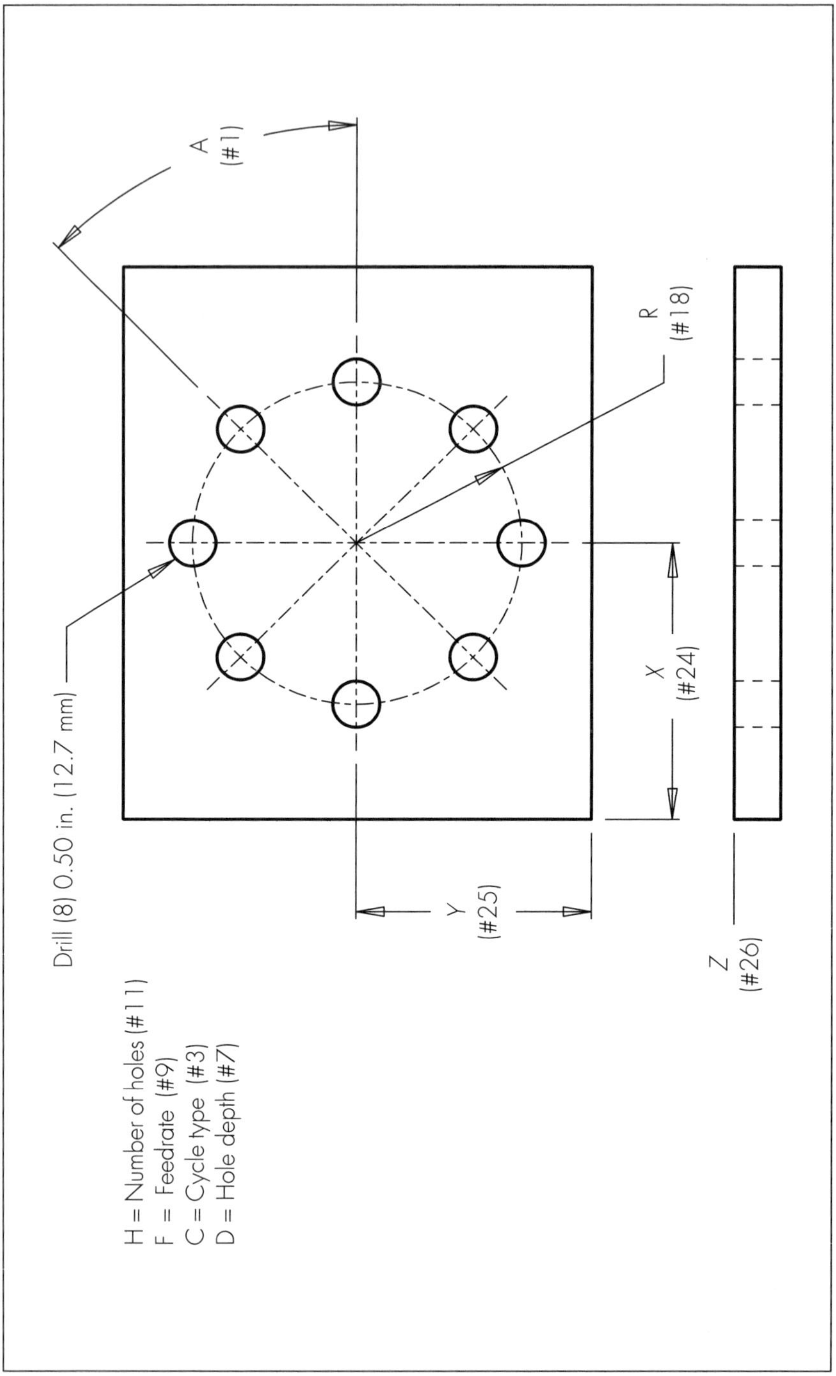

Figure 4-8. Arguments for custom macro B version of parametric program for drilling bolt holes.

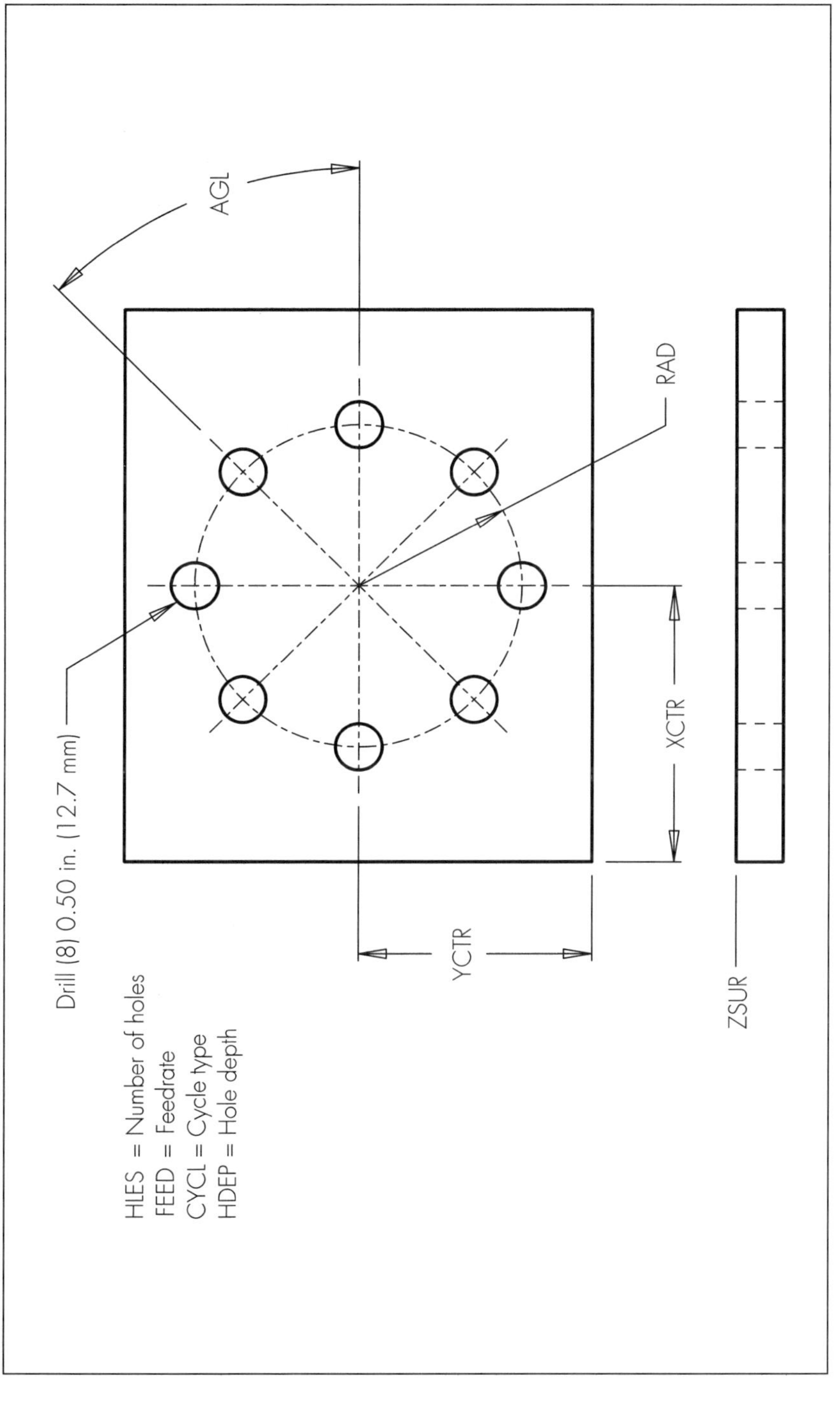

Figure 4-9. Arguments for user task 2 version of parametric program for drilling bolt holes.

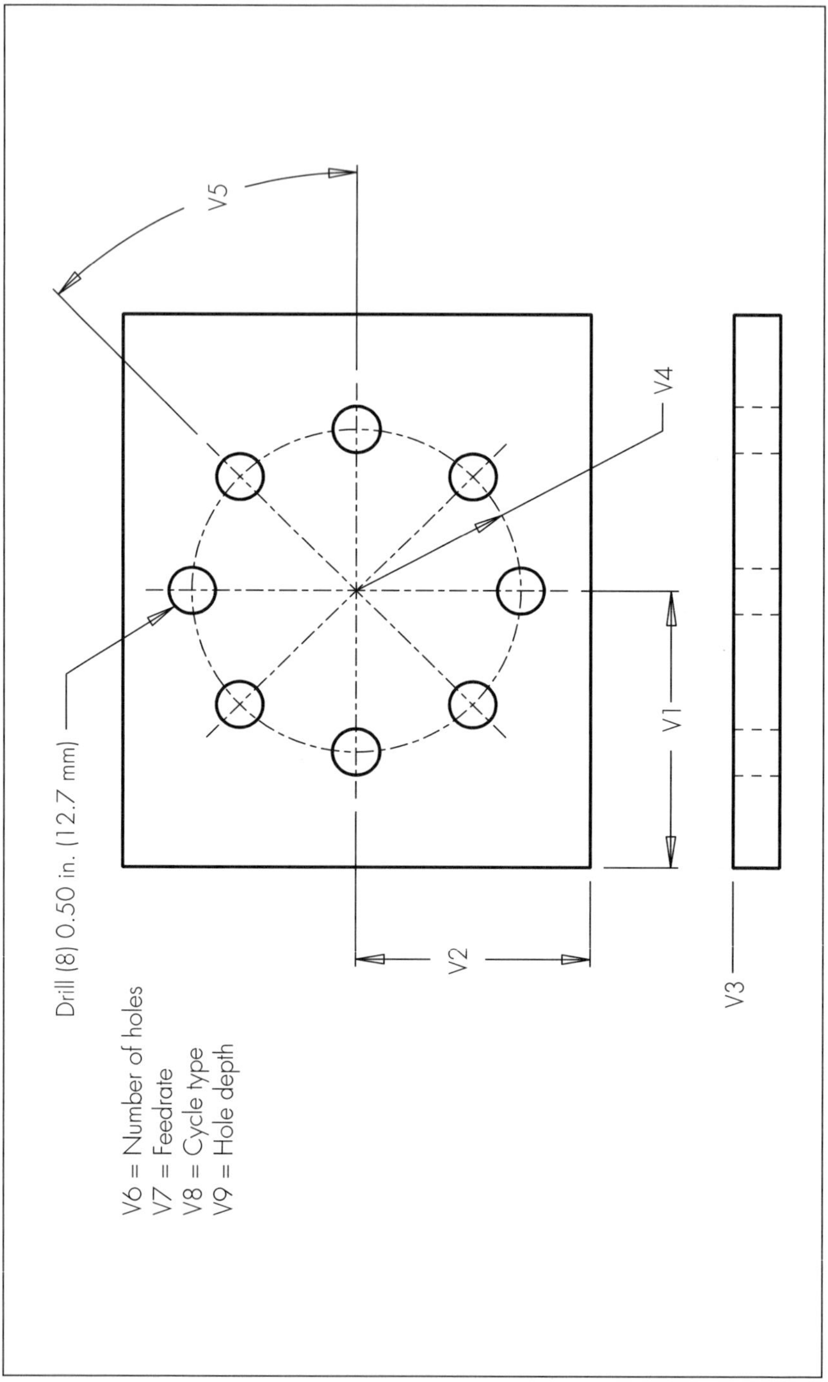

Figure 4-10. Arguments for Fadal's macro version of parametric program for drilling bolt holes.

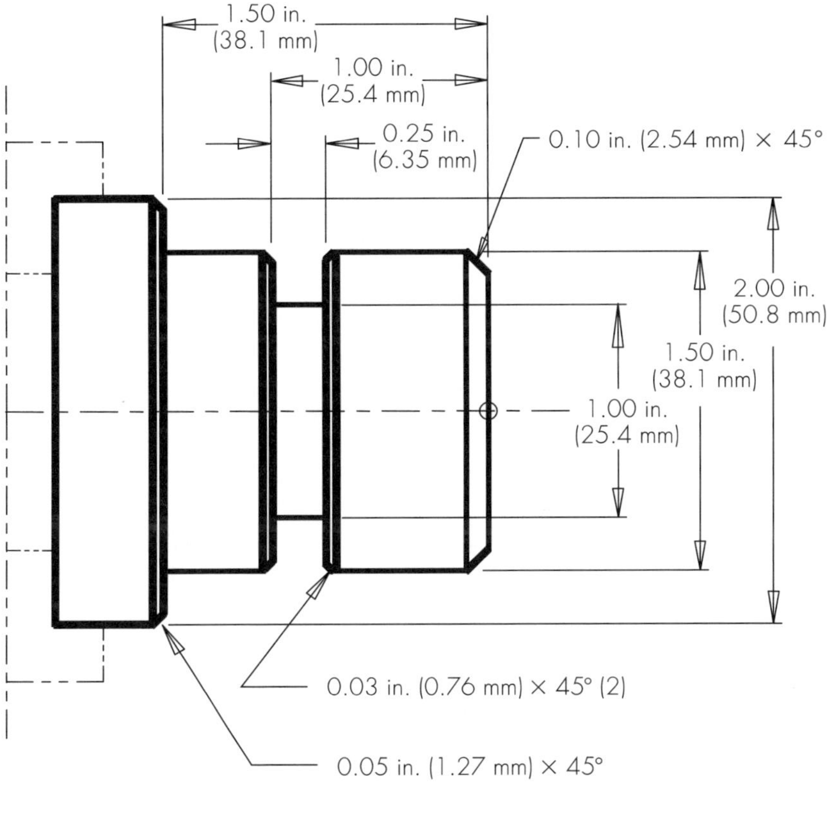

Figure 4-11. A common groove to be machined in a variety of workpieces. If multiple tool passes are to be made, a looping operation will be executed.

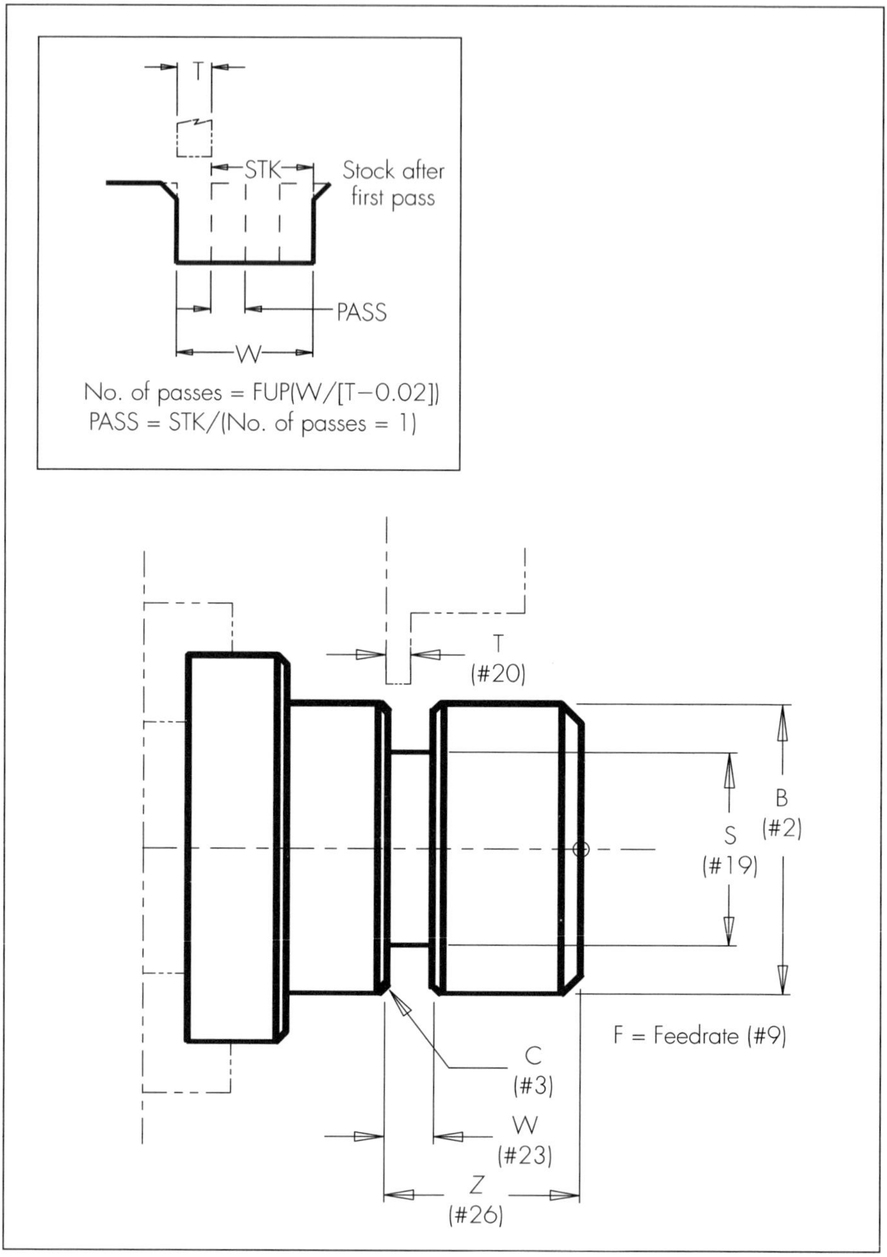

Figure 4-12. Custom macro B arguments for stepping the grooving tool. Inset shows calculations needed to determine the number of passes and move-over amount between passes when the groove is wider than the grooving tool.

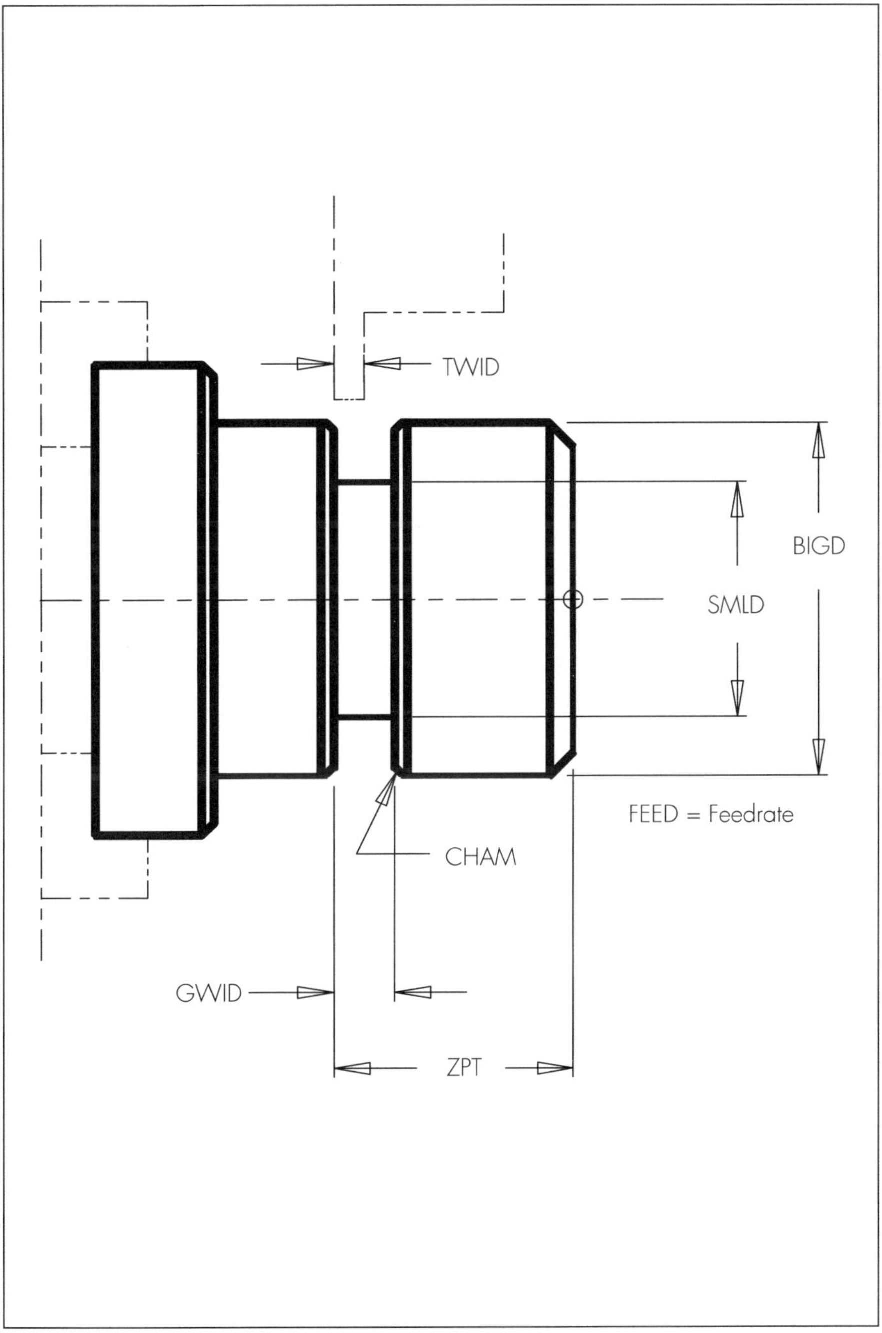

Figure 4-13. User task 2 arguments for stepping the grooving tool.

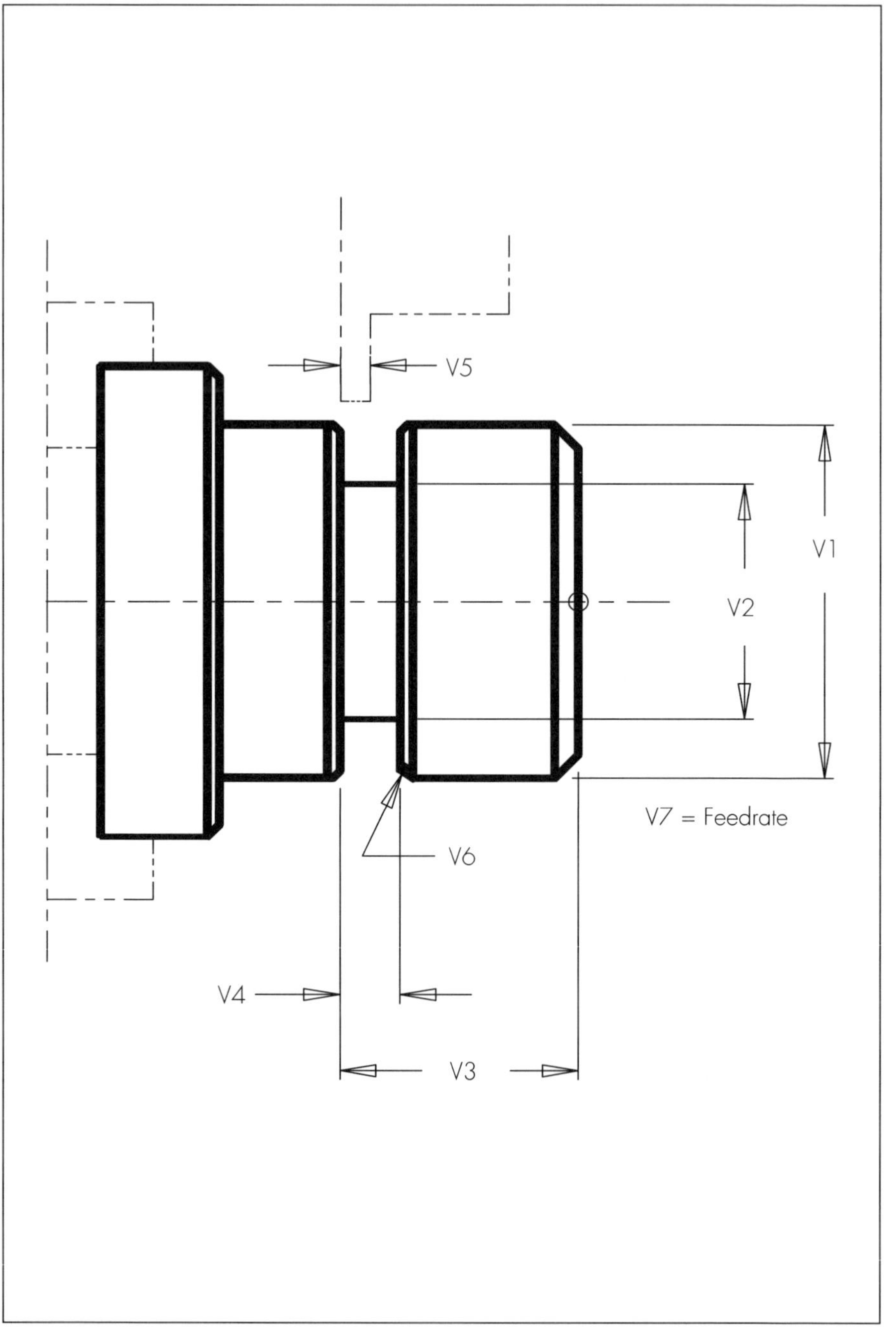

Figure 4-14. Fadal's macro arguments for stepping the grooving tool.

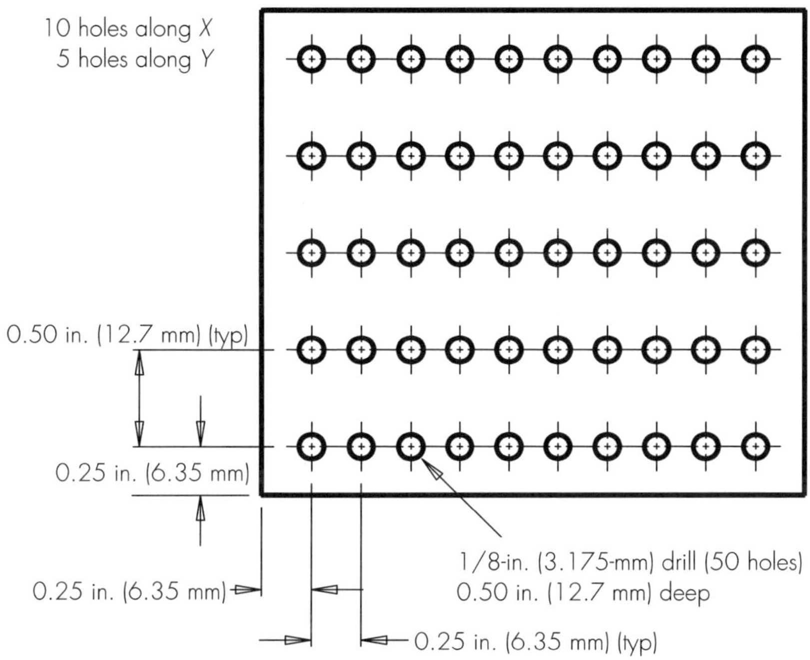

Figure 4-15. In grid-pattern machining such as required in this workpiece, a loop within a loop is a powerful and efficient tool.

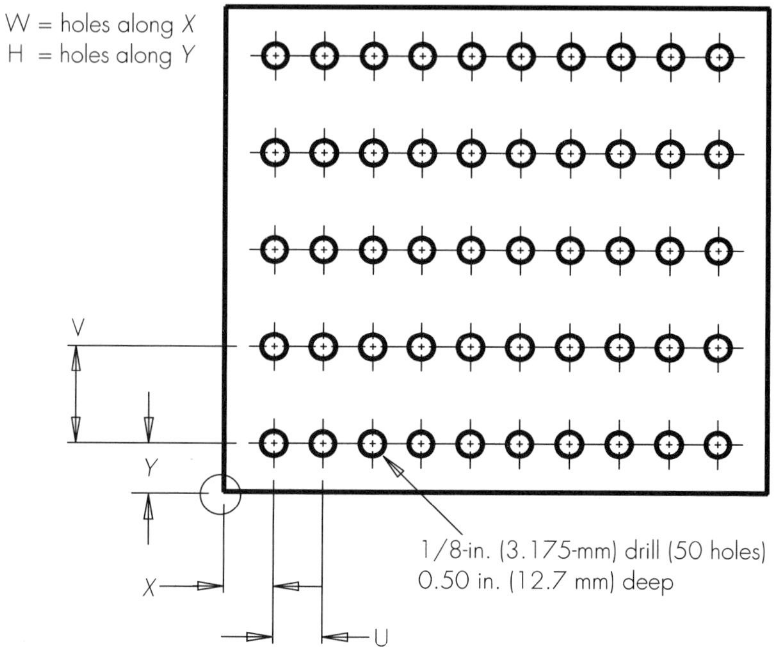

Figure 4-16. Custom macro B arguments for nesting loops.

Chapter 5

CNC Features of Parametric Programming

You have now been exposed to the most common computer-related features of parametric programming. Be sure to reference your control manufacturer's programming manual to see if you have other computer-related features not addressed in this text. Armed with your current knowledge, you should have no problem understanding the information presented in your control manuals.

Computer-related features by themselves give the parametric programmer a formidable programming tool, especially for user-created canned-cycle and family-of-parts applications. These computer-related features are extremely important and will be used throughout this chapter to build on what you now know and to introduce the CNC-related features of parametric programming.

CNC-related features enable functions that you would never consider possible with standard CNC programming. They offer access to many control functions that are not available with standard G-code-level commands. Of course, the first step to incorporating any new function or technique is knowing it is possible to do so. At the very least, this chapter should open your eyes to a world of new possibilities.

For each CNC feature, we first introduce the feature and show the syntax of how it is handled in each version of parametric programming. We then show applications for the feature. Though you may not have immediate need for some of the applications we show, our intention is as much to get you thinking about what can be done as to show you specific applications. With a little ingenuity, almost anything is possible.

It's important to understand at the outset that compared to computer-related features, CNC-related features vary much more dramatically from one version of parametric programming to another. For example, certain CNC-related features that are available in one version of parametric programming may not be available in another, and even the most common CNC-related features are handled quite differently among the different versions of parametric programming. For this reason, this chapter begins with the most common CNC-related features and works toward the more obscure. And since some of these features are very obscure, we describe some instances when these unusual CNC features can be helpful.

Aside from parametric programming, CNC controls vary with regard to their general capabilities and *user-friendliness*. A task that may require parametric programming capabilities on one control may be part of the standard functions of another. This means some of the applications in this chapter may not even be applicable in some versions of parametric programming, especially if the control manufacturer offers other (easier) ways to handle the problem without parametric programming techniques. For this reason, some examples in this chapter are given only for the version(s) of parametric programming for controls that most require their use.

Almost all example programs to this point have fallen into the category of either user-created canned cycles or family-of-parts. These two categories make it very easy to stress applications for the computer-related features. CNC-related features on the other hand, are more commonly related to the *utilities* and *driving accessory devices* categories, and many of the applications shown in this chapter reflect these categories.

SYSTEM VARIABLES

Most CNC-related features of parametric programming are handled with *system variables*. Like any other variable in parametric programming, most system variables contain numeric values. However, each has a fixed function and can only be used to reference its related function.

Generally speaking, system variables allow the parametric programmer to access many machine functions right from within the parametric program. Offsets, machine position, alarm generation, and current machine states are among the accessible functions. With custom macro B, system variables range in the #1000 through #6000 series. With user task 2 and Fadal's macro, system variables are addressed with special system variable names.

For two fundamental reasons this text cannot completely address the specific techniques needed to utilize every system variable for every version of parametric programming. First, the actual names of system variables vary from one version to the next, as does the actual usage for given system variables. This makes it difficult to give strong examples for every version. Second, and more importantly, names and usages for certain system variables vary even among control *models* within a given control manufacturer's product line. With Fanuc's custom macro B for example, the control model and even how the control is equipped with certain control options determine the actual names of system variables. This makes it virtually impossible to show an example for every possibility.

For this reason, you must be prepared to reference your control manufacturer's programming manual to determine which system variables are available with your particular CNC control(s). Despite this limitation, however, you should be able to fully understand the applications for the various system variables we show. It will be left up to you to relate these application examples to your particular control(s). This should require little more than a cross-check of your control manufacturer's manual to determine a pertinent system variable's name.

ACCESS TO TOOL OFFSETS

Almost all versions of parametric programming give you access to the tool offsets within your control. Though different versions of parametric programming utilize different system variable names to access specific tool offsets, most provide you access to *every kind* of tool offset within the control. In essence, this gives you the ability to read and write values to and from tool offsets.

You likely already know that most controls give you the ability to *write to* the tool offsets with standard G-code-level programming techniques. Fanuc, for example, uses a G10 command for this purpose. However, there is no standard CNC function on most controls (without parametric programming) that allows you to *read* the value stored within a tool offset and use it within your CNC programs. This new ability has some fairly sophisticated implications. But before we show applications, we first show the related system variables in each version of parametric programming addressed in this text.

Fanuc's custom macro B

As stated earlier, it will be impossible for this text to address all specific offset-related system variables since their names vary from one control model within the Fanuc product line to another. Additionally, an entirely different set of system variable names exists for machining center controls than for turning center controls. To complicate the issue even more, almost all models in the Fanuc product line can be equipped with a variety of offset-related alternatives, further changing the system variable naming structure related to tool offsets.

For this reason, all of our system-variable-related presentations will be directed at one specific machining center control (the 15M) and one specific turning center control (the 15T) in Fanuc's product line. Additionally, our discussion will be limited to showing each control's most basic offset configuration. We emphasize, however, that all techniques we show can be accomplished with any Fanuc (or Fanuc-compatible) control equipped with custom macro B. If you understand our presentations, it will be a relatively simple matter to cross-reference the related system variable names based on your own particular control model(s) and offset configuration.

Tool-length and cutter-radius compensation offsets. With the most basic tool offset configuration of the 15M machining center control, there is one offset value per tool offset. The CNC program determines how the offset value is used (as a tool-length and cutter-radius compensation offset). For this offset configuration, you will be using a series of system variables in the #2000 series. In this series #2001 is the value of offset number one, #2002 is the value of offset number 2. This continues until the total number of offsets in the control is reached (if the control has 199 offsets, #2199 is the value of offset number 199).

Remember, with parametric programming you can read and write to and from your tool offsets. The command

```
#100 = #2001
```

for example, places the current value of offset number one in common variable #100. On the other hand, you also can write to your offsets. The command

```
#2001 = #100
```

enters the current value of common variable #100 into offset number one. You must be careful, of course, not to overwrite a needed offset.

Fixture offsets. Another series of system variables is related to fixture offsets (also called work-shift offsets). For the 15M control, these system variables range from #5201 through #5335. This series of system variables assumes your control has only six fixture offsets. If you have purchased the option for more, a completely different set of system variables is involved.

As you know, each fixture offset has at least three values (one for *X*, one for *Y*, and one for *Z*). If your machine has additional axes, there will be more values in each fixture offset. In Chart 5-1 the fixture offset system variables are shown for a three-axis machining center controlled by a 15M control equipped with six fixture offsets.

As with tool offsets, you can read or write to or from any fixture offset. The commands

```
#5221 = –12.1263
#5222 = –12.2245
#5223 = –13.2247
```

enter the value of fixture offset number one's *X* as –12.1263, fixture offset number one's *Y* as –12.2245, and fixture offset number one's *Z* as –13.2247. On the other hand, the commands

```
#100 = #5221
#101 = #5222
#102 = #5223
```

store the current *X*, *Y*, and *Z* values of fixture offset number one into common variables #100, #101, and #102, respectively.

Chart 5-1. Machining center fixture offset system variables.

Fixture offset #0 (common)	*Fixture offset #4 (G57)*
X — #5201	*X* — #5281
Y — #5202	*Y* — #5282
Z — #5203	*Z* — #5283
Fixture offset #1 (G54)	*Fixture offset #4 (G58)*
X — #5221	*X* — #5301
Y — #5222	*Y* — #5302
Z — #5223	*Z* — #5303
Fixture offset #2 (G55)	*Fixture offset #4 (G59)*
X — #5241	*X* — #5321
Y — #5242	*Y* — #5322
Z — #5243	*Z* — #5322
Fixture offset #3 (G56)	
X — #5261	
Y — #5262	
Z — #5263	

Turning center wear offsets. Each turning center wear offset contains four values (*X*, *Z*, R, and T). A different series of system variables gives you access to each type. Chart 5-2 shows the wear offset related system variables for a 15T control.

As with any other offset, you can read and write to and from any turning center wear offset. The command

#100 = #2101

stores the current *Z* value of wear offset number one in common variable #100. On the other hand, the command

#2101 = #100

enters the current value of #100 into the *Z* value of wear offset number one.

Turning center geometry offsets. Turning center geometry offsets also contain four values (*X*, *Z*, R, and T). A different series of system variables gives you access to each type. Chart 5-3 shows the geometry offset-related system variables for a 15T control.

Chart 5-2. Turning center wear offset-related system variables.

Offset	*X*	*Z*	*R*	*T*
1	#2001	#2101	#2201	#2301
2	#2002	#2102	#2202	#2302
3	#2003	#2103	#2203	#2303
.	.		.	.
.	.		.	.
.	.		.	.
99	#2099	#2199	#2299	#2399

Chart 5-3. Turning center geometry offset-related system variables.

Offset	*X*	*Z*	*R*	*T*
1	#2401	#2501	#2601	#2701
2	#2402	#2502	#2602	#2702
3	#2403	#2503	#2603	#2703
.	.		.	.
.	.		.	.
.	.		.	.
99	#2499	#2599	#2699	#2799

As with any other offset, you can read and write to and from any turning center geometry offset. The command

#100 = #2501

stores the current *Z* value of geometry offset number one in common variable #100. On the other hand, the command

#2501 = #100

enters the current value of #100 into the *Z* value of geometry offset number one.

A trick with system variable numbers. It is often necessary to reference a given offset value by its offset number. But as you can see from the set of system variables related to offsets, a rather inconvenient system variable number must be used instead of the offset number itself. For this reason, it can be helpful to reference a variable (local or common) as part of the system variable's designation for a given tool offset. This variable will contain the offset number to be referenced. Consider these commands for a 15M machining center control.

```
N1 #100 = 1
N2 #[2000 + #100] = 6.5573
```

In line N1, the value of common variable #100 is being set to one. In line two, this value (one) is being used as part of the system variable number for a specific tool offset. With this technique, #100 is actually the tool-length compensation offset number being given a value. In line N2, the control will first add 2000 to #100, resulting in a value of 2001. The pound sign outside the brackets specifies that *system variable* #2001 (the value of offset number one) is being referenced.

In Chapter 2, an application is shown that can take advantage of this technique. Here is the call statement to the right-side milling macro shown in Chapter 2.

N025 G65 P1000 X0. Y0. Z0. W5.0 H3.0 T0.5 **D1.0** F5.0 (Mill right side)

In this call statement, letter address argument D is being used to specify the actual *diameter* of the milling cutter. However, if you use cutter-radius compensation on a regular basis, it may be more convenient to let D represent the tool *offset number* in which the milling cutter's radius is stored. This will more consistently reflect the method by which your operators normally use cutter-radius compensation offsets. Here is the new call statement, with the modified D word.

N025 G65 P1000 X0. Y0. Z0. W5.0 H3.0 T0.5 **D31.0** F5.0 (Mill right side)

Note that the value of D is now 31. The operator must, of course, store the radius of the milling cutter in offset number 31 (0.500 if a 1-in. [25.4-mm]-diameter cutter is used) before this program can be run. Here is the modified custom macro program which takes advantage of this new capability.

O1000 (Custom macro to mill right side)
N1 #100 = #[2000 + #7] (Copy cutter radius from offset to common variable #100)
N2 G00 X[#24 + #23 + **#100**] Y[#25 – 0.1 + **#100**] (Rapid to starting position)
N3 G01 Z[#26 – #20 – 0.1] F50. (Fast feed to work surface)
N4 Y[#25 + #11 + **#100**] F#9 (Mill right side)
N5 G00 Z[#26 + 0.1] (Rapid to above workpiece)
N6 M99 (End of custom macro)

In line N1, the value stored in the offset (number 31 in our case) is placed in common variable #100. This common variable is referenced in every motion command that requires the cutter radius.

This technique also is quite helpful when an entire series of offsets must be specifically arranged in some manner. Using a simple loop, any number of offsets can be manipulated with a very short series of commands. Here is an example loop that clears (sets to zero) 99 tool-length and cutter-radius compensation offsets. It requires just five commands.

```
.
.
.
#100 = 1 (Initialize counter for loop [to first offset number] to one)
N1 IF[#100 GT 99] GOTO 5 (Test if finished with loop)
#[2000 + #100] = 0 (Set current offset value to zero)
#100 = #100 + 1 (Step counter)
GOTO 1 (Go back to test to see if loop is finished)
N5 . . .
.
.
.
```

Note that similar techniques can be used with system variables for other offset types (fixture offsets and all kinds of turning center offsets).

Okuma's user task 2

Names for system variables related to offsets remain much the same among the entire Okuma control product line. About the only major difference you will notice is between machining center and turning center system variable names. That notwithstanding, we still recommend that you reference your programming manual from Okuma to confirm the names of offset-related system variables.

Tool-length and cutter-radius compensation. The system variable name for tool-length compensation offsets is VTOFH. A number in brackets to the right of the system variable name specifies the tool-length compensation offset number. As with any offset, you can read and write to and from the offsets with this variable. For example, the command

```
VTOFH[1] = V1
```

enters the current value of common variable V1 into tool-length compensation offset number one. On the other hand, the command

```
V1 = VTOFH[1]
```

stores the current value of tool-length compensation offset number one in common variable V1.

In similar fashion, another system variable (VTOFD) handles the cutter-radius compensation offsets. The command

VTOFD[1] = V1

enters the current value of common variable V1 into cutter-radius compensation offset number one. The command

V1 = VTOFD[1]

stores the current value of cutter-radius compensation offset number one in common variable V1. You must, of course, be careful not to overwrite a needed offset.

Fixture offsets. Three more system variables (VZOFX, VZOFY, and VZOFZ) give you access to your fixture offsets. (Note: If your machine has more than three axes, more than three system variables are involved.) The commands

VZOFX[1] = V1
VZOFY[1] = V2
VZOFZ[1] = V3

enter the current values of common variables V1, V2, and V3, respectively, into the *X*, *Y*, and *Z* positions of fixture offset number one. On the other hand, the commands

V1 = VZOFX[1]
V2 = VZOFY[1]
V3 = VZOFZ[1]

store the current *X*, *Y*, and *Z* values of fixture offset number one into V1, V2, and V3, respectively.

Turning center offsets. With turning centers, the *X* and *Z* value of each offset is available with system variables VTOFX and VTOFZ. The command

VTOFX[1] = V1

enters the current value of common variable V1 into the *X* position of offset number one. The command

V1 = VTOFX[1]

stores the current *X* value of offset number one in common variable V1.

In similar fashion, the tool-nose radius compensation values of each offset are available through system variables VNSRX and VNSRZ.

A trick with system variable numbers. You will often need to manipulate the value in brackets (the offset number) when working with offset-related system variables. Note that this number can be referenced with a variable. Consider these commands:

```
N1 TLGT = 1
N2 VTOFH[TLGT] = 6.5573
```

In line N1, the value of local variable TLGT is being set to one. In line two, this value (one) is being used as part of the system variable for a specific tool offset. With this technique, TLGT is actually the tool-length compensation offset number. In line N2, TLGT is used to determine which offset is being referenced.

Here is the call statement to the right-side milling macro shown in our example application in Chapter 2.

N025 CALL O1000 XZER = 0 YZER = 0 ZZER = 0 WIDT = 5.0 HGHT = 3.0 THK = 0.5 **TDIA = 1.0** FEED = 5.0 (Mill right side)

In this call statement, local variable TDIA is being used to specify the diameter of the milling cutter. However, if you use cutter-radius compensation on a regular basis, it may be more convenient to let a local variable, TLFF for example, represent the tool *offset number* in which the milling cutter's radius is stored. This will more consistently reflect the method by which your operators normally use tool offsets. Here is the new call statement, modified to use local variable TLFF as the offset number in which the cutter's radius is entered.

N025 CALL O1000 XZER = 0 YZER = 0 ZZER = 0 WIDT = 5.0 HGHT = 3.0 THK = 0.5 **TLFF = 5.0** FEED = 5.0 (Mill right side)

Note that the value of TLFF is 5.0 (in bold). This indicates that the radius position of *offset number five* is being used to store the cutter's radius. The operator must, of course, store the radius of the milling cutter in offset number five (0.500 if a 1-in. [25.4-mm]-diameter cutter is used) before this program is run. Here is the modified user task 2 program which takes advantage of this new capability.

O1000 (User task program to mill right side)
N1 TRAD = VTOFD[TLFF]
N2 G00 *X* = XZER + WIDT + **TRAD** *Y* = YZER – 0.1 + **TRAD** (Rapid to starting position)
N3 G01 *Z* = ZZER – THK – 0.1 F50. (Fast feed to work surface)
N4 *Y* = YZER + HGHT + **TRAD** F = FEED (Mill right side)

```
N5 G00 Z = ZZER + 0.1 (Rapid to above workpiece)
N6 RTS (End of user task program)
```

In line N1, the value stored in the offset (number 5 in this case) is placed in local variable TRAD. This local variable is then referenced in every motion command that requires the cutter radius.

As in custom macro B, this technique is helpful when an entire series of offsets must be manipulated in some manner. Using a simple loop, any number of offsets can be manipulated with a very short series of commands. Here is an example loop that clears (sets to zero) 99 tool-length and cutter-radius compensation offsets. It requires only six commands.

```
.
.
.
TLFF = 1 (Initialize counter for loop [to first offset number] to one)
N1 IF [TLFF GT 99] GOTO N5 (Test if finished with loop)
VTOFH[TLFF] = 0 (Set current tool length offset value to zero)
VTOFD[TLFF] = 0 (Set current cutter radius offset to zero)
TLFF = TLFF + 1 (Step counter)
GOTO N1 (Go back to test to see if loop is finished)
N5 . . .
.
.
.
```

Here, again, similar techniques can be used with system variables for other offset types (fixture offsets and all kinds of turning center offsets).

Fadal's macro

Fadal also uses a series of system variables to allow access to offset data.

Tool-length and cutter-radius compensation. For tool-length compensation offsets, you use a series of H codes that correspond to the offset number. For example, the command

```
#H01 = V1
```

enters the current value of V-series variable V1 into tool-length compensation offset number one. On the other hand, the command

```
#V1 = H01
```

enters the value of tool-length compensation offset number one into the macro's V-series variable V1.

In similar fashion a series of D codes allow access to cutter-radius compensation offsets. The command

#D01 = V1

enters the current value of V-series variable V1 into cutter-radius compensation offset number one. On the other hand, the command

#V1 = D01

enters the value of cutter-radius compensation offset number one into V series variable V1. Be careful not to overwrite a needed offset.

Fixture offsets. Another series of system variables is related to fixture offsets (also called work-shift offsets). As you know, each fixture offset has at least three values (one for *X*, one for *Y*, and one for *Z*). If your machine has additional axes, there will be more values in each fixture offset.

FX1 through FX48 allow access to the *X* value of each fixture offset. FY1 through FY48 allow access to the *Y* value for each fixture offset. FZ1 through FZ48 allow access to the *Z* value for each fixture offset. If your machine has a rotary axis, FA1 through FA48 or FB1 through FB48 allow access to the rotary axis value of each fixture offset. For example, the commands

#FX1 = V1
#FY1 = V2
#FZ1 = V3

enter the values of V-series variables V1, V2, and V3 into fixture offset number one's *X*, *Y*, and *Z* positions respectively. On the other hand, the commands

#V1 = FX1
#V2 = FY1
#V3 = FZ1

enter the values fixture offset number one's *X*, *Y*, and *Z* positions into V-series variables V1, V2, and V3, respectively.

A trick with system variable numbers. As with the other versions, you will often need to manipulate the system variable number when working with offset-related system variables. Note that the number of the system variable can be referenced with a variable. Consider these commands:

N1 #V1 = 1
N2 #H(V1) = 6.5573

In line N1, the value of V-series variable V1 is being set to one. In line two, this value (one) is being used as part of the system variable name for a specific tool offset. With this technique, V1 is actually the tool-length compensation offset number being set. Take note that this variable (or arithmetic expression) must be placed within parentheses. In line N2, V1 is used to determine which offset is being referenced.

Here are the calling commands to the right-side milling macro for the example shown in Chapter 2.

```
.
.
.
N025 #V1 = 0 'Assign argument V1
N030 #V2 = 0 'Assign argument V2
N035 #V3 = 0 'Assign argument V3
N040 #V4 = 5.0 'Assign argument V4
N045 #V5 = 3.0 'Assign argument V5
N050 #V6 = 0.5 'Assign argument V6
N055 #V7 = 1.0 'Assign argument V7
N060 #V8 = 5.0 'Assign argument V8
N065 M98 P1000 (Mill right side)
.
.
.
```

In line N055, V-series variable V7 is being used to specify the *diameter* of the milling cutter. However, if you use cutter-radius compensation on a regular basis, you may find it more convenient to let V7 represent the *tool offset number* in which the milling cutter's radius is stored. This will more consistently reflect the method by which your operators normally use tool offsets. Following are the new calling commands, modified to use V7 as the offset number in which the cutter's radius is stored.

```
.
.
.
N025 #V1 = 0 'Assign argument V1
N030 #V2 = 0 'Assign argument V2
N035 #V3 = 0 'Assign argument V3
N040 #V4 = 5.0 'Assign argument V4
N045 #V5 = 3.0 'Assign argument V5
N050 #V6 = 0.5 'Assign argument V6
N055 #V7 = 5.0 'Assign argument V7
```

N060 #V8 = 5.0 'Assign argument V8
N065 M98 P1000 (Mill right side)

.

.

.

Note that the value of V7 is now 5. This indicates that the radius position of *offset number five* is being used to store the cutter's radius. The operator must store the radius of the milling cutter in offset number five (0.500 if a 1-in. [25.4-mm]-diameter cutter is used) before this program is run. Here is the modified macro program which takes advantage of this new capability.

O1000 (Macro program to mill right side)
N1 **#V21 = D(V7)** 'Place cutter radius in V21
N2 #R0 = V1 + V4 + **V21** 'Calculate *X* position for cutter
N3 #R1 = V2 – 0.1 – **V21** 'Calculate *Y* starting position
N4 #R2 = V3 V6 0.1 'Calculate work surface position in *Z*
N5 #R3 = V2 + V5 + 0.1 + **V21** 'Calculate *Y* ending position
N6 #R4 = V8 'Transfer feedrate into R4
N7 #R5 = V3 + 0.1 'Calculate clearance position in *Z*
N8 G00 X + R0 Y + R1 (Rapid to starting position)
N9 G01 Z + R2 F50.0 (Fast feed to work surface)
N10 Y + R3 F + R4 (Mill right side)
N11 G00 Z + R5 (Rapid to above workpiece)
N12 M99 (End of parametric program)

In line N1, the value stored in the offset (number five) is placed in V-series variable V21. This V-series variable is then referenced every time the cutter radius is required.

This technique, in macro, as in other versions, also is helpful when an entire *series* of offsets must be manipulated in some manner. Using a simple loop, any number of offsets can be manipulated with a very short series of commands. This example loop clears (sets to zero) 99 tool-length and cutter-radius compensation offsets with only eight commands.

.

.

.

#V1 = 1 'Initialize counter for loop to one
#:LOOPBEGIN 'Beginning of loop
#IF V1>99 THEN GOTO :EXITLOOP 'Test if finished with loop
#H(V1) = 0 'Set current tool length offset value to zero
#D(V1) = 0 'Set current cutter radius offset to zero
#V1 = V1 + 1 'Step counter

```
#GOTO :LOOPBEGIN 'Go back to test to see if loop is finished
#:EXITLOOP 'End of loop
.
.
.
```

Note that similar techniques can be used with system variables for other offset types (fixture offsets and all kinds of turning center offsets).

Do you have more offsets than discussed here?

Again, we have only introduced the most common offset types in their most basic configurations. If your particular control(s) have more or different offset specifications (like zero-shift offsets), it is very likely that you have access to them through the use of (other) system variables. Reference your programming manual to determine which system variables are involved.

Applications for offset access

While the actual method of offset access varies from one version of parametric programming to another, the reasons why offset access is helpful remains quite consistent. Though this new capability may not at first glance seem at all earth-shattering, having the ability to read and write values to and from your offset tables has important implications.

As extra variables. The first application we present has little to do with actually needing offsets for a special purpose. We mention it only because if you should run out of variables for your parametric programming applications, your offset table makes an excellent place to find more. As long as you do not overwrite a needed offset, you can use your offsets as a kind of permanent common variable (since offset values are not lost when you turn the power off). Depending on the control model and type, you may have many unused offsets.

Improving compensation types. Offsets are normally used with some form of *tool compensation*. Machining centers require offsets for tool-length compensation, cutter-radius compensation, and fixture offsets. Turning centers require them for wear offsets, (possibly) geometry offsets, and tool-nose radius compensation. While all of these compensation types are very important and most are relatively easy to use, all are quite inflexible in the *way* they must be used. In certain applications, you may not agree with the method by which they must be used. Or it simply may not be convenient to use the compensation type in its intended manner. With parametric programming, *you actually have the ability to improve your current compensation types* with utility programs.

Here is a specific example: suppose to improve your current tool-wear offset compensation type, you determine that offset values should never be over a given amount, say 0.150 in. (3.81 mm) (in your own opinion). Suppose, also, that your turning center operators are constantly making decimal-point entry mistakes when entering turning center wear offsets (for example, an operator determines a wear offset entry should be 0.002 in. [0.051 mm], and mistakenly enters the value as 0.200 in. [5.08 mm] instead). If any wear offset is larger than 0.150 in. (3.81 mm), it is quite likely that an entry error has been made.

With the availability for testing of wear offset values, you can, with parametric programming, actually develop a series of tests to determine if any of the current offset values are greater than 0.150 in. If they are, an alarm can be sounded.

In custom macro B. Let us say that this turning center has 12 turret stations and that wear offsets 1 through 12 must be tested in this manner (both *X* and *Z*). Here is the series of commands that do the testing. Note that a loop is used to minimize the number of required commands.

```
.
.
.
#100 = 1 (Counter for loop)
N1 IF [#100 GT 12] GOTO 5 (Test if loop is finished)
IF [#[2000 + #100] LT 0.15] GOTO 2 (Test if X offset is less than 0.15 in. [3.81
   mm])
#3000 = 100 (X OFFSET TOO BIG)
N2 IF [#[2100 + #100] LT 0.15] GOTO 3 (Test if Z offset is less than 0.15 in.)
#3000 = 101 (Z OFFSET TOO BIG)
N3 #100 = #100 + 1 (Step counter)
GOTO 1 (Go back to see if loop is finished)
N5 . . .
.
.
.
```

In the second and third IF statements, notice again how the system variable related to each wear offset is referenced using a common variable (#100). The expression #[2000 + #100] first adds the value of #100 to 2000 (result: 2001, 2002, 2003, and so on). The pound sign placed outside the brackets allows the system variable number itself to be a changing value, based on the value of common variable #100.

In user task 2. Again we'll use the example of the turning center with 12 turret stations and in which wear offsets 1 through 12 must be tested in this manner (both *X* and *Z*). Following is the series of commands that do the testing. Note that a loop is used to minimize the number of required commands.

```
.
.
.
TLFF = 1 (Counter for loop)
```

```
N1 IF [TLFF GT 12] GOTO N5 (Test if loop is finished)
IF [VTOFX[TLFF] LT 0.15] GOTO N2 (Test if X offset is under 0.15 in. [3.81
   mm])
VUACM[1] = 'X OFFSET TOO BIG'
VDOUT[991] = 100
N2 IF [VTOFZ[TLFF] LT 0.15] GOTO N3 (Test if Z offset is under 0.15 in.)
VUACM[1] = 'Z OFFSET TOO BIG'
VDOUT[991] = 101
N3 TLFF = TLFF + 1 (Step counter)
GOTO N1 (Go back to see if loop is finished)
N5 . . .
.
.
.
```

In the second and third IF statements, notice how the system variable related to each wear offset is referenced using the local variable TLFF. The coding *VTOFX[TLFF]* renders the value of the offset number equal to the current value of TLFF (offset 1, 2, 3, and so on).

Any other ideas? Remember that almost any compensation type can be improved, based on specific problems you (or your operators) are having. Many machining center controls, for example, force operators to work with very large, cumbersome tool-length compensation values. If workpiece sizing must be done on a regular basis (possibly to bring a critical pocket to its correct depth), it can be difficult for operators to calculate the direction (plus or minus) and amount of offset change. And once they do, they must then add or subtract this value from a very large tool-length value to come up with the correct value of the offset. This task takes time and is error-prone, even for experienced operators. For those tools that require this kind of critical and cumbersome sizing, you could easily add a special sizing offset. The operator will simply enter the small deviation separating the desired and actual pocket depth in this offset, making it much easier to size the pocket. In the positioning command that brings the tool to the bottom of this critical pocket, you can easily reference the value of this secondary offset, making the tool go precisely to its correct depth. The same techniques can be used for cutter-radius compensation when critical tolerances must be held during peripheral milling.

User-defined compensation types. There are certain tooling, setup, and programming problems that no standard compensation functions can handle. Yet with the ability to access offset values from within parametric programs, *you can actually develop your own special compensation utilities* and solve previously impossible problems. Truly, you are limited only by your imagination in terms of what can be done.

Figure 5-1 shows a workpiece that illustrates the need for a user-defined compensation type.

The turning operations for this workpiece are done by a screw machine. When set up, the screw machine will run about 500-600 workpieces before tool maintenance is required. Note the rather open tolerance on the overall length of the workpiece (plus or minus 0.002 in. [0.051 mm]). During tool maintenance on the screw machine, the overall length of the workpiece will change slightly. Though the change in workpiece length is very small and the overall length is still well within the tolerance band, this change in workpiece length can wreak havoc with subsequent CNC machining center operation, owing to the nature of the setup being used. Figure 5-2 shows the setup.

Note that eight workpieces are sandwiched and clamped against one another in the fixture. When machining the hole in the outside diameter of each workpiece, any change in workpiece length will cause an accumulation of error in the hole locations being machined on the machining center.

Remember, about 500-600 blank workpieces are machined by the screw machine with no deviation in workpiece length. It is only when a new batch of blanks comes through (about once per shift) that the CNC machining center operator must deal with this problem. Admittedly, this may fall more into the realm of a fundamental fixture design problem than a programming one, and your tool designer may be able to design a fixture that eliminates the problem altogether.

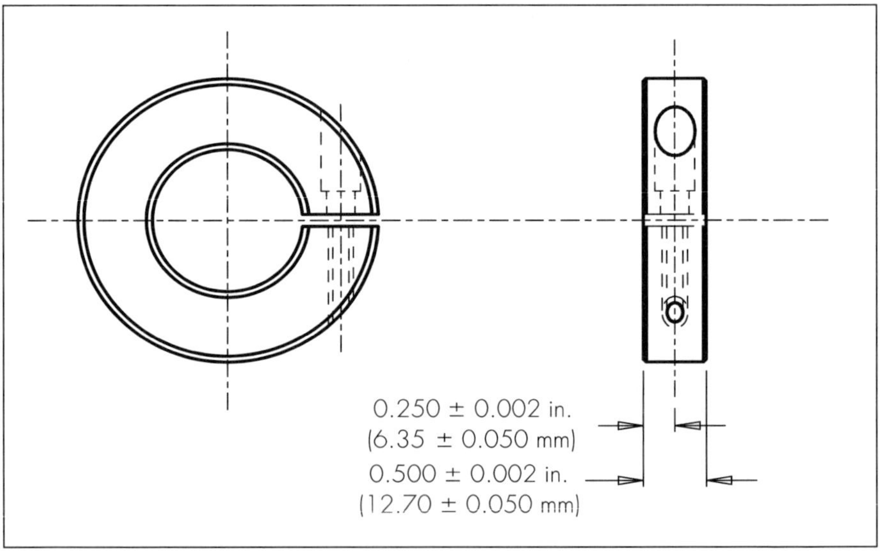

Figure 5-1. Access to offset values enable the parametric programmer to develop custom compensation utilities that solve previously impossible problems.

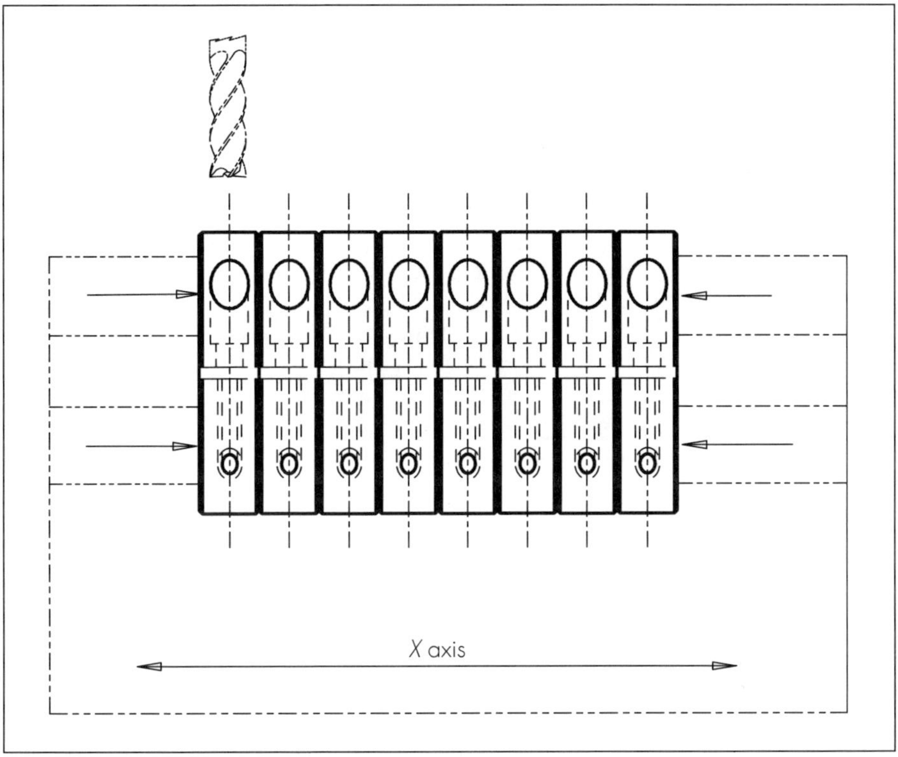

Figure 5-2. Machining center setup for the workpiece shown in Figure 5-1 requiring a user-defined compensation utility.

However, this is a real-world example of the kind of problem a CNC person faces on a regular basis—one that no standard compensation type is designed to solve.

With your new capability to access offset values from within a CNC program, however, you can easily solve this problem. A special tool offset (we'll use offset number 99) will be used in which to store the deviation between the nominal workpiece size (0.500 in. [12.7 mm]) and the actual workpiece size. If, for example, a batch of workpiece blanks comes through as 0.5007 in. (12.718 mm) long, the operator will simply store a value of plus 0.0007 in offset number 99. If the blanks coming through are 0.4994 in. (12.685 mm) long, the operator stores a value of minus 0.0006 in offset 99.

Here is a portion of the program shown in custom macro B format that counterbores the eight holes (you should now be able to easily convert this program to any version of parametric programming). Program zero is the left-most end of the stack in *X* and the center of the workpieces in *Y* and *Z*. Notice how every hole location in *X* references the value stored in offset 99.

```
.
.
.
N225 T03 M06 (Place counterbore in spindle)
N230 G54 G90 S700 M03 T04 (Select coordinate system, absolute mode, turn
   spindle on, get next tool ready)
N235 G00 X[0.25 + #2099/2] Y – 1.0 (Move into position for first hole)
N240 G43 H03 Z1.25 (Instate tool-length compensation, rapid up close in Z)
N245 G82 R1.25 Z0.25 F4.5 (Counterbore first hole)
N250 X[0.75 + #2099 + #2099/2] (Second hole)
N255 X[1.25 + #2099 * 2 + #2099/2] (Third hole)
N260 X[1.75 + #2099 * 3 + #2099/2] (Fourth hole)
N265 X[2.25 + #2099 * 4 + #2099/2] (Fifth hole)
N270 X[2.75 + #2099 * 5 + #2099/2] (Sixth hole)
N275 X[3.25 + #2099 * 6 + #2099/2] (Seventh hole)
N280 X[3.75 + #2099 * 7 + #2099/2] (Eighth hole)
N285 G80 (Cancel cycle)
.
.
.
```

You can easily determine whether enhancing current compensation types or developing new ones can help by watching and talking to your CNC operators. What kinds of problems are they having? Are they making the same mistakes over and over again when adjusting offsets for sizing purposes? Are they taking additional time performing (error-prone) calculations before they can even enter the offset adjustment? Are they having problems understanding the rationale behind how certain offset values must be entered? Do they have to measure one attribute in order to determine the offset setting of another (measuring over the pins after a threading operation, for instance). If the answer to any of these questions is yes, you can likely simplify the value and method of offset entry. As stated, only your own ingenuity will limit what can be done.

Referencing offsets in program zero assigning commands. If you use offsets to assign program zero (fixture offsets on machining centers or geometry offsets on turning centers), you can skip this section. On the other hand, if you are assigning program zero in the program (with G92 on machining centers or G50/G92 on turning centers), you will find this section quite helpful indeed.

One of the advantages of using offsets to assign program zero is that the program zero specification is kept separate from the program. Program zero assigning values are entered into offsets by the operator during setup. This is especially helpful on turning centers when the same cutting tools are commonly used from one setup to the next. For tools that remain in the turret from job to job, program zero assigning values need not be changed.

While there are exceptions to this depending on your methods of assigning program zero, if you are assigning program zero in the program, it is likely that the setup person must re-enter the program zero setting numbers for *every tool in every setup* (even for tools used from one production run to the next). This takes time and is quite error-prone since four values per tool must be changed. Here is a portion of a turning center program (in Fanuc format) that uses G50 to assign program zero from within the program. Note that this is a very common technique with Fanuc-controlled turning centers that do not have geometry offsets.

O0013 (Program number)
N005 G50 X10.2356 Z6.8277 (Assign program zero for tool one)
N010 T0101 (Index to station one)
N015 G96 S500 M03 (Start spindle)
N020 G00 X4.2 Z0.005 (Rapid to approach position)
N025 G01 X – 0.06 F0.012 (Rough face workpiece)
N030 G00 Z0.1 (Rapid away from workpiece)
N040 G00 X10.2356 Z6.8277 T0100 (Rapid to tool-change position, and cancel offset)
N045 M01 (Optional stop)
.
.
.

Note that line N005 assigns program zero and in line N040, the tool is sent back to precisely this starting position. With this format, again, four values per tool must be changed within the CNC program for every program run.

With your new ability to reference offset values from within your CNC program, you can easily modify the G50 command that assigns program zero and the G00 command that returns the machine to its tool-change position for each tool to include a reference to an offset. In effect, you gain one of the main benefits of geometry offsets, the ability to keep program zero assigning values separate from the program.

Picture a turning center with 12 tool stations and 32 offsets (a common scenario). The operator will still use offsets 1 through 12 as wear offsets. However, now he or she can use offsets 21 through 32 to store the program zero assigning values for tools 1 through 12. Consider the same program, modified to utilize offsets as program zero assigning values:

O0013 (Program number)
N005 G50 X[#2021] Z[#2121] (Assign program zero for tool one)
N010 T0101 (Index to station one)
N015 G96 S500 M03 (Start spindle)
N020 G00 X4.2 Z0.005 (Rapid to approach position)

```
N025 G01 X – 0.06 F0.012 (Rough face wc kpiece)
N030 G00 Z0.1 (Rapid away from workpiece)
N040 G00 X[#2021] Z[#2121] T0100 (Rapid to tool-change position, cancel
   offset)
N045 M01 (Optional stop)
.
.
.
```

Now, instead of entering the values X10.2356 and Z6.8277 in the program (in two places each), the setup person will enter the value 10.2356 in the *X* register of offset number 21 and 6.8277 in the *Z* register of offset number 21. With this technique, only two entries (one for *X* and one for *Z*) are required per tool. And when going from one production run to the next, these values need not be reentered for tools that stay in the turret.

Note that this technique also can be used on machining centers if you are using G92 to assign program zero in the program (it is especially helpful if you are repeating the G92 command at the beginning of each tool). The command (given in Fanuc format)

```
N005 G92 X[#2097] Y[#2098] Z[#2099]
```

retrieves the value in offset number 97 and uses it for the *X* value of the G92 command. The value in offset number 98 is used for *Y*, and the value in offset 99 is used for *Z*.

Remember that this technique overcomes only one of the limitations of assigning program zero in the program, and that the most severe limitation still exists. If the machine is not in its proper starting position when the program is activated (the machine is out of position), the control will utilize an incorrect program zero assigning command—and the results can be disastrous. Later, as we discuss access to the machine's position, we show how to refine this technique and minimize the potential for disaster.

ACCESS TO CURRENT POSITION

This function of parametric programming gives you the ability to determine the machine's position in any axis at any time during the execution of a program. In some versions of parametric programming, determining machine position can be done in any of several ways, and almost all allow, at the very least, the ability to access the machine's current position in the absolute mode relative to program zero in the current coordinate system.

Again, we show how this is done in each version of parametric programming before we discuss the implications of when it can be helpful.

In custom macro B

Custom macro B allows you to access the machine position in several ways. The three most important are (1) relative to the machine's zero return (reference) position, (2) relative to the program zero point of the current coordinate system (absolute position), and (3) relative to the program zero point of the current coordinate system at the instant of probe stylus contact. Here are the system variables.

In a three-axis machining center:

#5001 — Current *X* position relative to program zero in the current coordinate system
#5002 — Current *Y* position relative to program zero in the current coordinate system
#5003 — Current *Z* position relative to program zero in the current coordinate system

#5021 — Current *X* position relative to the machine's zero return position
#5022 — Current *Y* position relative to the machine's zero return position
#5023 — Current *Z* position relative to the machine's zero return position

#5061 — Current *X* position relative to the program zero point in the current coordinate system at the instant of probe stylus contact
#5062 — Current *Y* position relative to the program zero point in the current coordinate system at the instant of probe stylus contact
#5063 — Current *Z* position relative to the program zero point in the current coordinate system at the instant of probe stylus contact

In a two-axis turning center:

#5001 — Current *X* position relative to program zero in the current coordinate system
#5002 — Current *Z* position relative to program zero in the current coordinate system

#5021 — Current *X* position relative to the machine's zero return position
#5022 — Current *Z* position relative to the machine's zero return position

#5061 — Current *X* position relative to the program zero point in the current coordinate system at the instant of probe stylus contact
#5062 — Current *Z* position relative to the program zero point in the current coordinate system at the instant of probe stylus contact

In user task 2

User task 2 allows you to access the current absolute position relative to the program zero point in the current coordinate system (the current absolute position) in each axis. For a three-axis machine tool, three system variables are involved.

VSIOX — Current *X* position relative to the program zero point in the current coordinate system
VSIOY — Current *Y* position relative to the program zero point in the current coordinate system
VSIOZ — Current *Z* position relative to the program zero point in the current coordinate system

In Fadal's macro

Macro also allows you to access the current absolute position relative to the program zero point in the current coordinate system (the current absolute position) in each axis. For a three-axis machine tool, three system variables are involved.

AX — Current *X* position relative to the program zero point in the current coordinate system
AY — Current *Y* position relative to the program zero point in the current coordinate system
AZ — Current *Z* position relative to the program zero point in the current coordinate system

Applications for axis position access

There are numerous instances when it can be helpful to find the machine's current position from within a parametric program. Here we offer a few.

With user-created canned cycles. In the preceding three chapters we offered several user-created canned-cycle examples. With most machining operations, it is necessary to specify (with arguments) the location of the machining operation to be performed. In the case of the user-created canned cycle for thread milling, for example, it is necessary to specify the location of the hole to be thread milled in *X* and *Y*. With the grooving example, it is necessary to specify the location of the groove to be machined in *X* and *Z*.

With certain applications, you may elect to allow your user-created canned-cycle parametric program to *assume* that the tool is in certain relationship with the machining operation being performed when the parametric program is activated. This will minimize the number of arguments that must be specified in the call statement. With the user-created canned cycle for thread milling, for example, you can let the parametric program assume that the milling cutter is positioned right in the center of the hole to be thread milled in *X* and *Y*. This will eliminate the need to specify the *X* and *Y* position to the parametric program, which in turn shortens the call statement.

The best way to achieve this goal is to access the machine's current absolute position in *X*, *Y*, and *Z* at the very beginning of the parametric program (before any motion is generated). This will allow you to work in the absolute mode during the positioning movements of the user-created canned-cycle parametric program. The following example for thread milling (shown in user task 2 format) is identical to that shown at the end of Chapter 3, with one exception. Notice that the specification of the hole center in *X* and *Y* (XCTR and YCTR) are omitted from the call statement. In line N160 of the main program, the tool is sent to the hole center in *X* and *Y* (2.0 in *X* and 2.0 in *Y*). The user task 2 program will access this position and use it as a reference point for motion commands.

O0006 (Program number)

.

.

(Machine hole to get ready for thread milling)

.

.

N150 T04 M06 (Place thread milling cutter in spindle)
N155 G90 G54 S400 M03 T05 (Select absolute mode, coordinate system, start spindle, get next tool ready)
N160 G00 X2.0 Y2.0 (Rapid over to center of hole)
N165 G43 H04 Z.1 (Rapid up to workpiece in *Z*)
N170 CALL O1006 THK = 0.75 PICH = 0.125 TDIA = 1.0 MDIA = 3.0 ARAD = 1.0 FEED = 5.0 (Call user task program and mill thread)

.

.

.

This is the user task program to mill threads:

O1006 (Program number)
XCTR = VSIOX (Store *X* position in local variable XCTR)
YCTR = VSIOY (Store *Y* position in local variable YCTR)
TRAD = TDIA/2 (Set TRAD to the radius of the cutter)
CURZ = – [THK + 0.1] (Initial *Z* position)
G00 X = XCTR Y = YCTR + MAJR/2 – ARAD (Rapid to center of approach radius, point 1)
Z = CURZ (Rapid to beginning *Z* position)
CURZ = CURZ – PICH/4 (Calculate *Z* position at end of approach radius)
G01 X = XCTR – ARAD + TRAD F = FEED * 5 (Fast feed to beginning of approach radius, point 2)
G02 X = XCTR Y = YCTR + MAJR/2 – TRAD Z = CURZ R = ARAD – TRAD F = FEED (Circular approach to beginning of thread, point 3)
CURZ = CURZ – PICH/2 (Calculate *Z* position half way around thread)
G02 Y = YCTR – MAJR/2 + TRAD Z = CURZ R = MAJR/2 – TRAD (Mill half way around thread, point 4)
CURZ = CURZ – PICH/2 (Calculate *Z* position at end of thread)
G02 Y = YCTR + MAJR/2 – TRAD Z = CURZ R = MAJR/2 – TRAD (Mill balance of thread, point 5)

```
CURZ = CURZ – PICH/4 (Calculate Z position at end of retract radius)
G02 X = XCTR + ARAD – TRAD Y = YCTR + MAJR/2 – ARAD Z = CURZ R
   = ARAD – TRAD (Retract to point 6)
G00 X = XCTR (Return to center of approach radius)
Z0.1 (Rapid to clearance position above workpiece)
RTS (End of user task program)
```

While we only show one version of this program, you can easily apply the same techniques to the custom macro B and macro versions of this program.

A tool-length measuring program. While many companies have moved the setup-related task of measuring tool lengths for machining center setups off line (measuring tool lengths for the next job during the current production run), many others still measure tool lengths on line during setup. The reasons vary with the type of operation: it could be that no one is available to measure tool lengths during the production run; or maybe the lot size is so small that there is no time to set up and measure tool lengths off line.

Regardless of the reason, if tool lengths must be measured on line, you must make it as easy as possible for your setup people to perform the tool-length measuring function. *Tool-length measuring probes* make this task about as easy as it can be. However, many companies do not have them and must rely solely on the setup person to perform rather tedious and error-prone measurements for each tool while the machine is down between production runs.

If your setup people must take manual measurements on the machine to measure each tool's length (and, of course, enter each tool-length value into the corresponding offset), you can dramatically expedite the task with a utility parametric program designed to help measure tool lengths. This program will take advantage of the ability to access the machine's current position from within a program and the ability to enter values into tool-length offsets.

We illustrate with an example, and to make it simple to follow, we make five assumptions. First, we assume that the length of each tool is the tool-length compensation offset value. Second, that the offset number for each tool corresponds to the tool station number (offset one for station number one, offset two for station number two, and so on). Third, that all tools are to be measured, assembled, and loaded into the machine's tool magazine prior to running this utility program. Fourth, that the setup person uses a 3-in. (76.2-mm) block to touch each tool and that this block is placed directly on the table top prior to tool measurement. And finally, that the setup people will always have a series of tools in consecutive tool stations to measure.

In custom macro B format. Before first using this program, you must measure the distance from the table top to the machine spindle nose. This value will be placed in permanent common variable #500. Figure 5-3 shows the value of #500, as well as the 3-in. (76.2-mm) block used to measure each tool. Note that this measurement is taken with the machine resting at its zero return position.

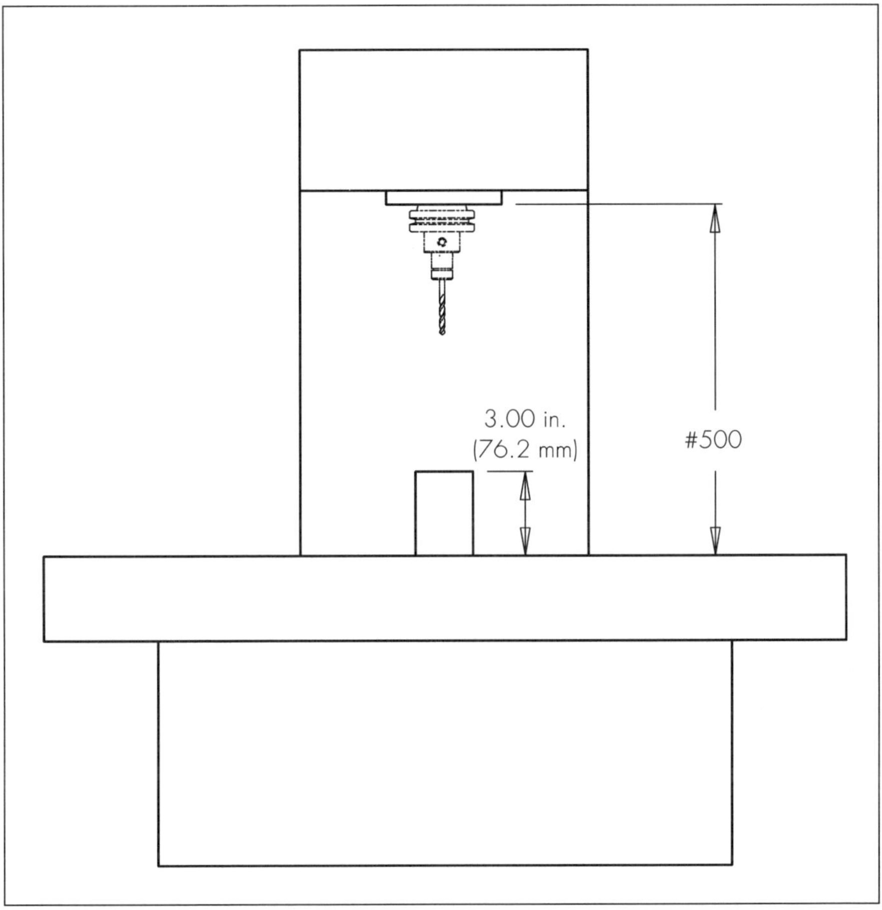

Figure 5-3. With a tool-length measuring parametric program, certain values are needed before the program can be used, in this case, the tool-measuring block and the value of #500.

```
O0014 (Program number)
#100 = 1 (First tool to measure)
#101 = 15 (Last tool to measure)
#102 = #100 (Initialize counter for loop)
N1 IF [#102 GT #101] GOTO 99 (Test if finished)
G91 G28 Z0 M19 (Return to tool-change position, orient spindle)
T#102 M06 (Place current tool in spindle)
#3006 = 101 (TOUCH TOOL TIP TO BLOCK) (Stop with message command)
#[2000 + #102] = #500 – 3 = ABS[#5023] (Set offset)
G91 G01 Z0.5 F30.0 (Move away from block)
```

```
#102 = #102 + 1 (Step counter)
GOTO 1 (Go back to test)
N99 M30 (End of loop and program)
```

While you should be able to follow most of this program, it does introduce some new functions. First, we are using #500 as a *system constant* (introduced in Chapter 2). This value (the distance from the table top to the spindle nose at the machine's reference position) will be measured and entered only once. The control will retain the value of #500 from day to day (it is a permanent common variable) and it will be available whenever this tool-length measuring utility program is used.

The setup person will initially edit the first two commands of this program to specify the current first tool and last tool number for tools to be measured. He or she will then place the 3-in. (76.2-mm) block on the table under the spindle nose and activate the program. After the loop is initialized and the first execution of the IF statement is found to be false, the G28 command will send the machine to its tool-change position. The first tool to be measured will be placed in the spindle and the program will halt due to the *stop with message* command.

Though it has not been introduced (yet), system variable #3006 is very similar to the #3000 alarm-generation system variable. There is one notable difference, however. Instead of placing the machine in a true alarm state, it simply places the machine in a program stop state, just like M00. But unlike M00, the display screen will automatically change and the message TOUCH TOOL TIP TO BLOCK will be placed on the screen. An M00 command would work just as well as #3006, but the setup person must be monitoring the program page of the display screen in order to see any message.

At this point, the setup person places the machine in a manual mode (jog, manual, handwheel, etc.) and manually brings the tool tip for the first tool down to touch the 3-in. (76.2-mm) block. When the tool tip touches the block, the setup person places the machine back in the automatic mode and reactivates the cycle. The command

```
#[2000 + #102] = #500 – 3 – ABS[#5023]
```

correctly calculates the tool-length value and places it into the tool offset corresponding to the current tool station number. Remember that system variable #5023 is the distance in *Z* from the machine's zero return position to its current position. Since this value will be a negative number (in our case), the absolute function is being used to reverse its sign. When this value plus an additional 3 in. (76.2 mm) (for the block height) is subtracted from permanent common variable #500, the result will render the tool's length. This value is then placed into the current offset specified as #[2000 + #102].

The program will then back the tool off slightly, step the counter to get to the next tool, and go back to the test. This process will then be repeated for each tool to be measured.

With the exception of the *stop with message* command, the techniques used in this program can be easily adapted to any version of parametric programming (use an M00 to replace the *stop with message* command). However, both Okuma and Fadal have very helpful built-in functions that adequately help the operator measure and enter tool lengths. For this reason, we do not show this example in user task 2 or macro.

Measurement of program zero on machining centers. Just as we can access machine position for the purpose of measuring tool lengths, so can we for the purpose of measuring the location of the program zero point. For machining center applications, nothing beats a spindle probe for this function. However, if you do not have a spindle probe on your machine, these functions of parametric programming even allow you to use a standard conductivity-type edge finder to handle just about any application a true spindle probe is designed to handle. (By *conductivity-type*, we mean an edge finder that incorporates a light that will illuminate the instant it comes into contact with any metallic surface.) The only major limitation is that the actual task of touching the edge finder to each surface must be performed manually. This capability of parametric programming opens the door to developing a series of very helpful utility parametric programs to facilitate the measurement of program zero based on your own specific program zero point specifications. Again, anything that can be done with a true spindle probe can be done with an edge finder.

A corner pickup utility in custom macro B. This example locates the lower left corner of a rectangular workpiece in *X* and *Y* and the top surface of the workpiece in *Z*. It assumes the use of an easy-to-use conductivity-type edge finder. Since this program is to be used on a regular basis, we will utilize two *system constants*. The radius of the edge finder (0.100 in. [2.54 mm]) is stored in permanent common variable #500 and the length of this edge finder is stored in #501. Once each surface location is found, the distance from the machine's reference point to the program zero surface will be stored in fixture offset number one. Prior to running this program, the setup person will manually position the edge finder about 0.5 in. (12.7 mm) in each direction from the corner to be picked up (0.5 in. to the left of the corner in *X*, 0.5 in. below the corner in *Y*, and 0.5 in. above the corner in *Z*).

```
O0015 (Program to pick up lower left corner)
G91 G01 Y0.75 Z – 0.75 F30.0 (Move to first touch position)
#3006 = 101 (TOUCH LEFT SIDE IN X)
#5221 = #5021 + #500 (Set X of fixture offset number one)
G91 G01 X – 0.2 (Move away in X)
Y – 0.75 (Move down in Y)
X0.55 (Move to second touch position)
```

```
#3006 = 102 (TOUCH BOTTOM SURFACE IN Y)
#5222 = #5022 + #500 (Set Y of fixture offset number one)
G91 Y – 0.2 (Move away in Y)
Z0.75 (Move above top surface in Z)
Y – 0.55 (Move to third touch position)
#3006 = 103 (TOUCH TOP SURFACE IN Z)
#5223 = #5023 – #501 (Set Z of fixture offset number one)
G91 G01 Z0.5 (Move away in Z)
G54 X0 Y0 (Instate coordinate system, move over corner)
M30 (End of program)
```

Using these techniques, you can develop any pickup routine that a true probing system is designed to do (any other corner, center of hole, boss or pin, center of slot, etc.).

A corner pickup utility in user task 2. This example also locates the lower left corner of a rectangular workpiece in *X* and *Y* and the top surface of the workpiece in *Z*. It assumes the use of an easy-to-use conductivity-type edge finder. Since this program is to be used on a regular basis, we will utilize two *system constants*. The radius of the edge finder (0.100 in., [2.54 mm]) is stored in V-series variable V31, and the length of this edge finder is stored in V32. Once each surface location is found, the distance from the machine's reference point to the program zero surface will be stored in fixture offset number one. Note that since the machine's reference position must be the point of reference for fixture offset entry, this program begins by setting the reference position as the program zero point in coordinate system number one (specified by G54 H01). Prior to running this program, the setup person will manually position the edge finder about 0.5 in. (12.7 mm) in each direction from the corner to be picked up (0.5 in. to the left of the corner in *X*, 0.5 in. below the corner in *Y*, and 0.5 in. above the corner in *Z*).

```
O0015 (Program to pick up lower left corner)
VTOFX[1] = 0 (Set machine reference position in X as origin of fixture offset
   number one)
VTOFY[1] = 0 (Set machine reference position in Y as origin of fixture offset
   number one)
VTOFZ[1] = 0 (Set machine reference position in Z as origin of fixture offset
   number one)
G54 H01 (Instate fixture offset number one)
G91 G01 Y0.75 Z – 0.75 F30.0 (Move to first touch position)
M00 (TOUCH LEFT SIDE IN X)
VTOFX[1] = VSIOX + V31 (Set X of fixture offset number one)
G91 G01 X – 0.2 (Move away in X)
Y – 0.75 (Move down in Y)
X0.55 (Move to second touch position)
```

```
M00 (TOUCH BOTTOM SURFACE IN Y )
VTOFY[1] = VSIOY + V31 (Set Y of fixture offset number one)
G91 Y – 0.2 (Move away in Y )
Z0.75 (Move above top surface in Z )
Y – 0.55 (Move to third touch position)
M00 (TOUCH TOP SURFACE IN Z )
VTOFZ[1] = VSIOZ – V32 (Set Z of fixture offset number one)
G91 G01 Z0.5 (Move away in Z )
G54 H01 X0 Y0 (Instate coordinate system, move over corner)
M02 (End of program)
```

Using these techniques, you can develop any pickup routine that a true probing system is designed to do (any other corner, center of hole or boss, center of slot, etc.).

A corner pickup utility in Fadal's macro. As with the other two versions of parametric programming we address in this text, this example locates the lower left corner of a rectangular workpiece in *X* and *Y*, and the top surface of the workpiece in *Z*. It assumes the use of an easy-to-use conductivity-type edge finder. Since this program is to be used on a regular basis, we will utilize two *system constants*. The radius of the edge finder (0.100 in. [2.54 mm]) is stored in V-series variable V98, and the length of the edge finder is stored in V99. Once each surface location is found, the distance from the machine's reference point to the program zero surface will be stored in fixture offset number one. Note that since the machine's reference position must be the point of reference, this program begins by setting the reference position as the program zero point in coordinate system number one (specified by G54). Prior to running this program, the setup person will manually position the edge finder about 0.5 in. (12.7 mm) in each direction from the corner to be picked up (0.5 in. to the left of the corner in *X*, 0.5 in. below the corner in *Y*, and 0.5 in. above the corner in *Z*).

```
O0015 (Program to pick up lower left corner)
#FX1 = 0 'Set machine reference position in X as origin of fixture offset number one
#FY1 = 0 'Set machine reference position in Y as origin of fixture offset number one
#FZ1 = 0 'Set machine reference position in Z as origin of fixture offset number one
G54 (Instate fixture offset number one)
G91 G01 Y0.75 Z – 0.75 F30.0 (Move to first touch position)
#PRINT "TOUCH LEFT SIDE IN X"
M00 (Program stop)
#FX1 = AX + V98 'Set X of fixture offset number one
G91 G01 X – 0.2 (Move away in X)
```

```
Y – 0.75 (Move down in Y )
X0.55 (Move to second touch position)
#PRINT "TOUCH BOTTOM SURFACE IN Y"
M00 (Program stop)
#FY1 = AY + V98 'Set Y of fixture offset number one
G91 Y – 0.2 (Move away in Y )
Z0.75 (Move above top surface in Z )
Y – 0.55 (Move to third touch position)
#PRINT "TOUCH TOP SURFACE IN Z"
M00 (Program stop)
FZ1 = AZ – V99 'Set Z of fixture offset number one
G91 G01 Z0.5 (Move away in Z )
G54 X0 Y0 (Instate coordinate system, move over corner)
M02 (End of program)
```

Using these techniques, you can develop any pickup routine that a true probing system is designed to do (any other corner, center of hole or boss, center of slot, etc.).

Assigning program zero in the program. If you are assigning program zero with offsets (fixture offsets on machining centers or geometry offsets on turning centers), you can skip this section. However, if you are assigning program zero in the program (with G92 on machining centers or G50/G92 on turning centers) as is required on older Fanuc controls, you must read this section. Applying this technique can make your machine much safer to utilize.

Earlier in this chapter we described how to reference offset values in conjunction with program zero assignment. However, this technique did not eliminate the single largest cause of machine crashes with machines requiring program zero to be assigned in the program. As you well know, if the machine is not in its planned starting position when the program is activated, the control will "believe" the program zero setting command in the program, and program zero will not be assigned correctly. It is quite likely that this will result in a crash.

By having the ability to access the machine's current position relative to the zero return position from within your program, you can actually use the machine's current position as your program zero assigning command. The control will take into consideration the distance from the zero return position when program zero is assigned. In effect, the machine will never be out of position again. Consider this portion of a turning center program:

O0013 (Program number)
N005 G50 X[#2021 + #5021] Z[#2121 + #5022] (Assign program zero for tool number one)
N010 T0101 (Index to station one)
N015 G96 S500 M03 (Start spindle)

```
N020 G00 X4.2 Z0.005 (Rapid to approach position)
N025 G01 X – 0.06 F0.012 (Rough face workpiece)
N030 G00 Z0.1 (Rapid away from workpiece)
N040 G00 X5.0 Z6.0 T0100 (Rapid to tool-change position, cancel offset)
N045 M01 (Optional stop)
.
.
.
```

Notice in line N005 that offset number 21 is being used to store the program zero assigning values (referenced by system variables #2021 and #2121). Remember, these offset values are equal to the distance from program zero to the tip of each tool while the machine is resting at its zero return position. To these values we add whatever distance the machine may be from the zero return position. If the machine happens to be at the zero return position, the values of #5021 and #5022 will be zero. If the machine is to the negative side of the zero return position, the values of these system variables will be negative; if to the positive side, the values will be positive. So regardless of the machine's position, the program zero assigning command will be correct. This technique allows still another benefit of geometry offsets: in line N040, notice that you no longer have to send the machine back to a special tool-change position. Instead, you can cut cycle time by moving the tool to any location at which the turret can safely index.

ACCESS TO ALARM GENERATION

Almost every version of parametric programming allows the programmer to perform tasks related to *error trapping*. Error trapping is the process of finding conditions that require the stoppage of a program's execution. During our discussion of conditional branching in Chapter 4, we introduced the alarm-generation functions of each version of parametric programming addressed in this text. Here we show another example in each version and review the syntax of alarm generation.

A part-counting parametric program

While most CNC controls have some form of part counter, few control-based part counters have the ability to halt the program after a specified number of workpieces has been run.

In custom macro B format. This part counter uses a *global flag* (we will use permanent common variable #500) as the actual counter. This allows the part-counting utility program to keep track of the current part count even after the power has been turned off. Prior to using this part counter for the very first time, the setup person will confirm that the value of permanent common variable #500 is zero (this value can be displayed and entered through the display screen and control panel).

At the end of any program that uses the part counter utility, a simple call statement will specify the number of workpieces to be run before the alarm is sounded (with letter address argument C, which is represented by local variable #3). An optional letter address argument (M for multiple) can be included in the call statement to specify how many workpieces are being machined per cycle. If M (represented by local variable #13) is left out of the call statement, the custom macro will assume that one workpiece is run per cycle. Here is a portion of a main program that uses the part counter utility.

.
.
.

N350 G65 P9101 C1500. (Call part counter utility, specify that 1500 workpieces are to be run, one per cycle)
M30 (End of main program)

Part counting custom macro utility program:

```
O9101 (Program number)
IF [#13 NE #0] GOTO 1 (If M is specified skip to N1)
#13 = 1 (M not specified, set default value to one)
N1 #500 = #500 + #13 (Step part count by the number of parts per cycle)
IF [#500 LT #3] GOTO 99 (If not done yet, exit program)
#500 = 0 (Reset counter to 0 for next production run)
#3000 = 100 (PART COUNT ACHIEVED)
N99 M99 (End of custom macro)
```

Of course, any time the operator wants to check the current part count, he or she can simply display permanent common variable #500. But with this utility program, the operator need not be present to stop the cycle when the part count is reached. This is especially handy with machines that can run several cycles without operator intervention (like bar-feeding turning centers).

To review the alarm-generation system variable (#3000), if it is executed, it will cause the machine to go into alarm state. The alarm will be very similar to any other program storage (PS) alarm. The program's execution will be halted and the operator will have to clear the alarm by pressing the reset button before anything else can be done.

The alarm message page will be automatically displayed and the message in parentheses (which must be in capital letters, up to 26 characters) will be displayed. The alarm number, which is chosen by the programmer and placed next to the equal sign must be in the range of 100-255 on most controls. This is simply a number the programmer can use to further document the alarm, though the message in parentheses is usually sufficient for this purpose. For the alarm generated in the part counter program, the message generated will appear like this:

MC-100 PART COUNT ACHIEVED

The MC in the message stands for MaCro alarm (as opposed to servo alarm, program storage alarm, overtravel alarm, etc.).

In user task 2 format. This part counter uses a *global flag* (we will use common variable V32) as the actual counter. Since global variables are nonvolatile, this again allows the part counting utility program to keep track of the current part count even after the power has been turned off. Prior to using this part counter for the very first time, the setup person will confirm that the value of common variable V32 is zero (this value can be displayed and entered through the display screen and control panel).

At the end of any program that uses the part counter utility, a simple call statement will specify the number of workpieces to be run before the alarm is sounded (with local variable argument TTLC for ToTaL Count). An optional local variable argument (MULT for MULTiple) can be included in the call statement to specify how many workpieces are being machined per cycle. If MULT is left out of the call statement, the custom macro will assume that one workpiece is run per cycle. Here is a portion of a main program that uses the part counter utility:

.
.
.

N350 CALL O9101 TTLC=1500. (Call part counter utility, specify that 1500 workpieces are to be run, one per cycle)
M02 (End of main program)

Part counting custom macro utility program:

O9101 (Program number)
IF MULT GOTO N1 (If argument MULT is specified skip to N1)
MULT = 1 (MULT is not specified, set default value to one)
N1 V32 = V32 + MULT (Step part count by the number of parts per cycle)
IF [V32 LT TTLC] GOTO N99 (If not done yet, exit program)
V32 = 0 (Reset counter to 0 for next production run)
VUACM[1] = 'RUN COMPLETED'
VDOUT[991] = 100
N99 RTS (End of user task program)

Of course, any time the operator wants to check the current part count, he or she can simply display common variable V32. But with this utility program, the operator need not be present to stop the cycle when the part count is reached. This is especially handy with machines that can run several cycles without operator intervention (like bar-feeding turning centers).

To review the alarm-generation system variables, VUACM (which stands for UserAlarm CoMment), places the message within single quotation marks (' ') on the display screen. The number in brackets (normally one) specifies the character number with which to begin the message. The message can be up to 16 characters long. System variable VDOUT with a value in brackets of 991 causes the machine to go into alarm state. The alarm will be quite similar to any other program storage alarm. The program's execution will be halted and the operator will have to clear the alarm by pressing the reset button before anything else can be done.

The alarm message page will be automatically displayed and the message in single quotes (in capital letters) will be displayed. The alarm number is chosen by the programmer and placed to the right of the equal sign in the VDOUT command. This is simply a number the programmer can use to further document the alarm, though the message in single quotes is usually sufficient for this purpose. For the alarm generated in the part-counter program, the message generated will appear like this:

3202 Alarm C User reserve code 100 RUN COMPLETED

In Fadal's macro format. This part counter uses a *global flag* (we will use common variable V99) as the actual counter. Since V-series variables are non-volatile, this allows the part-counting utility program to keep track of the current part count even after the power has been turned off. Prior to using this part counter for the very first time, the setup person will confirm that the value of V-series variable V99 is zero (this value can be displayed and entered through the display screen and control panel).

At the end of any program that uses the part-counter utility, simple calling commands specify the number of workpieces to be run before the alarm is sounded (with V-series variable argument V1). Another V-series variable (V2) will specify how many workpieces are being machined per cycle.

```
.
.
.
N350 #V1 = 1500.0 'Set number of parts to run to 1500
N355 #V2 = 1.0 'Set number parts per cycle
N360 M98 P9101 (Call part counter utility)
N365 M02 (End of main program)
```

Part counting macro utility program:

```
O9101 (Program number)
#V99 = V99 + V2 'Step part count by the number of parts per cycle
```

```
#IF V99 < V1 THEN GOTO :EXIT 'If not done yet, exit program
#V99 = 0 'Reset counter to 0 for next production run
#:REPEAT 'Statement label for message repeat
#PRINT "PART COUNT ACHIEVED" 'Print alarm message
#INPUT V20 'Halt program activation
#GOTO :REPEAT 'Repeat alarm if the cycle is reactivated
#:EXIT 'Statement label for end of program
M99 (End of custom macro)
```

Any time the operator wants to check the current part count, he or she can simply display common variable V99. But with this utility program, the operator need not be present to stop the cycle when the part count is reached. This is especially handy with machines that can run several cycles without operator intervention (like bar-feeding turning centers).

To review alarm generation, the PRINT command is used to print a message (of any kind) on the display screen. The message must be enclosed within standard quotation marks (" "). The PRINT command by itself does not actually stop the program. For this reason, we use an INPUT command that is normally used to request information from the operator. The V-series variable in the INPUT command can be any unused V-series variable.

Since the operator can still reactivate the program by simply pressing cycle start, logic is included after the INPUT command that keeps the program from going any further. The GOTO statement sends the control back to the print command, placing the machine in an *endless loop*. The operator could press cycle start any number of times, but the program will not continue.

Minimizing the time required for error trapping

During our discussion of conditional branching in Chapter 4, we offered an example of error trapping that used several IF statements to set default values, test for missing mandatory arguments, and test for conflicting arguments in the call statement. Here we show the program again (in custom macro B format) as part of our discussion of the impact of error trapping on cycle time.

```
O1006 (Program number)
(PRESET Z AND T)
IF [#26 NE #0] GOTO 5 (Test if Z is vacant)
#26 = 0.75 (Preset Z to 0.75)
N5 IF [#20 NE #0] GOTO 6 (Test if T is vacant)
#20 = 1.0 (Preset T to 1.0)
(TEST MANDATORY VARIABLES & GENERATE ALARM IF MISSING)
N6 IF [#24 EQ #0] GOTO 98 (If X is missing, generate alarm)
IF [#25 EQ #0] GOTO 98 (If Y is missing, generate alarm)
IF [#17 EQ #0] GOTO 98 (If Q is missing, generate alarm)
```

```
IF [#13 EQ #0] GOTO 98 (If M is missing, generate alarm)
IF [#1 EQ #0] GOTO 98 (If A is missing, generate alarm)
IF [#9 EQ #0] GOTO 98 (If F is missing, generate alarm)
IF [#2 EQ #0] GOTO 98 (If B is missing, generate alarm)
#20 = #20/2 (Set #20 to radius of tool instead of diameter)
#13 = #13/2 (Set #13 to radius of major diameter)
IF [#2 EQ 1.0] GOTO 20 (Test if blind hole)
(CALCULATIONS FOR TOP TO BOTTOM)
#101 = #24 – #1 + #20 (X position of tool center at approach position)
#102 = #24 + #1 – #20 (X position of tool center at retract position)
#103 = #25 + #13 – #20 (Y position of tool center at points 3 and 5)
#104 = #25 + #13 – #1 (Y position of tool center at points 2 and 6)
#105 = – [#26 + #17/4] (Z position of point 3)
#106 = #105 – #17/2 (Z position of point 4)
#107 = #106 – #17/2 (Z position of point 5)
#108 = #107 – #17/4 (Z position of point 6)
#109 = #1 – #20 (Radius of approach tool path)
#110 = #13 – #20 (Radius of tool path for milling thread)
#111 = #25 – #13 + #20 (Y position of tool center at point 4)
#112 = 2 (Set G-code number for circular motion [G02])
GOTO 60
(CALCULATIONS FOR BOTTOM TO TOP)
N20 #101 = #24 + #1 – #20 (X position of tool center at approach position)
#102 = #24 – #1 + #20 (X position of tool center at retract position)
#103 = #25 + #13 – #20 (Y position of tool center at points 3 and 5)
#104 = #25 + #13 – #1 (Y position of tool center at points 2 and 6)
#105 = –[#26  #17/4] (Z position of point 3)
#106 = #105 + #17/2 (Z position of point 4)
#107 = #106 + #17/2 (Z position of point 5)
#108 = #107 + #17/4 (Z position of point 2)
#109 = #1 – #20 (Radius of approach tool path)
#110 = #13 – #20 (Radius of tool path for milling thread)
#111 = #25 – #13 + #20 (Y position of tool center at point 4)
#112 = 3 (Set G-code number for circular motion [G03])
(MOTIONS)
N60 G00 X#101 Y#104 (Move in XY to approach point)
Z – #26 (Rapid to start of thread milling in Z)
G#112 X#24 Y#103 Z#105 R#109 F#9 (Approach to point 3)
Y#111 Z#106 R#110 (Mill half way around circle to point 4)
Y#103 Z#107 R#110 (Mill the rest of the way to point 5)
X#102 Y#104 Z#108 R#109 (Retract from thread)
G00 X#24 Y#25 (Move to hole center)
Z0.1 (Rapid out of hole)
```

```
GOTO 99
(ALARM IF DATA IS MISSING)
N98 #3000 = 100 (DATA MISSING IN CALL)
N99 M99 (End of custom macro)
```

The bulk of this program does nothing but test and calculate. There are actually only eight motion commands in the entire program. While the computers within CNC controls are getting faster and faster, even today's 32-bit controls will require some execution time to perform the calculations and tests. You will likely notice a short pause after activating this program before the machine actually starts moving.

If the lot size is relatively small, the extra processing time required to perform these non-motion commands will have only a small impact on the length of time it takes to complete the production run. However, as quantities grow, you will want to do something about this wasted cycle time.

While our example is related to a user-created canned-cycle application, and the techniques we show will work well, the impact of error trapping on cycle time is usually more critical in family-of-parts applications. Just remember that all techniques shown in this example can be applied to part family applications with equal ease and success.

Using optional block-skip. One (rather crude) way to solve the added cycle time problem is to use the optional block-skip function (slash code) in all calculation and testing commands. Assuming this is the only parametric program being used and the common variables needed by this program are not overwritten in some other program, optional block-skip will allow the control to skip redundant commands after the first workpiece is run. Consider this program:

```
O1006 (Program number)
IF [#101 NE #0] GOTO 1 (Test to ensure #101 has been set)
#3000 = 110 (TURN BLOCK SKIP OFF)
N1 (PRESET Z AND T)
/IF [#26 NE #0] GOTO 5 (Test if Z is vacant)
/#26 = 0.75 (Preset Z to 0.75)
/N5 IF [#20 NE #0] GOTO 6 (Test if T is vacant)
/#20 = 1.0 (Preset T to 1.0)
(TEST MANDATORY VARIABLES, GENERATE ALARM IF MISSING)
/N6 IF [#24 EQ #0] GOTO 98 (If X is missing, generate alarm)
/IF [#25 EQ #0] GOTO 98 (If Y is missing, generate alarm)
/IF [#17 EQ #0] GOTO 98 (If Q is missing, generate alarm)
/IF [#13 EQ #0] GOTO 98 (If M is missing, generate alarm)
/IF [#1 EQ #0] GOTO 98 (If A is missing, generate alarm)
/IF [#9 EQ #0] GOTO 98 (If F is missing, generate alarm)
/IF [#2 EQ #0] GOTO 98 (If B is missing, generate alarm)
```

```
/#20 = #20/2 (Set #20 to radius of tool instead of diameter)
/#13 = #13/2 (Set #13 to radius of major diameter)
/IF [#2 EQ 1.0] GOTO 20 (Test if blind hole)
(CALCULATIONS FOR TOP TO BOTTOM)
/#101 = #24 – #1 + #20 (X position of tool center at approach position)
/#102 = #24 + #1 – #20 (X position of tool center at retract position)
/#103 = #25 + #13 – #20 (Y position of tool center at points 3 and 5)
/#104 = #25 + #13 – #1 (Y position of tool center at points 2 and 6)
/#105 = –[#26 + #17/4] (Z position of point 3)
/#106 = #105 – #17/2 (Z position of point 4)
/#107 = #106 – #17/2 (Z position of point 5)
/#108 = #107 – #17/4 (Z position of point 6)
/#109 = #1 – #20 (Radius of approach tool path)
/#110 = #13 – #20 (Radius of tool path for milling thread)
/#111 = #25 – #13 + #20 (Y position of tool center at point 4)
/#112 = 2 (Set G-code number for circular motion [G02])
/GOTO 60
(CALCULATIONS FOR BOTTOM TO TOP)
/N20 #101 = #24 + #1 – #20 (X position of tool center at approach position)
/#102 = #24 – #1 + #20 (X position of tool center at retract position)
/#103 = #25 + #13 – #20 (Y position of tool center at points 3 and 5)
/#104 = #25 + #13 – #1 (Y position of tool center at points 2 and 6)
/#105 = –[#26 – #17/4] (Z position of point 3)
/#106 = #105 + #17/2 (Z position of point 4)
/#107 = #106 + #17/2 (Z position of point 5)
/#108 = #107 + #17/4 (Z position of point 2)
/#109 = #1 – #20 (Radius of approach tool path)
/#110 = #13 – #20 (Radius of tool path for milling thread)
/#111 = #25 – #13 + #20 (Y position of tool center at point 4)
/#112 = 3 (Set G-code number for circular motion [G03])
(MOTIONS)
N60 G00 X#101 Y#104 (Move in XY to approach point)
Z – #26 (Rapid to start of thread milling in Z)
G#112 X#24 Y#103 Z#105 R#109 F#9 (Approach to point 3)
Y#111 Z#106 R#110 (Mill half way around circle to point 4)
Y#103 Z#107 R#110 (Mill the rest of the way to point 5)
X#102 Y#104 Z#108 R#109 (Retract from thread)
G00 X#24 Y#25 (Move to hole center)
Z0.1 (Rapid out of hole)
/GOTO 99
(ALARM IF DATA IS MISSING)
/N98 #3000 = 100 (DATA MISSING IN CALL)
N99 M99 (End of custom macro)
```

At the very beginning of this program, notice the test to ensure that #101 has been set. If the operator forgets to turn the optional block-skip switch off for the first workpiece after turning on the power, the machine will go into alarm state. When the first workpiece is run, the optional block-skip switch can be turned on. From this point, the control will ignore the calculations and tests, minimizing cycle time.

A check-sum test. Though optional block-skip is effective in cutting cycle time, using it for that purpose eliminates the possibility of using it for any other purpose since most machines have only one optional block-skip function. Another way to minimize wasted calculation and testing time is to use a *check-sum test*. Here's how it works.

Two *global flags* are used. Since our example is in custom macro B, we use permanent common variables #500 and #501. In permanent common variable #500, we store the total of all values that might be changed that will affect the accuracy of the testing and calculations previously done. Note that in our case, only the letter address arguments are involved with testing and calculations. Therefore, only their corresponding local variables will be involved with the check-sum test. However, it is possible that other control functions, like tool offsets, are involved with your testing and calculations. For the check-sum test to be effective, *everything* that will affect the accuracy of the previous tests and calculations must be included in the check-sum test. Here is the program for the same thread milling application, modified to use the check-sum test technique:

O1006 (Program number)
(PRESET Z AND T)
IF [#26 NE #0] GOTO 5 (Test if *Z* is vacant)
#26 = 0.75 (Preset *Z* to 0.75)
N5 IF [#20 NE #0] GOTO 6 (Test if T is vacant)
#20 = 1.0 (Preset T to 1.0)
(CHECK-SUM TEST)
#500 = #26 + #20 + #24 + #25 + #17 + #13 + #1 + #9 + #2
IF [#500 EQ #501] GOTO 60 (If check-sum test passes, go to motion commands)
#501 = #500 (Set #501 equal to #500 so if nothing changes, check-sum test will pass next time)
(TEST MANDATORY VARIABLES, GENERATE ALARM IF MISSING)
N6 IF [#24 EQ #0] GOTO 98 (If *X* is missing, generate alarm)
IF [#25 EQ #0] GOTO 98 (If *Y* is missing, generate alarm)
IF [#17 EQ #0] GOTO 98 (If Q is missing, generate alarm)
IF [#13 EQ #0] GOTO 98 (If M is missing, generate alarm)
IF [#1 EQ #0] GOTO 98 (If A is missing, generate alarm)
IF [#9 EQ #0] GOTO 98 (If F is missing, generate alarm)
IF [#2 EQ #0] GOTO 98 (If B is missing, generate alarm)

```
#20 = #20/2 (Set #20 to radius of tool instead of diameter)
#13 = #13/2 (Set #13 to radius of major diameter)
IF [#2 EQ 1.0] GOTO 20 (Test if blind hole)
(CALCULATIONS FOR TOP TO BOTTOM)
#101 = #24 – #1 + #20 (X position of tool center at approach position)
#102 = #24 + #1 – #20 (X position of tool center at retract position)
#103 = #25 + #13 – #20 (Y position of tool center at points 3 and 5)
#104 = #25 + #13 – #1 (Y position of tool center at points 2 and 6)
#105 = –[#26 + #17/4] (Z position of point 3)
#106 = #105 – #17/2 (Z position of point 4)
#107 = #106 – #17/2 (Z position of point 5)
#108 = #107 – #17/4 (Z position of point 6)
#109 = #1 – #20 (Radius of approach tool path)
#110 = #13 – #20 (Radius of tool path for milling thread)
#111 = #25 – #13 + #20 (Y position of tool center at point 4)
#112 = 2 (Set G-code number for circular motion [G02])
GOTO 60
(CALCULATIONS FOR BOTTOM TO TOP)
N20 #101 = #24 + #1 – #20 (X position of tool center at approach position)
#102 = #24 – #1 + #20 (X position of tool center at retract position)
#103 = #25 + #13 – #20 (Y position of tool center at points 3 and 5)
#104 = #25 + #13 – #1 (Y position of tool center at points 2 and 6)
#105 = –[#26 – #17/4] (Z position of point 3)
#106 = #105 + #17/2 (Z position of point 4)
#107 = #106 + #17/2 (Z position of point 5)
#108 = #107 + #17/4 (Z position of point 2)
#109 = #1 – #20 (Radius of approach tool path)
#110 = #13 – #20 (Radius of tool path for milling thread)
#111 = #25 – #13 + #20 (Y position of tool center at point 4)
#112 = 3 (Set G-code number for circular motion [G03])
(MOTIONS)
N60 G00 X#101 Y#104 (Move in XY to approach point)
Z – #26 (Rapid to start of thread milling in Z)
G#112 X#24 Y#103 Z#105 R#109 F#9 (Approach to point 3)
Y#111 Z#106 R#110 (Mill half way around circle to point 4)
Y#103 Z#107 R#110 (Mill the rest of the way to point 5)
X#102 Y#104 Z#108 R#109 (Retract from thread)
G00 X#24 Y#25 (Move to hole center)
Z0.1 (Rapid out of hole)
GOTO 99
(ALARM IF DATA IS MISSING)
N98 #3000 = 100 (DATA MISSING IN CALL)
N99 M99 (End of custom macro)
```

Close to the beginning of this program (just after the default settings) is the command that sets permanent common variable #500 to the total of all local variables representing letter address arguments in the call statement. Then comes the check-sum test. For the very first workpiece in the production run, this test will be evaluated as false, since there is no way the total of all local variables involved with this application will equal the most recent setting of #501. Because this test will render a false result, all of the testing and calculation commands will be performed. Just after the check-sum test, #501 is being set equal to #500. As long as none of the letter address arguments are changed, the next time the check-sum test is executed, it will render a *true* result. In this case, the control will skip all tests and calculations and proceed directly to the motion commands, saving program execution time. The tests and calculations will be performed again only in the event that any of the letter address arguments are changed by the operator.

Again, this example involves only the letter address arguments. Keep in mind that any value that, if changed, would affect the correctness of calculations and tests included in the check-sum test.

OTHER SYSTEM VARIABLES OF INTEREST

As stated at the beginning of this chapter, versions of parametric programming vary with regard to how many CNC-related features are available. From this point forward, we isolate those system variables that are available in just one of the versions addressed in this text, custom macro B. Be sure to check in your own control manufacturer's programming manual to determine if you have other system variable functions not addressed in this text.

#1000 series—Access to terminal locations

Custom macro B allows you to access certain terminal locations on a printed circuit board right from within a program. For the purposes of outputting signals to some outside device, you can set any one of 32 terminal locations high (on) or low (off). When setting it high, you are sending a 24-volt signal to the terminal location, which is usually done to activate some outside device. When setting it low, you are removing the 24-volt signal.

Another set of 32 terminals is used as input signals. These are commonly used to attain confirmation that the outside device has completed its function. If an input terminal is high, the outside device has sent the signal back telling the control it has completed its function. If the input terminal is low, there is no 24-volt signal currently at the location, meaning the device is still in the process of completing its function.

System variables #1000 through #1031 are used to monitor the input terminals, while system variables #1100 through #1131 are used to send the 24-volt signal to the output terminals. Input terminal locations are marked on the schematic diagram as UI000 through UI031. Output terminal locations are marked as UO000 through UO031.

Consider the following program example that could be used to activate any M-code-type device, such as an indexer, hydraulic clamp, or automatic door.

```
O9001 (Program number)
#1100 = 1 (Activate outside device)
N1 IF [#1000 EQ 1] GOTO 99 (Test if device is finished)
GOTO 1 (Go back and test again)
N99 M99 (End of custom macro)
```

The first command sends the 24-volt signal to output terminal location UO000. This activates the indexer, hydraulic clamp, automatic door, or any other device. The IF statement tests to see if a confirmation signal has yet been received from the outside device. That is, has input terminal location UI000 received a 24-volt signal yet? If it has not, which is likely for the first few times the IF statement is executed, the control proceeds to the GOTO statement and the test is repeated. In reality, the control is stuck in a loop until the confirmation signal is read. When it is received, the control exits the custom macro.

Note that custom macro even allows the programmer to activate this program by specifying a user-defined M-code, making it appear as if a true M-code is being used to activate the outside device. There is more on user-defined M-codes later in this chapter.

Access to timers

System variables #3001 and #3002 allow you to access the control's internal clock/timer. Variable #3001 is a timer incremented in milliseconds, while #3002 is incremented in hours. For newer Fanuc controls, you also have access to the current time and date, giving you even more control of what you can do with timer-related functions.

These timers within the control are always running. However, you can reset the milliseconds timer and hour timer by setting them to a value of zero. For example, the command #3001 = 0 resets the milliseconds timer; #3002 = 0 resets the hour timer.

A thousand milliseconds, of course, equals one second. Here is a simple example showing how to use the milliseconds timer to make the control pause for 2 seconds (just as the G04 dwell command would):

```
#3001 = 0 (Reset milliseconds timer)
N1 IF [#3000 GT 2000] GOTO 99 (If finished exit loop)
GOTO 1 (Go back to test)
N99 . . .
.
.
.
```

These timers will be helpful whenever you need your custom macro to keep track of time. A case in point is when trying to monitor cycle time for a given tool, as would be needed with any kind of tool-life monitoring. With this technique, you can come up with a limited tool-life management system. But keep in mind that most controls offer excellent tool-life management systems that can outperform much of what you can do with parametric programming techniques. If you need tool-life management on a regular basis, look into the specific optional feature for tool-life management.

Timing events. You can actually time the period that a given cutting tool is machining. When its total time is exhausted, you can command the control change to a fresh tool or generate an alarm telling the operator that it is time to perform tool maintenance.

Older Fanuc controls with custom macro B, for example, have two timers, one for milliseconds and one for hours. Newer Fanuc controls have a clock that offers time in a similar format to the time/date function of personal computers.

Consider the following commands that accumulate the time tool number one is cutting in permanent common variable #500.

O0001 (Program number)
N005 T0101 (Index to tool number one)
N010 G96 S500 M03 (Start spindle)
N015 G00 X3.0 Z0.1 (Rapid into position)
N020 #3001 = 0 (Reset milliseconds timer)
N025 G71 P030 Q050 U0.040 W0.005 D0.125 F0.012 (Rough turn)
N030 G00 X1.0 (Beginning of finish pass definition)
N035 G01 Z – 1.0 (Turn first diameter)
N040 X2.0 (Feed up first face)
N045 Z – 2.0 (Turn second diameter)
N050 X3.0 (Feed up second face)
N055 #500 = #500 + #3001 (Accumulate time in #500)
N060 G00 X4.5 Z4.0 (Rapid to tool-change position)
.
.
.
N305 IF [#500 LT 40000] GOTO 320 (If time is less than 40 minutes, go to end of program)
N310 #500 = 0 (Reset accumulated time)
N315 #3000 = 100 (TOOL ONE DULL)
N320 M30

Notice that line N020, the milliseconds timer, is reset to zero. From the instant this command is executed, #3001 will accumulate time. After the tool is finished, #500 is stepped by the amount of time that has passed in line N055. At the end of the program, the test in line N305 determines if the tool has been used for more than 40 minutes (40000 milliseconds). When it has, #500 is set back to zero (ready for the next insert) and the alarm in line N315 is given, telling the operator to change the inserts.

Suppressing single block and M-code await

System variable #3003 allows you to take control of two important control functions, single block and whether M-codes are awaited. If #3003 is set to 0 (the initialized state), single block is not suppressed and the control will wait for auxiliary functions (M-codes) to be completed before going on to the next command. If #3003 is set to one, single block will be suppressed and the control will wait for the completion of M-codes to continue with the program. If #3003 is set to two, single block is not suppressed, and the control will not wait for the completion of auxiliary functions. If set to four, single block is suppressed and M-codes will be awaited.

An application for suppressing single block. If single block is on, there may be times when detrimental things could happen. For example, when tapping a hole, the machine must flow through the entire tapping operation. For machining centers and many turning centers, this is simply part of the G84 tapping canned-cycle function. Even if single block is on, the control will treat the entire tapping operation as one block.

However, many Fanuc-controlled turning center controls do not have a tapping cycle. They force the programmer to write the tapping commands longhand. When tapping longhand, there is no way of inhibiting single block with normal G-code-level programming. But with custom macro, it is possible to suppress single block. Here is an example that shows how. (Letter address *Z* is represented in the custom macro by local variable #26, R by #18, and F by #9.)

Main program:
O0003 (Program number)
N005 G00 T0101 M41 (Select tool, offset, and spindle range)
N010 G97 S500 M03 (Start spindle cw at 500 rpm)
N015 G65 P1003 Z – 1.0 R.25 F.0625 (Tap hole)
N020 G00 X5.0 Z4.0 T0100 (Move to tool-change position, cancel offset)
N025 M30 (End of program)

In custom macro B:

O1003 (Program number)
G00 X0 Z#18 (Rapid to approach position)
#3003 = 1 (Suppress single block)
G32 Z#26 F#9 (Feed into hole)
M04 (Reverse spindle)

G32 Z#18 (Feed back out of hole)
#3003 = 0 (Reactivate single block)
M99 (End of custom macro)

Prior to entering the hole, the command #3003 = 1 makes the control ignore the single block function (even if it is on). After coming out of the hole the command #3003 = 0 reactivates the single block function.

An application for not waiting for M-codes. For many M-codes, the machine-tool builder determines whether the confirmation signal will be waited before the control can go on to the next command. In most cases, you will agree with the machine-tool builder's decision as to whether a given M-code should be awaited or not. However, when time savings is an issue, there may be instances when you may not agree and wish to take control of whether the control waits for a given M-code to be completed.

For example, some machining centers require that a door to the automatic tool changer be completely closed at the completion of each tool change. This is quite common for machines with totally enclosed guarding. In some cases, an M-code is used to close the door. In this example, the control should not have to wait for the door to completely close before going on to the next command. The control should be allowed to simply continue with the program, since the door can be closing without causing any danger.

Whether or not the control waits for the completion of the door closing M-code can be controlled by system variable #3003. If your particular machine does wait (wasting time), simply set #3003 = 2 prior to activating the M-code in question. Of course, you must set #3003 = 0 when finished with the M-code in order to get the control back to its normal state for other M-codes that do require the control to wait.

Control of feed hold, feedrate override, and exact stop check

In similar manner, system variable #3004 gives you control of three more important CNC functions. Table 5-1 shows the meaning of each value of #3004.

For all intents and purposes, exact stop check can be controlled by other means (G09 and G61), so you typically do not have to use system variable #3004 for this purpose. But, as with suppressing single block, #3004 can be very helpful when you need to control feed hold and feedrate override. If you have applications that require that feed hold and feedrate override be suppressed (like tapping with G01), you can give the command #3004 = 3 just prior to the commands in question.

After giving the motion commands, don't forget to set #3004 = 0 to reinstate control of feed hold and feedrate override.

Stop with message

Earlier in this chapter we discussed the #3000 alarm-generating system variable. As you know, this system variable, if read, will put the control into alarm state

Table 5-1. Defining the values of #3004.

Value	*Feed hold*	*Feedrate override*	*Exact stop check*
0	Effective	Effective	Effective
1	Suppressed	Effective	Effective
2	Effective	Suppressed	Effective
3	Suppressed	Suppressed	Effective
4	Effective	Effective	Suppressed
5	Suppressed	Effective	Suppressed
6	Effective	Suppressed	Suppressed
7	Suppressed	Suppressed	Suppressed

and display a message on the screen. To continue with the program, the operator must reset the alarm.

There are times when you will want to stop the machine and give the operator a message, but you want the operator to be able to continue with the program by simply pressing cycle start. Most programmers use the simple M00 program stop for this purpose. If, for example, you wish to stop the machine so the operator can clear chips and add tapping compound, you could give this command:

```
N050 M00 (BLOW CHIPS AWAY AND ADD TAPPING COMPOUND)
```

However, if the operator is not monitoring the program mode while running the program, the message in parentheses will not be seen. The operator will have to switch display screen modes in order to see the message. The programmer can force the message to appear with system variable #3006. With #3006, the operator will be able to continue with the cycle by simply pressing the cycle start button. Here is an example of the syntax for the stop with message system variable. Note how similar it is to the alarm-generation system variable (#3000).

```
#3006 = 100 (BLOW CHIPS AWAY)
```

As with the #3000 alarm generation system variable, #3006 uses a message number (100 in this example). The message in parentheses will be displayed next to the message number in the format:

```
MS-100 BLOW CHIPS AWAY
```

Access to modal information

The #4000 series system variables give you access to currently instated modal information.

Variables #4001 through #4021 specify the currently active G-code within each G-code group. The various G-code groups are listed in the beginning of your programming manual. For machining centers, G00, G01, G02, and G03 are in G-code group number one. To attain the currently instated G-code in this group, you could give this command:

```
#101 = #4001
```

Group number two is plane selection (G17, G18, and G19). Group number three is absolute and incremental modes (G90 and G91). Again, to see the complete list of G-code groups, consult your programming manual.

Here are other system variables in the #4000 series that allow access to modal information:

#4102 — Currently instated B word
#4107 — Currently instated D word
#4108 — Currently instated E word
#4109 — Currently instated F word
#4111 — Currently instated H word
#4113 — Currently instated M word
#4114 — Currently instated N word
#4115 — Currently instated O word
#4119 — Currently instated S word
#4120 — Currently instated T word

An application. For the most part, these system variables are not commonly required by the typical parametric programmer. However, there is at least one instance in which their use may be helpful. Suppose you wish to use the incremental mode for motions within a custom macro, yet you are unsure of the machine's current state relative to incremental and absolute mode when the custom macro is activated. At the end of your custom macro, you wish to leave the machine in the state you found it. Consider these commands:

```
O1054 (Program number)
#101 = #4003 (Determine the current setting of G90 or G91)
G91 (Set incremental mode)
(Motions under the influence of incremental mode)
.
.
.
#4003 = #101 (Reset mode to original state)
M99 (End of custom macro)
```

Notice that in the command #101 = #4003, we are storing the current value of G-code group number three (which happens to be the group related to G90 or G91) in common variable #101. Then the G91 mode is selected to perform the machining operation. Near the end of the custom macro, we have the command #4003 = #101. This command is reselecting the mode (stored earlier) by reassigning system variable #4003.

Admittedly, this may be a far-fetched application for the typical parametric programmer. In most cases, only the machine-tool builder or aftermarket supplier would have an application for this kind of technique.

OTHER CNC FEATURES OF PARAMETRIC PROGRAMMING

As you have seen, this chapter has been working from common CNC features toward the less known. In this discussion, we simply mention CNC features that are equipped with some versions of parametric programming that have nothing to do with system variables. For a fuller understanding of the strengths of your CNC system, you should reference your control manufacturer's programming manual to determine whether or not you have one or more of these capabilities in your particular version of parametric programming.

PROTECTION FOR IMPORTANT PROGRAMS

If you have important parametric programs that you do not want accidentally deleted or modified, it's best to protect them. This protection capability is especially useful with parametric programs related to the family-of-parts, user-created canned-cycle, and utility application categories. Fanuc, for example, allows you to protect the entire series of programs numbered in the 9000 series. Unfortunately, the parameters that are related to this function vary even among different Fanuc control *models*, meaning you will have to reference your programming manual to determine the parameters involved.

LABELING VARIABLES ON THE DISPLAY SCREEN PAGE

Some versions allow you to label variables on the display screen page. This can be helpful to alert the operator that certain variables should not be changed. When using global variables and system constants, for example, you wouldn't want an operator accidentally changing an important variable. This feature is also quite helpful for variables used as calibration values needed by probing systems.

In custom macro B, for example, permanent common variables (the #500 series) can be labeled with a SETVN command (SETVN stands for SET Variable Name). The command

```
SETVN = 500 (COUNTER)
```

for example, places the message COUNTER next to permanent common variable #500. In custom macro B, this message is limited to eight characters in length.

USER-DEFINED G- AND M-CODES

Some versions of parametric programming allow you to actually invoke your custom macros with G- and M-codes. This can make your parametric programs appear to be true G- and M-code functions. Since this can be a very helpful function (unique to custom macro B), we discuss it here in detail.

Defining your own G-codes

As you know, you can easily command a custom macro to be executed by giving the G65 command. Included in the G65 command is a P word that specifies the program number to be executed. Though this is not at all a bad way to call a custom macro, there is an easier way.

When you have an often-used custom macro, for example, it is sometimes helpful to assign it its own G-code. You can actually have a G-code of your choosing call the custom macro. To do this, you will first need to make sure that you do not overwrite a currently-used G-code. Look in the list of G-codes included in the programming manual to find an unused G-code number. Also, keep in mind that you can use G-codes for calling custom macros that range from 100 to 255 (not just up to 99). This should help you avoid conflicts.

There are certain parameters related to how G-codes are created. The actual parameter numbers will vary even from one control *model* to another, so you will have to look in the custom macro section of your programming manual to find the exact parameter numbers.

To execute a custom macro by a user-defined G-code, the custom macro program number *must* be between O9010 and O9019. For 10-, 11-, and 15-series Fanuc controls, parameters 7050-7059 allow the user to specify which G-code numbers are related to G-code calling program numbers. If, for example, parameter number 7050 is set to 12, whenever a G12 is executed, the control will execute program O9010. The control will execute program O9011 whenever the G-code specified by the value of parameter 7051 is read. The list of letter address arguments must still be specified.

Creating G-codes simply allows you to specify a G-code, say G12, instead of G65 P1004. However, this can sometimes be helpful to solve certain G-code numbering compatibility problems. For example, suppose you have a Yasnac control that has two circle milling G-codes (G12 for clockwise and G13 for counterclockwise). With custom macro B, you intend to write a user-created canned cycle for circle milling on your Fanuc control that duplicates the Yasnac style. If you use the same letter address arguments as are used by Yasnac; if you name your programs as O9010 and O9011; and if you set parameters 7050 and 7051 (10-, 11-, or 15-series controls only) to 12 and 13, programming for circle milling operations will be consistent between Fanuc and Yasnac.

Defining your own M-codes

With custom macro B, you also have the ability to create new and modify the function of current M-codes. This is done in much the same way as you create user-defined G-codes. For most controls, programs O9001 through O9009 are used as user-defined M-codes. Parameters 7071 through 7079 are used to specify the M-code numbers involved (on 10-, 11-, and 15-series controls). If for example, parameter 7071 is set to 14, program number O9001 is called whenever M14 is executed. By comparison with user-defined G-codes, most control models do not allow a list of arguments to be passed with user-defined M-codes. This means, of course, that you will only be using user-defined M-codes to activate custom macros that do not require arguments to be passed.

In addition to creating M-codes, you have the ability to change the function of any M-code your machine is currently using. You can even activate the original M-code from within the custom macro program the M-code is calling. This is especially helpful with the machining center M06 command.

Suppose you have a vertical machining center that uses M06 to make tool changes. However, the normal function of M06 with most machines is only to make tool changes. If the machine is not currently resting at the tool-change position (normally the *Z* axis zero-return position for vertical machining centers) when the M06 is commanded, the machine will go into alarm state.

With custom macro B, you can change the function of M06 so that whenever it is activated, the machine will first go to the zero-return position (even incorporating the spindle orient M19 on the way) and then make the actual tool change. Here's how (for a 10-, 11-, or 15-series Fanuc control): First, set parameter number 7071 to 6. This will cause the control to execute program number O9001 whenever M06 is executed. Next, load this program into the control:

```
O9001 (Program number)
G91 G28 Z0 M19 (Send the machine to tool-change position, orient spindle)
M06 (Change tools)
G90 (Reselect absolute mode)
M99 (End of custom macro)
```

With this program, the machine will first be sent to its tool-change position. In the next command, the M06 will actually make the tool change. Note that an M-code of the same number specified in a user-defined M-code program will cause the control to activate its original function. (A custom macro cannot call itself.) Finally, we save some work in the main program by reselecting the absolute mode.

As with G-codes, you can also add new M-codes. As stated during the discussion of the #1000 series system variables, you can even access outside devices with user-defined M-codes. To complete the example given during the discussions of the #1000 series system variables, here we show how you could make M13 activate the outside device.

First, set the first available M-code parameter (let's say number 7071) to 13. From this point, any time the control reads an M13, it will activate custom macro O9001. Next, the outside device activating custom macro is loaded into the control as program number O9001.

```
O9001 (Program number)
#1100 = 1 (Activate indexer)
N1 IF [#1000 EQ 1] GOTO 99
GOTO 1
N99 M99
```

As stated earlier, the #1100 = 1 statement applies a 24-volt signal to terminal location UO000 (to which the indexer is connected). Then the custom macro program goes into a waiting loop, looking for the confirmation signal from the outside device. As soon as UI000 is closed (by the outside device), system variable #1000 will be equal to one, ending the loop and letting the control continue.

Another time that user-defined M-codes can really help is when you have similar machines made by different machine-tool builders. For example, if you have two similar turning centers, each having the same approximate horsepower, chuck size, and range of travels, it is quite likely that you will want to run programs on either machine.

Unfortunately, machine-tool builders have not standardized M-codes, even some rather common ones. For example, one machine may require an M23 to select the spindle's low range and M25 to select high range. Another may use M41 for low range and M42 for high range. Yet another may require no M-code (only one spindle range) but the control for this machine may hang up if any unrecognizable M-code is read.

Rather than forcing the operator or programmer to constantly deal with this incompatibility problem (constantly changing programs), you can create user-defined M-codes for one of the machines involved. In the spindle range example, let's say you want to use M41 and M42 for low and high spindle range on both machines. On the machine that currently uses M23 and M25, set parameter 7071 to 41 and parameter 7072 to 42 (for 10-, 11-, and 15-series Fanuc controls). Then load these two simple programs into the control's memory.

```
O9001 (Program number)
M23 (Select low spindle range)
M99 (End of custom macro)
```

```
O9002 (Program number)
M25 (Select high spindle range)
M99 (End of custom macro)
```

Once this is done, whenever this control reads an M41, it will execute program O9001, which of course selects the low spindle range. When it reads an M42, program O9002 is executed, selecting the high spindle range.

If you happen to have a machine that has only one spindle range, simply add two user-defined M-codes (for M41 and M42) that execute empty custom macro programs (programs with a program number followed immediately by M99). Though these custom macros will do absolutely nothing, at least the control will not hang up when an M41 or M42 is read.

Keep in mind that the spindle range selection for turning centers is only one example of incompatibility problems with M-codes. Others include bar-feed activating M-codes, part-catcher M-codes, tailstock M-codes, indexer M-codes, pallet-changer M-codes, and tool-changer activation M-codes.

COMMUNICATING DATA THROUGH THE RS-232C PORT

Some versions of parametric programming allow *external data output* commands designed to send data through the communications (RS-232C) port. Some even allow data transfers in both directions (*to* and *from* the CNC control). These functions are especially helpful when you wish to transmit production control data, such as workpiece count, time in cycle, and statistical process control data (when probing systems are used).

Fanuc's custom macro B is limited to sending data out, from the CNC control to some receiving device. The receiving device usually takes the form of a (serial) printer or personal computer. Four commands are related.

POPEN — Tells the control to open the serial port for outputting data
BPRNT — Outputs data (without decimal points)
DPRNT — Outputs data (with decimal points)
PCLOS — Closes the serial port at the end of a transmission

Since most values being communicated require decimal point placement, the DPRNT command is the more popular of the printing commands.

Suppose you have a custom macro program that uses a probe to measure the width of a slot. After machining, the probe measures the slot; however, you want the slot width to be automatically reported to your statistical process control (SPC) computer. You also want your parametric program to report the current number of the workpiece in the production run, as well as the slot width. You use permanent common variable #500 to track the current part count, and common variable #100 to store the slot width. The following series of commands will send this data through the communications port. Of course, the outside device must be connected and ready to receive this data at the time these commands are read.

.
.
.

```
POPEN (Open the serial port)
DPRNT PART NUMBER #500 40 SLOT WIDTH #100 24
PCLOS (Close the serial port)
.
.
.
```

The POPEN command opens the serial port. In the DPRNT command, the words PART NUMBER and SLOT WIDTH simply specify notes that will appear next to the values stored in #500 and #100. The numbers 40 and 24 specify the decimal point format for the numbers being printed. The format "40" is telling the control to print up to four digits to the left of the decimal point and no (0) digits to the right of the decimal point. Since #500 is an integer (whole number), there will be no portion of a whole number involved. This format (40) allows the printing of four digit numbers (up to 9999).

In similar fashion, the 24 after #100 will have the control print two digits to the left of the decimal point and four places to the right. This will print in xx.xxxx format for the slot width. The PCLOS command closes the serial port after the transmission.

If this is part number 45 and the slot is measured as 2.2018 in. (55.9257 mm) wide, the actual message, if printed, will appear as follows:

```
PART NUMBER 0045 SLOT WIDTH 02.2018
```

CONCLUSION

Though we have dealt with many of the CNC features of parametric programming, it is likely that you have more available with your particular version. Reference your own control manufacturer's programming manual to determine what else may be available. And remember, some CNC features are very obscure, and the discussions given here may be somewhat difficult to follow. However, with study, you may find you have many more CNC features available to you.

Chapter 6

Approaching and Verifying Parametric Programs

You have now been exposed to the various tools available with parametric programming. And armed with your new knowledge, you are perhaps anxious to tackle your own applications. But before you start your own parametric programming, there is a bit more information you should know that will complement what we've already covered. First, there are suggestions for *how to approach applications* in each of the five application categories. These we provide in the form of step-by-step instructions. Second are *tips on how to verify parametric programs*. You will find that since parametric programming is much more flexible than conventional programming, you must be more cautious when verifying them. However, once verified, parametric programs can be treated like canned cycles, minimizing the long-term program verification effort. Finally, there is another form of argument assignment used with custom macro B called argument assignment number two.

APPROACHING APPLICATIONS

Though we have shown several example programs, and for the most part have followed the suggestions about to be made, we now offer a set of step-by-step instructions you can use to approach parametric programming for each application category. While there is some similarity in the approach to each category, if you follow them, each specific suggestion should simplify your first few programs. Our intention is to make it as simple as possible to develop parametric programs to handle your own applications.

APPROACHING FAMILY-OF-PARTS APPLICATIONS

As you know from Chapters 1 and 2, with the family-of-parts category the parametric program machines the entire workpiece. Though other programs may be involved, generally speaking, one program (the main program) does the bulk of the machining operations. The changing arguments are specified and well documented at the beginning of the program, and the operator or setup person will modify these arguments as needed to machine a specific workpiece in the family.

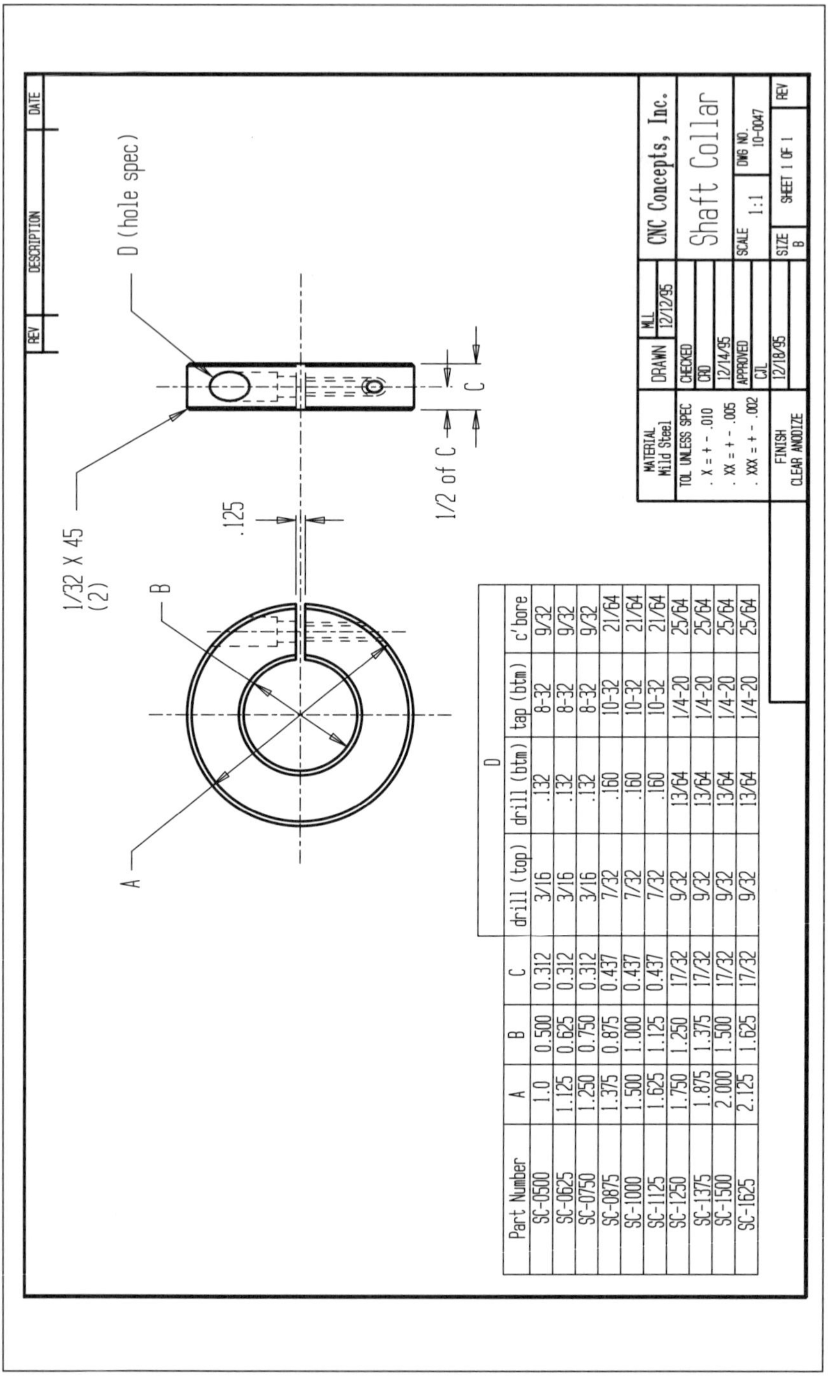

Part Number	A	B	C	D			
				drill (top)	drill (btm)	tap (btm)	c'bore
SC-0500	1.0	0.500	0.312	3/16	.132	8-32	9/32
SC-0625	1.125	0.625	0.312	3/16	.132	8-32	9/32
SC-0750	1.250	0.750	0.312	3/16	.132	8-32	9/32
SC-0875	1.375	0.875	0.437	7/32	.160	10-32	21/64
SC-1000	1.500	1.000	0.437	7/32	.160	10-32	21/64
SC-1125	1.625	1.125	0.437	7/32	.160	10-32	21/64
SC-1250	1.750	1.250	17/32	9/32	13/64	1/4-20	25/64
SC-1375	1.875	1.375	17/32	9/32	13/64	1/4-20	25/64
SC-1500	2.000	1.500	17/32	9/32	13/64	1/4-20	25/64
SC-1625	2.125	1.625	17/32	9/32	13/64	1/4-20	25/64

Figure 6-1. A typical method of dimensioning family-of-parts applications.

Figure 6-1 shows a common method of dimensioning with families-of-parts. Using this *shaft collar* part family, we will highlight the points we want to make throughout this section. Note that only the turned operations are going to be performed in this example. We show this application only in custom macro B format, since all points we make apply to every version of parametric programming. You should now be able to easily convert our example to your own version.

Step one—Develop a process to be used for the entire family

As you know, the *process* is of paramount importance to the success of any CNC program, and parametric programs are no exception. As is true in developing conventional CNC programs, your first step in developing parametric programs is to develop a workable process. In some cases, you may find it difficult to develop a universal process that works for all workpieces in the family. This is especially true in cases where there is a dramatic variance from the smallest workpiece in the family to the largest. (Any processing problems you encounter that are caused by workpiece variations should be taken as a sign that you may have to break your family-of-parts application into two or more parametric programs. The smallest workpieces in your family, for example, may require an entirely different process than the bulk of the workpieces in the family. In this case, it may be impossible to develop one parametric program to handle all workpieces.)

For the shaft collar family-of-parts application, the process is shown in Table 6-1. This process will work for all workpieces. (Note that only *turning* operations are performed.)

Table 6-1. Process sequencing for shaft collar machining.

Step	*Description*	*Tool*	*Station*
1	Rough face and turn	Turning tool	1
2	Drill	HSS drill	2
3	Rough bore	Boring bar	3
4	Finish bore	Boring bar	4
5	Finish turn	Turning tool	5
6	Cutoff	1/8-in. (3.175-mm) cutoff tool	6
7	Feed bar	Bar stop	7

Minimize cutting tools. Since one of the main reasons for using parametric programming with families-of-parts is to minimize change-over (setup) time, you will likely be focused on reducing the time it takes to change cutting tools during setup. When feasible, your goal should be to minimize variations in cutting tools. In our example, for instance, the same rough turning tool, finish turning tool, cutoff tool, and bar stop can be used for all workpieces in the family. It would be wise to maintain their tool station number from one setup to the next. Only the drill, rough boring bar, and finish boring bar will have to change (this, because of the hole size being machined).

As long as production quantities (lot sizes) are relatively low, we can sacrifice some cycle time to minimize setup time by choosing only two sets of internal tools for the entire family. For workpieces with a hole less than 1.0 in. (25.4 mm) in diameter, we'll use a 0.437-in. (11.10-mm) drill and 0.375-in. (9.525-mm)-diameter boring bars. For holes more than 1.0 in. in diameter, we'll use a 0.875-in. (22.23-mm)-diameter drill and 0.75-in. (19.05-mm)-diameter boring bars.

With this method, the parametric program can automatically determine the speed in rpm for the drill by testing against the inside diameter size. If the parametric program determines that the hole is less than 1.0 in. (25.4 mm) in diameter, it will base the rpm on a 0.437-in. (11.10-mm)-diameter drill. If 1.0 in. or more, it will base the rpm on a 0.875-in. (22.23-mm)-diameter drill.

Again, note that we could be sacrificing some rough boring time with larger holes to gain the benefit of minimizing the required cutting tools. This will allow us to tool the turret once, never having to change tools as long as we are running workpieces in this family. As production quantities grow, however, this may not be the wisest thing to do. With higher quantities, the time it takes to change tools in the turret during setup may be easily overcome with saved cycle time during the production run.

Step two—Determine your arguments

Early in your preparation to write the parametric program, you will need to isolate those things that change from one workpiece in the part family to the next. Of course, any dimension that changes will be specified as an argument. Any dimension that does not change (like the chamfer size in our example), can be handled as a constant value. But be careful with constant values. If there is any possibility that new parts in the family will require variations, write an argument for the workpiece attribute instead of using the constant value.

You also will need to document the argument assignments. Possibly you can simply mark up a print, showing the appropriate arguments. More likely, you will need to develop a special drawing and other support materials to help clarify how arguments are used by your parametric program. For our example, we will provide a simple list of all arguments and document the meaning of each right in the parametric program.

The changing workpiece attributes. When your design engineer uses variable dimensioning techniques as shown in the example, it will be quite easy to determine which dimensions must be specified as arguments. However, if your design engineer does not use variable dimensioning techniques, it can be quite difficult. In this case, you must study all workpiece drawings included in the part family. In our example, there are three obvious changing workpiece attributes (A, B, and C), which are specified right on the print. In our custom macro B version of this program, we use common variable #100 to represent the outside diameter (A dimension), #101 to represent the inside diameter (B dimension), and #102 to represent the workpiece length (C dimension).

The changing rough stock attributes. The condition of the rough stock prior to machining is equally as important as any finished workpiece attribute. When considering arguments, any changing value that will specify the size of your raw material must be specified as an argument. In our example, the outside diameter of the raw stock will vary with bar diameter. We'll use common variable #103 to specify the stock diameter. When it comes to removing stock on the face of the workpiece, we'll say this value remains a hard-and-fixed (constant) value of 0.050 in. (1.27 mm).

The changing cutting condition attributes. At first glance, you may feel that the cutting conditions (speeds, feeds, depth of cut, etc.) you use can be handled with constant values. Depending upon your application, you may be right, especially if there are absolutely no variations in your raw material. However, it is usually wise to enable your operator or setup person to easily modify the cutting conditions (at least speeds and feeds) for each operation performed. To allow this, you must assign a special argument for each cutting condition element. Here is how our example program will assign cutting condition-related arguments:

Argument	*Description*	*Initially set to*
#110	Speed for rough face and turn	500 sfm (152 smm)
#111	Feed for rough face and turn	0.012 ipr (0.305 mmpr)
#112	Depth of cut for roughing	0.125 in. (3.175 mm)
#113	Speed for drilling	80 sfm (24.4 smm)
#114	Feed for drilling	0.008 ipr (0.203 mmpr)
#115	Speed for rough boring	400 sfm (122 smm)
#116	Feed for rough boring	0.008 ipr (0.203 mmpr)
#117	Depth of cut for rough boring	0.080 in. (2.032 mm)
#118	Speed for finish boring	500 sfm (152 smm)
#119	Feed for finish boring	0.005 ipr (0.127 mmpr)
#120	Speed for finish turning	600 sfm (183 smm)
#121	Feed for finish turning	0.007 ipr (0.178 mmpr)
#122	Speed for cutoff tool	400 sfm (122 smm)
#123	Feed for cutoff tool	0.005 ipr (0.127 mmpr)

What about multiple materials? It is not uncommon for workpieces in part families to be machined from different materials. In this shaft collar application, for example, some collars may be made of mild steel while others are made of stainless steel. When going from one setup to the next, you will want to make it as easy as possible to specify the cutting conditions for the materials being machined. One easy way to do this is to use simple subprogramming techniques. One cutting conditions subprogram can be placed in the control's memory for each material in the family. Consider these two programs:

```
O2001 (Cutting conditions for mild steel)
#110 = 500.0 (Speed in sfm for rough face and turn)
```

```
#111 = 0.012 (Feed in ipr for rough face and turn)
#112 = 0.125 (Depth of cut for rough face and turn)
#113 = 80.0 (Speed in sfm for drilling)
#114 = 0.008 (Feed in ipr for drilling)
#115 = 400.0 (Speed in sfm for rough boring)
#116 = 0.008 (Feed in ipr for rough boring)
#117 = 0.080 (Depth of cut for rough boring)
#118 = 500.0 (Speed in sfm for finish boring)
#119 = 0.005 (Feed in ipr for finish boring)
#120 = 600.0 (Speed in sfm for finish turning)
#121 = 0.007 (Feed in ipr for finish turning)
#122 = 400.0 (Speed in sfm for cutoff tool)
#123 = 0.005 (Feed in ipr for cutoff tool)
M99 (End of subprogram)

O2002 (Cutting conditions for stainless steel)
#110 = 300.0 (Speed in sfm for rough face and turn)
#111 = 0.010 (Feed in ipr for rough face and turn)
#112 = 0.100 (Depth of cut for rough face and turn)
#113 = 60.0 (Speed in sfm for drilling)
#114 = 0.007 (Feed in ipr for drilling)
#115 = 2750.0 (Speed in sfm for rough boring)
#116 = 0.006 (Feed in ipr for rough boring)
#117 = 0.050 (Depth of cut for rough boring)
#118 = 350.0 (Speed in sfm for finish boring)
#119 = 0.004 (Feed in ipr for finish boring)
#120 = 400.0 (Speed in sfm for finish turning)
#121 = 0.005 (Feed in ipr for finish turning)
#122 = 275.0 (Speed in sfm for cutoff tool)
#123 = 0.004 (Feed in ipr for cutoff tool)
M99 (End of subprogram)
```

If machining a mild steel shaft collar, a call to program number O2001 will be made in the main program. When machining stainless steel, however, a call to O2002 will be made. With this technique, you can easily set up to run as many different materials as necessary.

Advanced technique—Minimizing cutting conditions arguments. As you can see from the list of arguments on cutting conditions, 14 arguments are required just to get cutting conditions assigned. This will substantially lengthen the arguments at the beginning of the parametric program, which may possibly confuse the operator. If cutting conditions must change on a regular basis, there may be no other choice. However, if cutting conditions remain more consistent, and if you still want to give the operator or setup person the ability to easily change them when required, you can utilize a special parametric program for the sole purpose of setting up cutting conditions. In essence, you can separate the selection of cutting conditions from your main program.

With custom macro B, you can use a G65 command (CALL command in user task 2) to activate a cutting conditions parametric program. If no arguments are specified, you can let the cutting conditions parametric program set default values for each cutting condition common variable (#110 through #123). If the operator or setup person wishes to change a cutting condition, an appropriate letter address argument can be specified. For our shaft collar example, the letter address arguments correspond to cutting conditions as shown in the following list. Of course, the person making cutting conditions changes must have access to this list to change the appropriate cutting condition.

Letter	*Description*	*Common Variable*
A (#1)	Speed for rough face and turn (500.0 sfm [152 smm])	#110
B (#2)	Feed for rough face and turn (0.012 ipr [0.305 mmpr])	#111
C (#3)	Depth of cut for roughing (0.125 in. [3.175 mm])	#112
D (#7)	Speed for drilling (80.0 sfm [24.4 smm])	#113
E (#8)	Feed for drilling (0.008 ipr [0.203 mmpr])	#114
F (#9)	Speed for rough boring (400.0 sfm [122 smm])	#115
H (#11)	Feed for rough boring (0.008 ipr [0.203 mmpr])	#116
I (#4)	Depth of cut for rough boring (0.080 in. [2.032 mm])	#117
J (#5)	Speed for finish boring (500.0 sfm [152 smm])	#118
K (#6)	Feed for finish boring (0.005 ipr [0.127 mmpr])	#119
M (#13)	Speed for finish turning (600.0 sfm [183 smm])	#120
Q (#17)	Feed for finish turning (0.007 ipr [0.178 mmpr])	#121
R (#18)	Speed for cutoff tool (400.0 sfm [122 smm])	#122
S (#19)	Feed for cutoff tool (0.005 ipr [0.127 mmpr])	#123

If the cutting conditions parametric program is program number O1001, the command

```
G65 P1001
```

will set up all cutting conditions with their default values. If the operator wishes to change only the speed for the cutoff tool to 300 sfm, the command

```
G65 P1001 R300.0
```

will be given. The cutting conditions parametric program will simply test each local variable corresponding to letter address arguments to see if they have been included in the G65 command. If none have been included, the default value of each corresponding common variable (#110 through #123) will be used. On the other hand, if one or more letter address argument has been set, its corresponding common variable will be set to the value specified in the G65 command. Though our example is specific to custom macro B, this technique can be applied to any version of parametric programming that incorporates the capability of testing whether a (local) variable has a value (user task 2 also has this capability). The cutting conditions program we will invoke from our family-of-parts parametric program follows. This program must, of course, be in the control's memory at the time the family-of-parts parametric program is run.

```
O1001 (Cutting conditions parametric program)
IF [#1 NE #0] GOTO 1 (Test if A is assigned)
#1 = 500.0 (If not set corresponding local variable to default value)
N1 IF [#2 NE #0] GOTO 2 (Test if B is assigned)
#2 = 0.012 (If not set corresponding local variable to default value)
N2 IF [#3 NE #0] GOTO 3 (Test if C is assigned)
#3 = 0.125 (If not set corresponding local variable to default value)
N3 IF [#7 NE #0] GOTO 4 (Test if D is assigned)
#7 = 80.0 (If not set corresponding local variable to default value)
N4 IF [#8 NE #0] GOTO 5 (Test if E is assigned)
#8 = 0.008 (If not set corresponding local variable to default value)
N5 IF [#9 NE #0] GOTO 6 (Test if F is assigned)
#9 = 400.0 (If not set corresponding local variable to default value)
N6 IF [#11 NE #0] GOTO 7 (Test if H is assigned)
#11 = 0.008 (If not set corresponding local variable to default value)
N7 IF [#4 NE #0] GOTO 8 (Test if I is assigned)
#4 = 0.080 (If not set corresponding local variable to default value)
N8 IF [#5 NE #0] GOTO 9 (Test if J is assigned)
#5 = 500.0 (If not set corresponding local variable to default value)
N9 IF [#6 NE #0] GOTO 10 (Test if K is assigned)
#6 = 0.005 (If not set corresponding local variable to default value)
N10 IF [#13 NE #0] GOTO 11 (Test if M is assigned)
#13 = 500.0 (If not set corresponding local variable to default value)
N11 IF [#17 NE #0] GOTO 12 (Test if Q is assigned)
#17 = 0.007 (If not set corresponding local variable to default value)
N12 IF [#18 NE #0] GOTO 13 (Test if R is assigned)
#18 = 400.0 (If not set corresponding local variable to default value)
N13 IF [#19 NE #0] GOTO 14 (Test if S is assigned)
#19 = 0.005 (If not set corresponding local variable to default value)
N14 #110 = #1 (Store value for use in parametric program)
#111 = #2 (Store value for use in parametric program)
#112 = #3 (Store value for use in parametric program)
#113 = #7 (Store value for use in parametric program)
```

```
#114 = #8 (Store value for use in parametric program)
#115 = #9 (Store value for use in parametric program)
#116 = #11 (Store value for use in parametric program)
#117 = #4 (Store value for use in parametric program)
#118 = #5 (Store value for use in parametric program)
#119 = #6 (Store value for use in parametric program)
#120 = #13 (Store value for use in parametric program)
#121 = #17 (Store value for use in parametric program)
#122 = #18 (Store value for use in parametric program)
#123 = #19 (Store value for use in parametric program)
M99 (End of custom macro)
```

Changing tool station numbers. It also may be helpful if your setup person can manipulate the tool station numbers used for cutting tools. In our example, the setup person may elect to permanently place both sets of internal tools (drill, rough boring bar, and finish bar) in the turret. The 0.437-in. (11.10-mm)-diameter drill may be in station two, and the 0.875-in. (22.225-mm)-drill may be in station seven. If this is the case, you can easily make the station number for each cutting tool an argument. For our custom macro B example, we will use these common variables for tool station numbers:

#131 — Station number for rough face and turn tool (1)
#132 — Station number for drill (2)
#133 — Station number for rough boring bar (3)
#134 — Station number for finish boring bar (4)
#135 — Station number for finish turning tool (5)
#136 — Station number for cutoff tool (6)
#137 — Station number for bar stop (7)

To minimize the number of arguments in the parametric program, you can use the same advanced technique shown for the cutting conditions parametric program. For the shaft collar example, we will use program O1002 for the tool station setting custom macro. As long as the tools are placed in the turret as shown in the foregoing sequence, the command

```
G65 P1002
```

will set the common variables accordingly. Here is the list of letter address arguments and their corresponding tool descriptions.

A (#1) — Rough turn and face tool (1)
B (#2) — Drill (2)
C (#3) — Rough boring bar (3)

D (#7) — Finish boring bar (4)
E (#8) — Finish turning tool (5)
F (#9) — Cutoff tool (6)
H (#11) — Bar stop (7)

The tool station setting parametric program is:

```
O1001 (Cutting conditions parametric program)
IF [#1 NE #0] GOTO 1 (Test if A is assigned)
#1 = 1 (If not set corresponding local variable to default value)
N1 IF [#2 NE #0] GOTO 2 (Test if B is assigned)
#2 = 2 (If not set corresponding local variable to default value)
N2 IF [#3 NE #0] GOTO 3 (Test if C is assigned)
#3 = 3 (If not set corresponding local variable to default value)
N3 IF [#7 NE #0] GOTO 4 (Test if D is assigned)
#7 = 4 (If not set corresponding local variable to default value)
N4 IF [#8 NE #0] GOTO 5 (Test if E is assigned)
#8 = 5 (If not set corresponding local variable to default value)
N5 IF [#9 NE #0] GOTO 6 (Test if F is assigned)
#9 = 6 (If not set corresponding local variable to default value)
N6 IF [#11 NE #0] GOTO 7 (Test if H is assigned)
#11 = 7 (If not set corresponding local variable to default value)
N7 #131 = #1 (Set common variable for use in another program)
#132 = #2 (Set common variable for use in another program)
#133 = #3 (Set common variable for use in another program)
#134 = #7 (Set common variable for use in another program)
#135 = #8 (Set common variable for use in another program)
#136 = #9 (Set common variable for use in another program)
#137 = #11 (Set common variable for use in another program)
M99 (End of custom macro)
```

Flag arguments. As mentioned in Chapter 4, flags are used to specify exactly how you want the parametric program to behave based on some special condition. As you evaluate the arguments to be supplied for family-of-parts applications, consider any condition that may change from one use of the program to the next.

In the case of the shaft collar, for example, there may be a need when the raw material to be machined is in the form of tubing instead of a solid bar. In this instance, the drilling operation will not be required. A flag can be used to specify not only if the raw material is in the form of tubing, but to specify the size of the hole. In the parametric program, a test can be made against this flag argument. If the argument is greater than zero, the drilling operation can be skipped and the rough boring can begin machining from the specified hole diameter.

Keep track of variables and document their default values. Even for seemingly simple applications, it is a good idea to track the variables you are using. This will help you avoid conflicts, eliminating the possibility of using the same variable twice. The following list tracks the variables used in the example application thus far (all of these variables are being used as arguments). For those values that have default settings, it also is wise to document their default values.

#100 — Outside diameter
#101 — Inside diameter
#102 — Part length
#103 — Stock diameter
#110 — Speed for rough face and turn (500.0 sfm [152 smm])
#111 — Feed for rough face and turn (0.012 ipm [0.305 mmpr])
#112 — Depth of cut for roughing (0.125 in. [3.175 mm])
#113 — Speed for drilling (80.0 sfm [24.4 smm])
#114 — Feed for drilling (0.008 ipr [0.203 mmpr])
#115 — Speed for rough boring (400.0 sfm [122 smm])
#116 — Feed for rough boring (0.008 ipr [0.203 mmpr])
#117 — Depth of cut for rough boring (0.080 in. [2.032 mm])
#118 — Speed for finish boring (500.0 sfm [152 smm])
#119 — Feed for finish boring (0.005 ipr [0.127 mmpr])
#120 — Speed for finish turning (600.0 sfm [183 smm])
#121 — Feed for finish turning (0.007 ipr [0.178 mmpr])
#122 — Speed for cutoff tool (400.0 sfm [122 smm])
#123 — Feed for cutoff tool (0.005 ipr [0.127 mmpr]
#131 — Station number for rough face and turn tool (1)
#132 — Station number for drill (2)
#133 — Station number for rough boring bar (3)
#134 — Station number for finish boring bar (4)
#135 — Station number for finish turning tool (5)
#136 — Station number for cutoff tool (6)
#137 — Station number for bar stop (7)

Step three—Be sure you have solved all problems related to the application

The more complicated the application, the more problems you will face. Before you begin writing the parametric program, be sure you have all the problems solved and that you have a clear understanding of how you will develop the parametric program. Solving problems before you begin writing the program is much simpler than having to backtrack during the writing of the program.

In the early planning stages, it is common to feel somewhat overwhelmed by the number of problems that must be solved. This may lead to a premature decision that the part family cannot be handled with parametric programming techniques. Do not be too quick to give up on your part family applications. As you document (write down) each problem and approach each one separately, you will likely find that no single problem is insurmountable. Use this divide-and-conquer approach to regain your confidence in the application.

Step four—Write the parametric program

With all the problems solved, and with a clear understanding of the process and values that change from one workpiece to the next, you should be ready to develop the parametric program. In many cases, your methods will be similar to those you use when developing any conventional CNC program. However, instead of using hard and fixed values for each CNC word, you will be using variables (and arithmetic expressions) to determine the values of CNC words. Though our example application does not require it, you also may need to develop loops and use other program flow logic (commonly with flags) to manipulate the method by which your parametric program is executed.

Here is the *first* attempt at the example program to machine the shaft collars. Note that the final steps will be to refine this program. Again, we show only the custom macro B version of this program and we assume the cutting conditions program (O1001) and tool station setting program (O1002) are in the control's memory. The arguments are currently set to machine part number SC-1250 shown in Figure 6-1.

```
O0001 (Program number)
(SET ARGUMENTS)
#100 = 1.75 (Outside diameter — A dimension)
#101 = 1.250 (Inside diameter — B dimension)
#102 = 0.5312 (Workpiece length — C dimension)
#103 = 2.0 (Rough stock diameter)
G65 P1001 (Call cutting conditions parametric program, use default values)
G65 P1002 (Call tool station setting parametric program, use default values)
(TEST FOR DRILL DIAMETER)
IF [#101 LT 1.0] GOTO 1 (Test if hole is under 1.0)
#145 = 0.875 (Set #145 to drill size)
GOTO 2 (Jump to line N2)
N1 #145 = 0.437 (Set #145 to drill size)
(START MACHINING OPERATIONS)
N2 T[#[#131 * 100 + #131]] (Index turret to rough face and turn tool)
G96 S#110 M03 (Start spindle)
G00 X[#103 + 0.2] Z0.005 (Rapid to facing position)
G01 X – 0.062 F#111 (Rough face)
G00 X#103 Z0.1 (Rapid to starting position for rough turn)
G71 P11 Q12 U0.040 W0.005 D#112 F#111 (Rough turn canned cycle)
N11 G00 X[#100 – 2 * 0.0312] (Rapid to bottom of chamfer)
```

```
G01 Z0 (Feed flush with face)
X#100 Z – 0.0312 (Machine chamfer on face)
Z – [#102 + 0.125 + 0.02] (Turn outside diameter)
N12 X#103 (Feed up to stock diameter)
G00 X5.0 Z5.0 (Rapid to tool-change position)
M01 (Optional stop)
T[#132 * 100 + #132] (Index turret to drill)
G97 S[3.82 * #113/#145] M03 (Start spindle)
G00 X0 Z0.1 (Rapid to approach position)
G01 Z – [#102 + 0.03 + 0.3 * #145] F#114 (Drill hole)
G00 Z0.1 (Rapid out of hole)
X5.0 Z5.0 (Rapid to tool-change position)
M01 (Optional stop)
T[#133 * 100 + #133] (Index turret to rough boring bar)
G96 S#115 M03 (Start spindle)
G00 X#145 Z0.1 (Rapid to approach position)
G71 P21 Q22 U – 0.04 W0.005 D#117 F#116 (Rough bore hole)
N21 G00 X[#101 + 2 * 0.0312] (Rapid to top of chamfer)
G01 Z0 (Feed flush with face)
X#101 Z – 0.0312 (Machine chamfer)
Z – [#102 + 0.02] (Bore through)
N22 X#145 (Feed down to drill diameter)
G00 X5.0 Z5.0 (Rapid to tool-change position)
M01 (Optional stop)
T[#134 * 100 + #134] (Index to finish boring bar)
G96 S#118 M03 (Start spindle)
G00 X#145 Z0.1 (Rapid to approach position)
G70 P21 Q22 F#119 (Finish bore)
G00 X5. Z5. (Rapid to tool-change position)
M01 (Optional stop)
T[#135 * 100 + #135] (Index to finish face and turn tool)
G96 S#120 M03 (Start spindle)
G00 X[#100 + 0.2] Z0 (Rapid to approach position)
G01 X#101 F#121 (Finish face)
G00 X#103 Z0.1 (Rapid to approach position for finish turning)
G70 P11 Q12 F#121 (Finish turn)
G00 X5.0 Z5.0 (Rapid to tool-change position)
M01 (Optional stop)
T[#136 * 100 + #136] (Index turret to cutoff tool)
G96 S#122 M03 (Start spindle)
G00 X[#103 + 0.2] Z – [#102 + 0.125] (Rapid to cutoff position)
G01 X[#100 – 2 * 0.0312] F#123 (Feed to chamfer bottom)
G00 X[#100 + 0.15] (Rapid out in X)
```

```
Z – [#102 + 0.125 – 0.0312] (Move over to form chamfer)
G01 X#100 (Feed flush with diameter)
X[#100 – 2 * 0.0312] Z[#102 + 0.125] (Form chamfer)
X[#101 – 0.1] (Cutoff workpiece)
G00 X#103 (Rapid out)
X5.0 Z5.0 (Rapid to tool-change position)
M01 (Optional stop)
T[#137 * 100 + #137] (Index to bar stop)
M05 (Stop spindle)
G00 X0 Z – [#102 + 0.125 – 0.1] (Rapid to approach position)
M14 (Open jaws)
G98 G01 Z0.05 F30.0 (Select feed per minute mode, fast feed bar out to position)
G99 M15 (Close jaws, re-select feed per revolution mode)
G00 X5.0 Z5.0 (Rapid to tool-change position)
M99 (End of bar feeding program)
```

Advanced technique—Understanding the turret index commands. You may be questioning each turret indexing command. As you know, most controls utilize a four-digit turret index command. The first two digits specify the tool station (and geometry offset) number, while the second two digits specify the wear offset number. The command T0101, for example, indexes the turret to station number one and selects wear offset number one. Remember that leading zeros can be omitted, meaning T101 will do the same thing as T0101. Knowing this, consider these commands:

```
#131 = 4 (Set #131 to desired turret station number)
T[#131 * 100 + #131] (Index turret and select wear offset)
```

As long as the wear offset number is to be the same number as the turret position (as is commonly the case), this T command will select the desired station and wear offset number. Note that #131 is currently set to four. The expression #131 * 100 + #131 will be evaluated as 404, which when specified with the T word, will index to station number four and invoke wear offset number four. Similar techniques can be used if you have a six-digit turret index command.

Step five—Error-trap potential mistakes

Before you release your program for production, you should do your best to trap potential mistakes. In our shaft collar application, for example, we know the range of each workpiece-defining argument because it is published on the engineering drawing. A series of tests can easily determine whether each argument falls within the range specified on the drawing. If any argument is not in the correct range, an alarm can be sounded.

Other areas that commonly require error trapping are functions related to offsets. You may wish to test the range of wear offsets to trap an incorrect entry. You may also wish to test the geometry offsets to ensure that they have all been set.

After you release your program and start production, look for mistakes your people are prone to making. With a little ingenuity, just about any mistake can be error-trapped. Here is a revised version of the shaft collar program which tests the range of each workpiece-related argument (dimension A, B, and C).

```
O0001 (Program number)
(SET ARGUMENTS)
#100 = 1.75 (Outside diameter — A dimension)
#101 = 1.250 (Inside diameter — B dimension)
#102 = 0.5312 (Workpiece length — C dimension)
#103 = 2.0 (Rough stock diameter)
G65 P1001 (Call cutting conditions parametric program, use default values)
G65 P1002 (Call tool station setting parametric program, use default values)
(ERROR TRAPPING)
IF [#100 GE 1.0] GOTO 41 (Test if A dimension is big enough)
#3000 = 100 (A DIMENSION TOO SMALL)
N41 IF [#100 LE 2.125] GOTO 42 (Test if A dimension is small enough)
#3000 = 101 (A DIMENSION TOO BIG)
N42 IF [#101 GE 0.5] GOTO 43 (Test if B dimension is big enough)
#3000 = 102 (B DIMENSION TOO SMALL)
N43 IF [#101 LE 1.625] GOTO 44 (Test if B dimension is small enough)
#3000 = 103 (B DIMENSION TOO BIG)
N44 IF [#102 GE 0.312] GOTO 45 (Test if C dimension is big enough)
#3000 = 104 (C DIMENSION TOO SMALL)
N45 IF [#102 LE 0.532] GOTO 46 (Test if C dimension is small enough)
#3000 = 105 (C DIMENSION TOO BIG)
N46 (TEST FOR DRILL DIAMETER)
IF [#101 LT 1.0] GOTO 1 (Test if hole is less than 1.0 in. [25.4 mm])
#145 = 0.875 (Set #145 to drill size)
GOTO 2 (Jump to line N2)
N1 #145 = 0.437 (Set #145 to drill size)
(START MACHINING OPERATIONS)
N2 T[#[#131 * 100 + #131]] (Index turret to rough face and turn tool)
G96 S#110 M03 (Start spindle)
G00 X[#103 + 0.2] Z0.005 (Rapid to facing position)
G01 X – 0.062 F#111 (Rough face)
G00 X#103 Z0.1 (Rapid to starting position for rough turn)
G71 P11 Q12 U0.040 W0.005 D#112 F#111 (Rough-turn canned-cycle)
N11 G00 X[#100 – 2 * 0.0312] (Rapid to bottom of chamfer)
G01 Z0 (Feed flush with face)
```

```
X#100 Z – 0.0312 (Machine chamfer on face)
Z – [#102 + 0.125 + 0.02] (Turn outside diameter)
N12 X#103 (Feed up to stock diameter)
G00 X5.0 Z5.0 (Rapid to tool-change position)
M01 (Optional stop)
T[#132 * 100 + #132] (Index turret to drill)
G97 S[3.82 * #113/#145] M03 (Start spindle)
G00 X0 Z0.1 (Rapid to approach position)
G01 Z – [#102 + 0.03 + 0.3 * #145] F#114 (Drill hole)
G00 Z0.1 (Rapid out of hole)
X5.0 Z5.0 (Rapid to tool-change position)
M01 (Optional stop)
T[#133 * 100 + #133] (Index turret to rough boring bar)
G96 S#115 M03 (Start spindle)
G00 X#145 Z0.1 (Rapid to approach position)
G71 P21 Q22 U – 0.04 W0.005 D#117 F#116 (Rough bore hole)
N21 G00 X[#101 + 2 * 0.0312] (Rapid to top of chamfer)
G01 Z0 (Feed flush with face)
X#101 Z – 0.0312 (Machine chamfer)
Z – [#102 + 0.02] (Bore through)
N22 X#145 (Feed down to drill diameter)
G00 X5.0 Z5.0 (Rapid to tool-change position)
M01 (Optional stop)
T[#134 * 100 + #134] (Index to finish boring bar)
G96 S#118 M03 (Start spindle)
G00 X#145 Z0.1 (Rapid to approach position)
G70 P21 Q22 F#119 (Finish bore)
G00 X5.0 Z5.0 (Rapid to tool-change position)
M01 (Optional stop)
T[#135 * 100 + #135] (Index to finish face and turn tool)
G96 S#120 M03 (Start spindle)
G00 X[#100 + 0.2] Z0 (Rapid to approach position)
G01 X#101 F#121 (Finish face)
G00 X#103 Z0.1 (Rapid to approach position for finish turning)
G70 P11 Q12 F#121 (Finish turn)
G00 X5.0 Z5.0 (Rapid to tool-change position)
M01 (Optional stop)
T[#136 * 100 + #136] (Index turret to cutoff tool)
G96 S#122 M03 (Start spindle)
G00 X[#103 + 0.2] Z – [#102 + 0.125] (Rapid to cutoff position)
G01 X[#100 – 2 * 0.0312] F#123 (Feed to chamfer bottom)
G00 X[#100 + 0.15] (Rapid out in X)
Z – [#102 + 0.125 – 0.0312] (Move over to form chamfer)
```

```
G01 X#100 (Feed flush with diameter)
X[#100 – 2 * 0.0312] Z[#102 + 0.125] (Form chamfer)
X[#101 – 0.1] (Cutoff workpiece)
G00 X#103 (Rapid out)
X5.0 Z5.0 (Rapid to tool-change position)
M01 (Optional stop)
T[#137 * 100 + #137] (Index to bar stop)
M05 (Stop spindle)
G00 X0 Z – [#102 + 0.125 – 0.1] (Rapid to approach position)
M14 (Open jaws)
G98 G01 Z0.05 F30.0 (Select feed per minute mode, fast feed bar out to position)
G99 M15 (Close jaws, re-select feed per revolution mode)
G00 X5.0 Z5.0 (Rapid to tool-change position)
M99 (End of bar feeding program)
```

Step six—Eliminate the execution of redundant commands

Notice that our last version of the parametric program begins with more than 80 noncutting commands (if you count the commands included in the cutting conditions and tool-station selecting programs). While newer 32-bit controls may be able to execute these commands quite rapidly, you will likely notice a pause in the machining cycle whenever they are executed. And notice that all of these commands must be executed *every time* the program is run (with the most recent version of the program). If production quantities are very low, repeating these redundant commands may not be terribly detrimental. However, in a bar-feeding application (or any application when the lot size is relatively high), repeating these redundant commands is very wasteful.

Keep in mind that all arguments in our application are stored in common variables, which are retained until the power is turned off. This means that the noncutting commands in our example program need to be executed only once (in the first cycle of the day). Once in production, these commands can be skipped until the power is turned off.

Using optional block-skip. As mentioned in Chapter 4, the easiest way to bypass redundant commands is with optional block-skip. Consider this revised program:

```
O0001 (Program number)
/(SET ARGUMENTS)
/#100 = 1.75 (Outside diameter — A dimension)
/#101 = 1.250 (Inside diameter — B dimension)
/#102 = 0.5312 (Workpiece length — C dimension)
/#103 = 2.0 (Rough stock diameter)
/G65 P1001 (Call cutting conditions parametric program, use default values)
/G65 P1002 (Call tool station setting parametric program, use default values)
```

```
/(ERROR TRAPPING)
/IF [#100 GE 1.0] GOTO 41 (Test if A dimension is big enough)
/#3000 = 100 (A DIMENSION TOO SMALL)
/N41 IF [#100 LE 2.125] GOTO 42 (Test if A dimension is small enough)
/#3000 = 101 (A DIMENSION TOO BIG)
/N42 IF [#101 GE 0.5] GOTO 43 (Test if B dimension is big enough)
/#3000 = 102 (B DIMENSION TOO SMALL)
/N43 IF [#101 LE 1.625] GOTO 44 (Test if B dimension is small enough)
/#3000 = 103 (B DIMENSION TOO BIG)
/N44 IF [#102 GE 0.312] GOTO 45 (Test if C dimension is big enough)
/#3000 = 104 (C DIMENSION TOO SMALL)
/N45 IF [#102 LE 0.532] GOTO 46 (Test if C dimension is small enough)
/#3000 = 105 (C DIMENSION TOO BIG)
/N46 (TEST FOR DRILL DIAMETER)
/IF [#101 LT 1.0] GOTO 1 (Test if hole is less than 1.0 in. [25.4 mm])
/#145 = 0.875 (Set #145 to drill size)
/GOTO 2 (Jump to line N2)
/N1 #145 = 0.437 (Set #145 to drill size)
(TEST TO BE SURE COMMON VARIABLES ARE ACTIVE)
IF [#100 NE #0] GOTO 61 (Test that #100 has a value)
#3000 = 110 (TURN OPTIONAL BLOCK SKIP OFF)
N61 (START MACHINING OPERATIONS)
N2 T[#[#131 * 100 + #131]] (Index turret to rough face and turn tool)
G96 S#110 M03 (Start spindle)
G00 X[#103 + 0.2] Z0.005 (Rapid to facing position)
G01 X – 0.062 F#111 (Rough face)
G00 X#103 Z0.1 (Rapid to starting position for rough turn)
G71 P11 Q12 U0.040 W0.005 D#112 F#111 (Rough turn canned cycle)
N11 G00 X[#100 – 2 * 0.0312] (Rapid to bottom of chamfer)
G01 Z0 (Feed flush with face)
X#100 Z – 0.0312 (Machine chamfer on face)
Z – [#102 + 0.125 + 0.02] (Turn outside diameter)
N12 X#103 (Feed up to stock diameter)
G00 X5.0 Z5.0 (Rapid to tool-change position)
M01 (Optional stop)
T[#132 * 100 + #132] (Index turret to drill)
G97 S[3.82 * #113/#145] M03 (Start spindle)
G00 X0 Z0.1 (Rapid to approach position)
G01 Z – [#102 + 0.03 + 0.3 * #145] F#114 (Drill hole)
G00 Z0.1 (Rapid out of hole)
X5.0 Z5.0 (Rapid to tool-change position)
M01 (Optional stop)
T[#133 * 100 + #133] (Index turret to rough boring bar)
```

```
G96 S#115 M03 (Start spindle)
G00 X#145 Z0.1 (Rapid to approach position)
G71 P21 Q22 U – 0.04 W0.005 D#117 F#116 (Rough bore hole)
N21 G00 X[#101 + 2 * 0.0312] (Rapid to top of chamfer)
G01 Z0 (Feed flush with face)
X#101 Z – 0.0312 (Machine chamfer)
Z – [#102 + 0.02] (Bore through)
N22 X#145 (Feed down to drill diameter)
G00 X5.0 Z5.0 (Rapid to tool-change position)
M01 (Optional stop)
T[#134 * 100 + #134] (Index to finish boring bar)
G96 S#118 M03 (Start spindle)
G00 X#145 Z0.1 (Rapid to approach position)
G70 P21 Q22 F#119 (Finish bore)
G00 X5.0 Z5.0 (Rapid to tool-change position)
M01 (Optional stop)
T[#135 * 100 + #135] (Index to finish face and turn tool)
G96 S#120 M03 (Start spindle)
G00 X[#100 + 0.2] Z0 (Rapid to approach position)
G01 X#101 F#121 (Finish face)
G00 X#103 Z0.1 (Rapid to approach position for finish turning)
G70 P11 Q12 F#121 (Finish turn)
G00 X5.0 Z5.0 (Rapid to tool-change position)
M01 (Optional stop)
T[#136 * 100 + #136] (Index turret to cutoff tool)
G96 S#122 M03 (Start spindle)
G00 X[#103 + 0.2] Z – [#102 + 0.125] (Rapid to cutoff position)
G01 X[#100 – 2 * 0.0312] F#123 (Feed to chamfer bottom)
G00 X[#100 + 0.15] (Rapid out in X)
Z – [#102 + 0.125 – 0.0312] (Move over to form chamfer)
G01 X#100 (Feed flush with diameter)
X[#100 – 2 * 0.0312] Z[#102 + 0.125] (Form chamfer)
X[#101 – 0.1] (Cutoff workpiece)
G00 X#103 (Rapid out)
X5.0 Z5.0 (Rapid to tool-change position)
M01 (Optional stop)
T[#137 * 100 + #137] (Index to bar stop)
M05 (Stop spindle)
G00 X0 Z – [#102 + 0.125 – 0.1] (Rapid to approach position)
M14 (Open jaws)
G98 G01 Z0.05 F30.0 (Select feed per minute mode, fast feed bar out to position)
G99 M15 (Close jaws, re-select feed per revolution mode)
G00 X5.0 Z5.0 (Rapid to tool-change position)
M99 (End of bar feeding program)
```

With this method, it's easy to see that all calculations and logic will be skipped if optional block-skip is turned on. Note that we still include one test in the program to ensure that optional block-skip is turned off for the first workpiece. If #100 (or any common variable used in this program) does not have a value, the operator has forgotten to turn off the optional block-skip function prior to running the first workpiece. Note that if any argument is changed during production, the operator must remember to turn off the optional block-skip function for the next cycle for the changed argument(s) to be used.

APPROACHING USER-CREATED CANNED CYCLES

Finding applications in this category is usually quite easy. Simply consider those machining operations you perform on a regular basis that are not supported with control-based canned cycles. It is likely that by this point in the text, you have several applications in mind.

Approaching user-created canned-cycle applications is quite similar to approaching family-of-parts applications. And generally speaking, since user-created canned cycles are commonly designed to perform a single machining operation (as opposed to machining the entire workpiece), they tend to be easier to handle than full part-family applications. That notwithstanding, some user-created canned-cycle applications can be quite challenging.

For illustration, we will use the drawing shown in Figure 6-2. Again, we show only one version of this program (custom macro B format) since all points made can be applied to any version of parametric programming. You should be able at this point to convert this version of the program to your version.

In Figure 6-2, notice the four large counterbored holes. These holes will be milled (using circular interpolation) with an endmill. We will presume that this is only one of many workpieces that requires counterboring (circle milling) with an endmill.

Step one—Determine what your user-created canned cycle must do

By knowing how your user-created canned cycle must be used among the various workpieces you machine, you should already have a pretty good idea of what it is you expect your parametric program to do. However, you'll want to ensure that the program you write will handle all of its requirements. The best way to do this is to compile a good collection of the drawings for the workpieces involved. As you study these, you may be surprised to find elements or conditions you did not consider.

Document what your parametric program is designed to do. You *must* document exactly what the program will do, especially if others in your company will be using your user-created canned cycle. Doing so will help to ensure that everyone reading your programs will understand how the program works, and if done

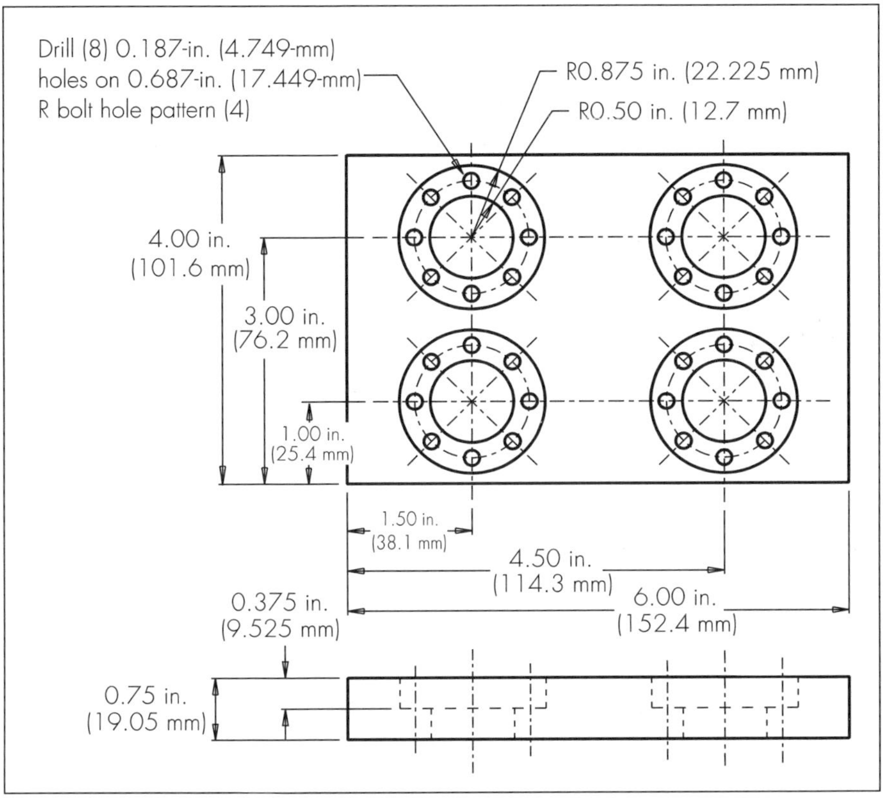

Figure 6-2. The dimensions needed to program user-created canned-cycle applications.

early in the preparation stage, this documentation will help you plan how your parametric program must function.

We cannot stress enough the importance of the documentation step. Before you think about ignoring it, consider the frustration involved in attempting to learn about certain control functions from your CNC control manufacturer's programming manuals. Many of these manuals are so poorly written, they make it nearly impossible to thoroughly understand even the most basic CNC functions. To avoid confusion among the people using your user-created canned-cycle parametric program, and to help you remember your methods in the future, be sure you make your program as easy as possible to understand. Good documentation will do that.

Describe functions and special considerations. Begin your documentation with a description of exactly what the program is intended to do. Following is a description of what our example circle milling user-created canned-cycle is designed to do, as well as some special criteria for its use.

This circle milling parametric program will completely machine circular counterbores (one per call) with an endmill. It assumes that a hole larger than the endmill diameter has been machined in the center of the counterbored hole, since the endmill will rapid into the hole in *Z* to get to its work surface. The endmill must be large enough in diameter to completely machine the counterbore in one *XY* pass. However, since many of our workpieces require rather deep counterbores, provisions are made for multiple *Z* passes within the calling command. Provisions also are made to allow both climb and conventional milling (with G03 or G02).

Make a sketch or drawing showing all arguments. To clearly illustrate the arguments of your application, make a drawing that shows them. Figure 6-3 shows such a drawing for the circle milling user-created canned cycle.

As you make this drawing, you are again forced to think through the things your program must know in order to function. In many cases, you will turn up

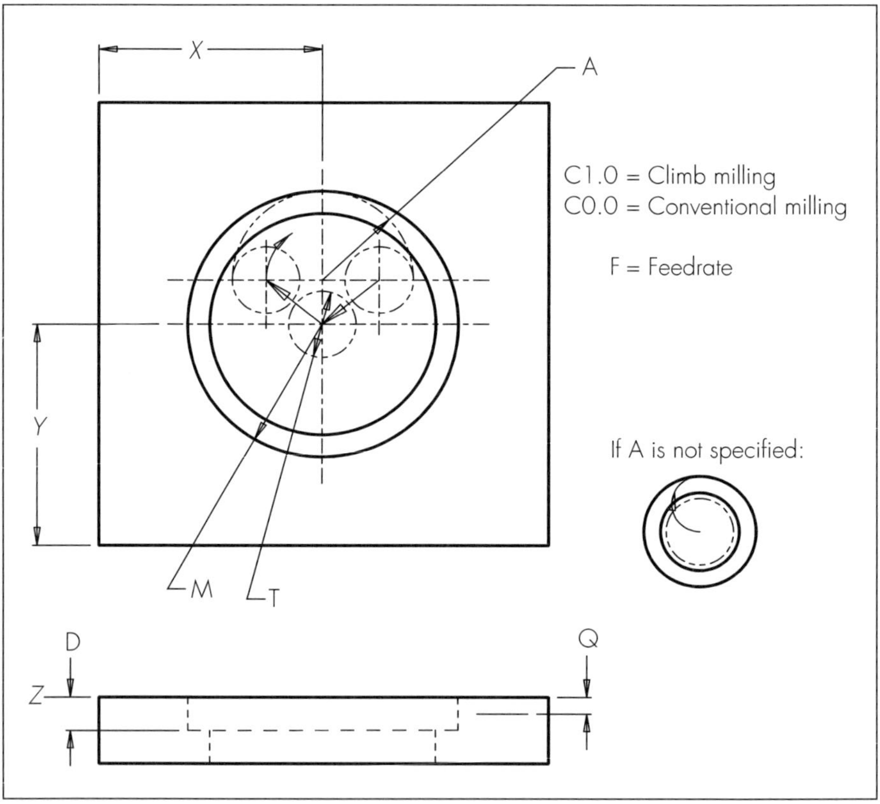

Figure 6-3. The arguments used in the user-created canned cycle for circle milling applied to the workpiece in Figure 6-2.

new conditions that require you to rethink your methods. For example, Figure 6-3 illustrates a situation with our circle milling user-created canned cycle that requires positioning movement to conserve time. However, the drawing shown in Figure 6-2 does not require such a movement. To handle this specific problem, we depend on the value of the approach radius (A in this case) to determine whether or not to make the prior approach movement. If A is not included in the call statement, the tool will move right from the center of the hole to the surface being milled (in a circular movement), also shown in Figure 6-3.

Write a test program that uses your user-created canned cycle. This program, if included with the documentation for your user-created canned cycle, will serve to show everyone using your program an example of its use. It also should help to motivate you to write the parametric program. As you write the example program, you quite likely will find yourself saying something like "Performing this machining operation has never been so easy!"

Here is the example program to machine the counterbored holes in Figure 6-2. Though not shown in this example program, note that the drilling operation must be done prior to using the circle milling parametric program.

```
O0021 (Test program)
N005 G54 G90 S500 M03 (Select coordinate system, absolute mode, start spindle)
N010 G00 X1.5 Y1.0 (Rapid to lower left hole)
N015 G43 H01 Z0.1 (Instate tool-length compensation, move up to work surface)
N020 G65 P1021 X1.5 Y1.0 Z0 D0.375 Q0.1 T0.875 M0.875 C1.0 F5.0 (Mill
    lower left counterbore)
N025 G65 P1021 X1.5 Y3.0 Z0 D0.375 Q0.1 T0.875 M0.875 C1.0 F5.0 (Mill
    upper left counterbore)
N030 G65 P1021 X4.5 Y3.0 Z0 D0.375 Q0.1 T0.875 M0.875 C1.0 F5.0 (Mill
    upper right counterbore)
N035 G65 P1021 X4.5 Y1.0 Z0 D0.375 Q0.1 T0.875 M0.875 C1.0 F5.0 (Mill
    lower right counterbore)
N040 G92 G28 Z0 (Return to tool-change position)
N045 M30 (End of program)
```

Notice that since letter address argument A is omitted from each calling command, the tool will feed from the center of the hole to the diameter being machined (in a circular movement). This example also assumes that there are no arguments set to default values in the parametric program.

Step two—Consider default values and flags

You can shorten the length of the typical call command by setting default values in the parametric program. This is quite helpful with user-created canned-cycle applications. In our circle milling example, it could be that the work surface is almost always program zero in Z. Maybe you almost always climb-mill. Maybe the depth of each *Z* pass is almost always 0.25 in. (6.35 mm). For any value that is commonly set a certain way, plan to preset the value in the parametric program to minimize the length of the call statement. Be sure you document which arguments are mandatory and which are set to default values.

In similar fashion, consider any argument you can use to flag the parametric program as to how you want it to behave. In our example, we are using the letter address C to specify whether or not to climb-mill. Be sure to consider your own special situation. You may wish to allow for the possibility that there is no hole machined in the workpiece prior to the circle milling operation. In this case, you could use a flag to specify whether the tool can be sent into the hole at rapid or whether it must be positioned at a cutting feedrate.

Step three—Consider error trapping

As with family-of-parts applications, be sure you consider the potential mistakes the user of your user-created canned cycle might make. At the very least, test to ensure that mandatory arguments are assigned. In our example, we might also need to test to ensure that the cutter radius specified (half of letter address argument T) is smaller than the radius to be milled (address M). As you release your parametric program for use in production, be on the lookout for mistakes others are making. Error-trapping tests can be added at any time.

Step four—Be sure you have solved all of the problems

Before you begin writing the parametric program, be sure you have a clear understanding of how you will handle all of the problems related to its application. With the circle milling program, for example, we have already mentioned (and solved) the problem of how the cutter approaches the surface being milled, based on the size of the hole currently in the workpiece. We have the additional problem of developing a loop to take multiple *Z* passes. For your own applications, if you are pondering how to handle any specific problem, it's good practice to solve the problem before beginning to write the program.

Step five—Write the parametric program

With a clear understanding of how you intend to meet the challenges of your application, you are truly ready to write the parametric program. However, be on the lookout for problems that crop up as you dig in. If you face new problems, you should stop your attempt and solve them first.

Here is the circle milling program:

```
O1021 (Circle milling parametric program)
(TEST MANDATORY ARGUMENTS)
IF [#1 EQ #0] GOTO 97 (Test argument A)
IF [#24 EQ #0] GOTO 97 (Test argument X)
IF [#25 EQ #0] GOTO 97 (Test argument Y)
IF [#26 EQ #0] GOTO 97 (Test argument Z)
IF [#7 EQ #0] GOTO 97 (Test argument D)
```

```
IF [#17 EQ #0] GOTO 97 (Test argument Q)
IF [#3 EQ #0] GOTO 97 (Test argument C)
IF [#9 EQ #0] GOTO 97 (Test argument F)
IF [#20 EQ #0] GOTO 97 (Test argument T)
IF [#13 EQ #0] GOTO 97 (Test argument M)
(TEST IF CLIMB OR CONVENTIONAL MILLING)
IF [#3 EQ 1] GOTO 5 (If climb milling, go to N5)
(XY CALCULATIONS FOR CONVENTIONAL MILLING)
#101 = #24 – #1 + #20/2 (X position at start of approach radius)
#102 = #24 + #1 – #20/2 (X position at end of escape radius)
#110 = 2 (G-code number for circular move)
GOTO 6
N5 (CALCULATIONS FOR CLIMB MILLING)
#101 = #24 + #1 – #20/2 (X position at start of approach radius)
#102 = #24 – #1 + #20/2 (X position at end of escape radius)
#110 = 3 (G-code number for circular move)
N6 (CALCULATIONS FOR BALANCE OF POSITIONS)
#103 = #25 + #13 – #1 (Y position at start and end of approach radius)
#104 = #1 – #20/2 (Radius of tool path center line during approach and escape)
#105 = #13 – #20/2 (Radius of tool path center line during full circle movement)
#106 = #25 + #13 – #20/2 (Y position at beginning of full circle movement)
(CALCULATIONS FOR NUMBER OF PASSES AND DEPTH OF EACH PASS
   FOR Z LOOP)
#111 = FUP[#7/#17] (Number of passes)
#112 = #7/#111 (Recalculated depth of each pass)
(INITIALIZE Z LOOP VARIABLES)
#121 = 1 (Counter)
#122 = #26 – #112 (Current Z position)
(MOVE TO STARTING POSITION)
G00 X#24 Y#25 (Move to hole center in XY)
Z[#26 + 0.1] (Rapid to just above work surface)
(BEGIN Z LOOP)
N40 IF [#121 GT #111] GOTO 95 (Test if finished)
G00 Z#122 (Move to current Z position for pass)
IF [#1 NE #0] GOTO 45 (If A is not vacant, go to N45)
G#110 Y#106 R[#13 – #20/2] F#9 (Form 1/2 circle approach radius)
GOTO 48 (Skip 1/4 circle approach)
N45 G01 X#101 Y#103 F40.0 (Fast feed to beginning of approach radius)
G#110 X#24 Y#106 R#104 F#9 (Form 1/4 circle approach)
N48 G#110 J – #105 (Form full circle movement)
IF [#1 NE #0] GOTO 50 (If A is not vacant, go to N50)
G#110 Y#25 R[#13 – #20/2] F#9 (Form 1/2 circle escape radius back to hole
   center)
```

```
GOTO 55 (Skip 1/4 circle approach)
N50 G#110 X#102 Y#103 F40.0 (Form 1/4 circle escape radius)
G01 X#24 Y#25 F40.0 (Fast feed back to hole center)
N55 (STEP CHANGING VALUES)
#121 = #121 + 1 (Step counter)
#122 = #122 – #112 (Step current Z position)
GOTO 40 (Go back to test if loop is finished)
N95 G00 Z[#26 + 0.1] (Rapid out of hole)
GOTO 99 (Go to end of custom macro)
N97 #3000 = 100 (DATA MISSING IN CALL)
N99 M99 (End of custom macro)
```

APPROACHING COMPLEX GEOMETRIC SHAPE APPLICATIONS

As stated in Chapter 1, just about any shape that can be defined with an arithmetic formula can be machined by a parametric program. This includes simple shapes like spheres, pyramids, and cones as well as more complicated shapes like those required when draft angles must be machined in plastic injection molds. However, remember that the complexity of developing programs in this category is directly related to the complexity of the shape. For extremely complex shapes, a good 3D computer-aided manufacturing (CAM) system can usually outperform anything you can do with parametric programming.

Applications in this category commonly require looping techniques. You initialize variables for the first pass (or motion), make the pass, step the counters, and repeat the loop. Figure 6-4 shows an example of a spherical shape that requires looping. For our example, we'll say this shape must be machined with a ball endmill on a vertical machining center.

Though there are many ways to approach this application, the example will have the endmill make a series of circular passes around the spherical shape, working from the lowest *Z* surface. The tool will begin the circular motion from the three-o'clock (*X*-plus) side of the sphere. For each circular pass, the tool will be raised slightly in *Z* and move slightly closer to the sphere center line in *X*.

Step one—Solve the arithmetic problems

Before you can begin writing the parametric program, you must have a clear understanding of the arithmetic calculations your program will use. In many cases, you will have several options for how the arithmetic can be done. Be sure to choose the best way. In our sphere-milling application, for example, after making the first circular pass, you could have the tool move up in *Z* a fixed amount, calculate a new *X* position, and make another pass. While this method would correctly machine the spherical shape, the witness marks left on the workpiece would not be evenly spaced. As the tool approaches the top of the sphere, the rings on the workpiece will be farther and farther apart. A better way (the way we

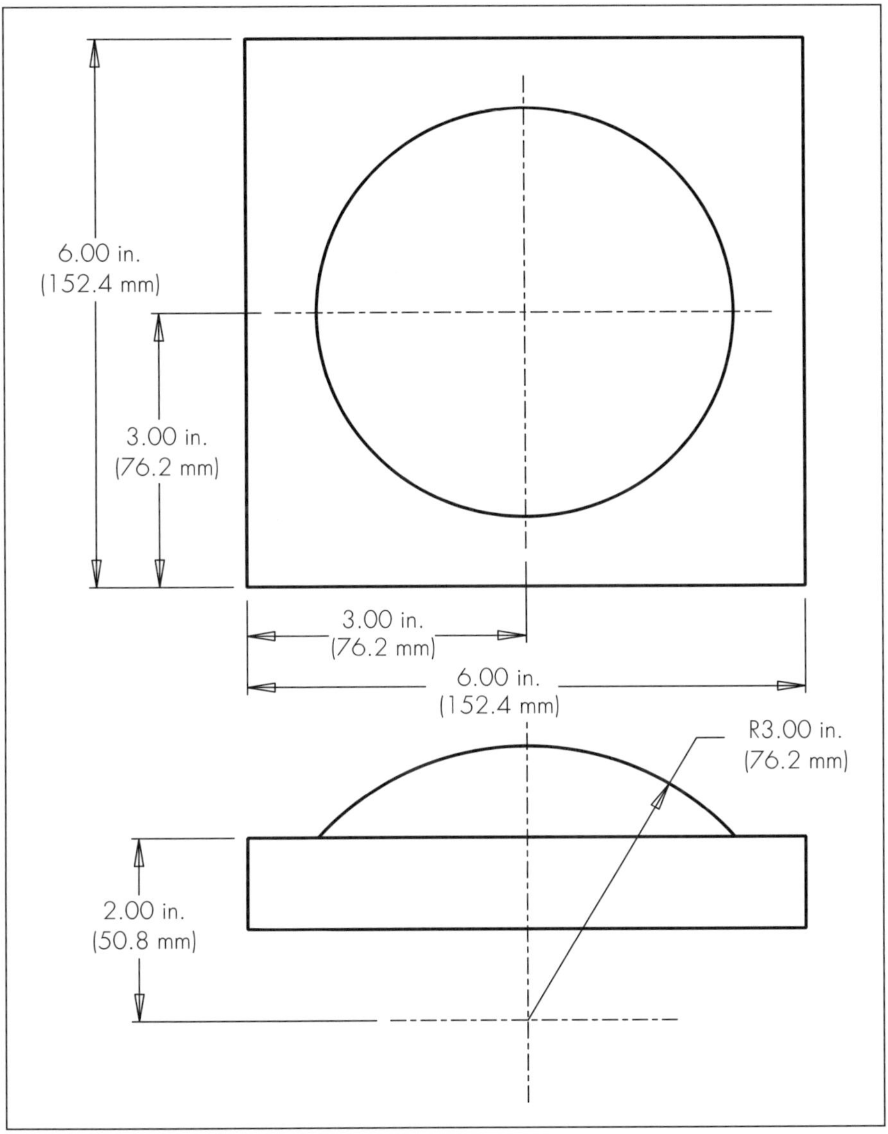

Figure 6-4. Though parametric programs can define virtually any workpiece, complex geometric shapes generally require looping.

will use) would be to recalculate the *XZ* position for each pass based on an incremental angle between passes.

Figure 6-5 shows the arithmetic needed to determine the initial tool position in *X* for making the first circular pass around the sphere. Note that this distance

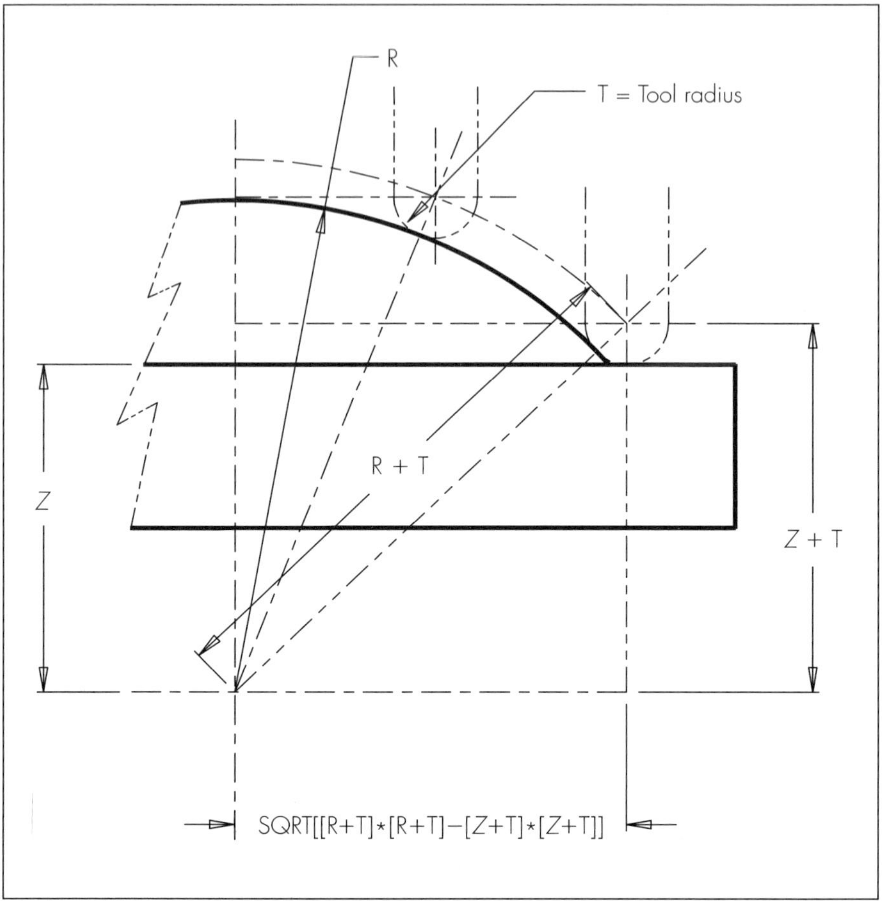

Figure 6-5. Arithmetic needed to calculate the initial XZ tool position for making the first circular pass around the workpiece shown in Figure 6-4.

from the center of the sphere to the center of the ball endmill in *X* is also the radius of the first full circle movement around the sphere.

While this works nicely to determine needed values for the first pass, subsequent passes require more. To use our planned solution, we have to determine the current angle from the sphere center line to the center of the ball endmill during the first pass. This is necessary so that we can make subsequent passes. Figure 6-6 shows how.

Note that this calculation is based on the arc tangent function. Once we determine this angle, each subsequent pass (requiring a new *X* and *Z* position) will be based on relatively simple sine and cosine functions.

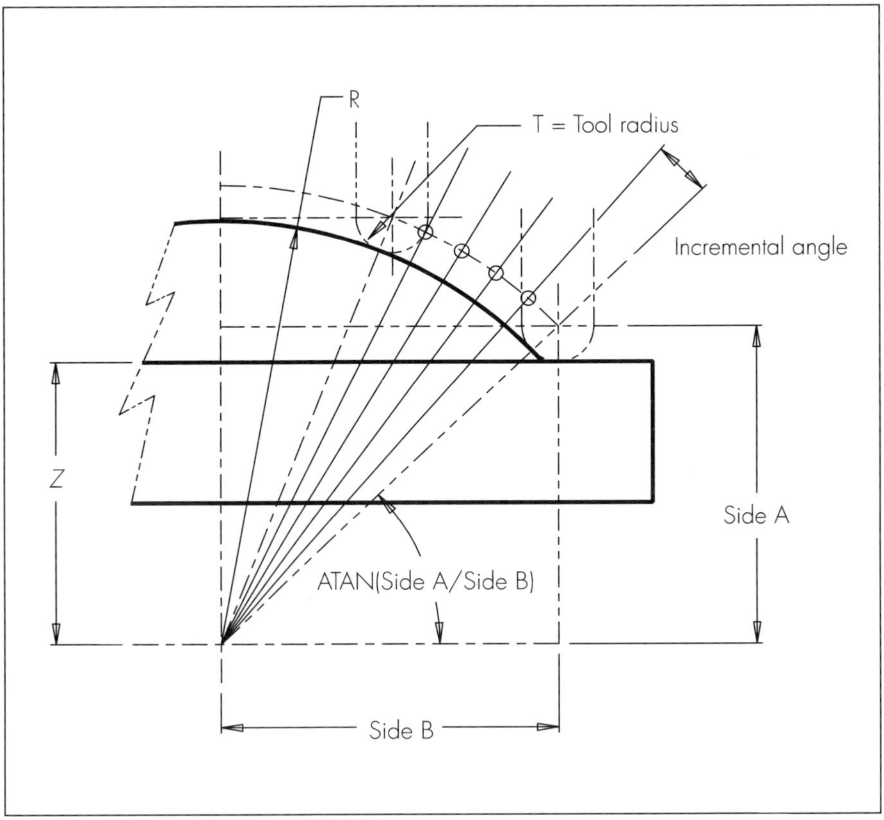

Figure 6-6. Arithmetic needed to calculate initial angle and subsequent XZ positions for the workpiece shown in Figure 6-4.

While you may be able to come up with another (and likely better) method of solving the arithmetic problems of this application, our main point is that regardless of what kinds of arithmetic problems you face, you must solve them before attempting to write your parametric program.

Step two—Determine your arguments

As with user-created canned-cycle applications, you must determine and document the arguments you intend to use with your parametric program. Since it is possible that other machining operations may be performed on this workpiece, we will invoke this parametric program as we would a user-created canned cycle (with a G65 command from the main program). If there are no other machining operations to perform, of course, you could include all arguments in the main program and treat the application as a part family. Figure 6-7 shows the arguments we have chosen.

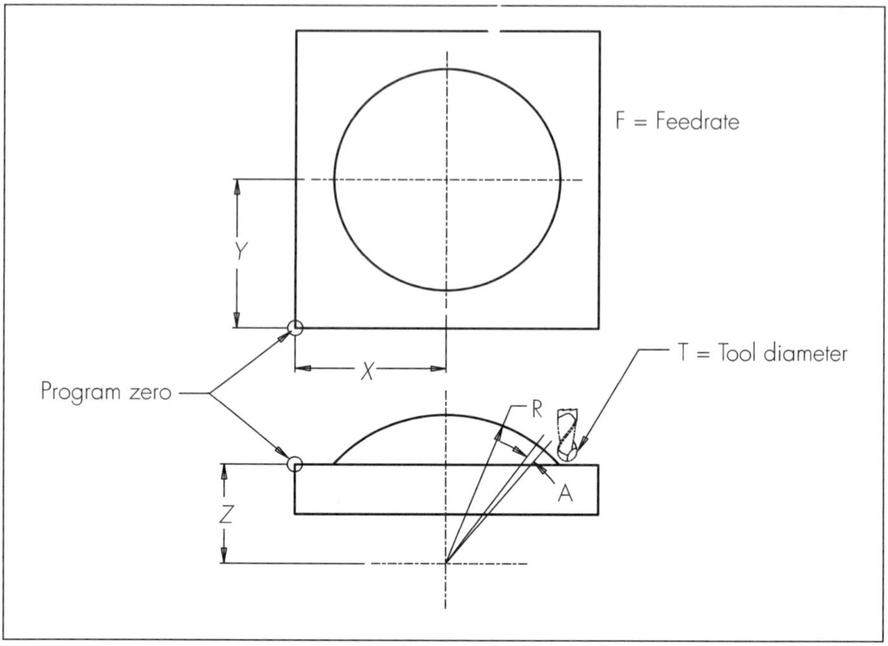

Figure 6-7. Arguments used by sphere-milling custom macro program for the example workpiece shown in Figure 6-4.

Step three—Write an example calling program

If you are handling the complex geometric shape as a kind of user-created canned cycle (as we are), write your test program before you attempt the parametric program. As with applications in the user-created canned-cycle category, this should give you the motivation you need to actually write the parametric program. Here is our test program.

O0022 (Main program)
N005 G54 G90 S800 M03 (Select coordinate system, absolute mode, start spindle)
N010 G00 X3.0 Y3.0 (Rapid to center of sphere)
N015 G43 H01 Z1.2 (Rapid to above sphere surface in *Z*)
N020 G65 P1022 X3.0 Y3.0 Z2.0 R3.0 A1.0 T0.5 F5.0 (Mill entire sphere)
N025 G91 G28 Z0 (Return to tool-change position)
N030 M30 (End of program)

Notice that line N020 commands the machining of the entire sphere based on a circular motion around the sphere every 1 degree (specified by A1.0). If better resolution is required, letter address argument A can be reduced.

Step four—Write the parametric program

With a clear understanding of how all problems are to be solved, you are ready to develop the parametric program. Here is the sphere milling parametric program in custom macro B format.

```
O1022 (Sphere milling custom macro program)
#20 = #20/2 (Make #20 the tool radius, not diameter)
#101 = #24 + SQRT [[#18 + #20] * [#18 + #20] – [#26 + #20] * [#26 + #20]]
   (Current X position)
#102 = ATAN [#26 + #20]/[#101] (Beginning angle)
#103 = #101 – #24 (Beginning distance from center of sphere to center of ball)
#110 = 1 (Counter)
#111 = FUP[ 90.0 – #102]/#1] (Number of passes)
#112 = [90.0 – #102]/#111 (Recalculate incremental angle between passes)
#102 = #102 + #112 (Current angle [note that first pass is made outside the loop])
G00 X#101 Y#25 (Move to first XY position)
G01 Z0 F#9/2 (Plunge to first Z position)
G02 I – #103 F#9 (Make first pass around sphere)
N1 IF [#110 GE #111] GOTO 99 (Test if finished)
#101 = #24 + [COS[#102] * [#18 + #20]] (Current X position)
#122 = [SIN[#102] * [#18 + #20]] – #26 – #20 (Current Z position)
#103 = #101 – #24 (Current distance from center of sphere to center of ball)
G01 Z#122 F[#9 * 5] (Move up in Z for next pass)
X#101 F#9 (Move over in X for next pass)
G02 I – #103 (Form current circular pass)
#110 = #110 + 1 (Step counter)
#102 = #102 + #112 (Step current angle)
GOTO 1 (Go back to test)
N99 G00 Z[#18 – #26 + 0.2] (Rapid out in Z)
M99 (End of custom macro)
```

APPROACHING UTILITY APPLICATIONS

The utility category for parametric programming is the most diversified application category. Generally speaking, you are only limited by your imagination as to what you can use utilities for.

Locate a problem area and develop a solution

The most important factor in approaching utility applications is recognizing cases in which the utilization of your CNC equipment could be improved. This may sound quite basic, but CNC machines are commonly so productive that we may be quite happy with our current methods even though there may be much room for improvement. Moreover, we as human beings have a propensity for *leaving well enough alone*.

To successfully approach utility applications, you must overcome this tendency. Be ever vigilant as to times when your CNC people have any kind of problem that leads to mistakes, wasted time, or in any other fashion keeps you from utilizing your machine tools at their peak levels. And once you find such a problem, muster the tenacity to stick with it until you have the solution. *Don't take no for an answer.* With the new tools you have at your disposal with parametric programming (and especially the CNC features), you have an arsenal of powerful new weapons with which to work.

To illustrate utility applications, we'll use a jaw boring operation. In this operation, we notice that our turning center setup people are taking a great deal of time setting up to bore soft jaws. They are performing this operation manually, using the machine's handwheel. As you well know, to bore jaws totally in manual mode can be tedious and time consuming. With a parametric program for jaw boring, the setup person can simply specify the size of the jaws to be bored and jaw boring will be completely automatic. This will even free the setup person to do other things as the jaws are bored. Figure 6-8 shows the application, including the common variable arguments to be used.

The parametric program for the jaw boring utility in custom macro B format follows. It assumes that the boring bar is in station number 12 and that program zero in *Z* is the face of the jaws.

```
O0023 (Jaw boring utility program)
#100 = 2.0 (Small diameter, boring starts from here)
#101 = 3.0 (Large diameter)
#102 = 0.5 (Depth of boring operation)
#103 = 300.0 (Speed for jaw boring)
#104 = 0.010 (Feedrate for jaw boring)
#105 = 0.1 (Depth of cut for jaw boring)
T1212 (Index to jaw boring bar)
G96 S#103 M03 (Start spindle)
G98 G01 X#100 Z0.1 F30.0 (Fast feed to position to begin jaw boring)
G99 (Re-select feed per revolution mode)
G71 P1 Q2 U – 0.04 W0.005 D#105 F#104 (Rough-boring canned cycle)
N1 G00 X#101 (Rapid up to large diameter)
G01 Z – #102 (Machine to bottom of jaw)
N2 X#100 (Feed down face to starting diameter)
G70 P1 Q2 F[#104 * 0.5] (Finish bore)
M30 (End of program)
```

APPROACHING APPLICATIONS TO DRIVE OPTION DEVICES

Parametric programming, as we learned in Chapter 1, commonly provides the necessary higher level programming capabilities required with certain accessory devices. Probes, post-process gaging systems, and digitizing systems are among

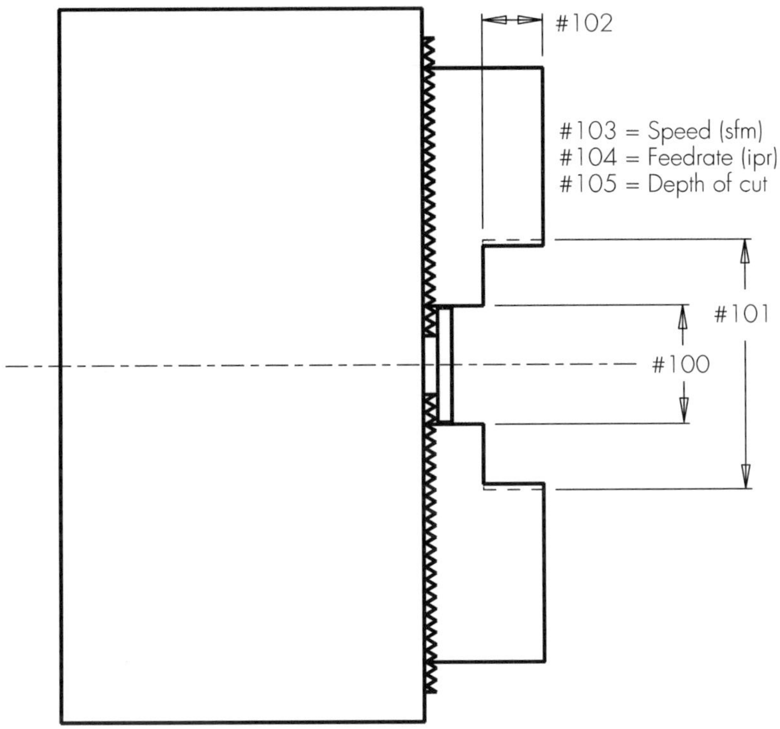

Figure 6-8. The arguments assigned for jaw-boring utility application.

the devices that require parametric programming capabilities. (We address probes in depth in Part Two of this text.)

As introduced in Chapter 5, many versions of parametric programming allow access to certain terminal locations that can be connected to outside devices. Since the supplier of the accessory device is commonly responsible for making the necessary connections, most CNC users do not have to be overly concerned with these terminals and how they are activated. For this reason, it is not necessary to dwell on approaching accessory device applications further than what has already been shown in Chapter 5.

VERIFYING PARAMETRIC PROGRAMS

Since parametric programs are commonly much more complicated than conventional CNC programs, you must verify them with care. Parametric programs falling into the categories of part families, user-created canned cycles, and complex geometric shapes are commonly designed to work under a wide variety of conditions, and you must verify them under all conditions before releasing them for production.

MISTAKES BEGINNERS ARE PRONE TO MAKING

Beginning parametric programmers are prone to making mistakes in three areas. Armed with the knowledge of what these pitfalls are should help you avoid them.

Syntax mistakes

As with standard G-code level programming, there are strict rules for parametric programming commands. If the CNC control cannot interpret a command, it will sound an alarm. Here are some examples of mistakes that will generate alarms.

Custom macro B mistakes:

```
GOTO N1 (N should not be this command)
IF #101 GT 5 GOTO 99 (Conditional expression should be in brackets)
#101 = SIN 30 (30 should be in brackets)
X#101 + 5 (Calculations within CNC words must be in brackets)
#101 = #24 + SIN[45.0] * #101] (Too many right-side brackets)
```

User task 2 mistakes:

```
GOTO 1 (1 must be replaced with N1)
IF PASS GT 5 GOTO N99 (Conditional expression should be in brackets)
CURZ = SIN 30 (30 should be in brackets)
CURX = XZER + SIN[45.0] * RAD] (Too many right-side brackets)
```

Fadal's macro mistakes:

```
#GOTO LOOPSTART 'colon (:) must precede LOOPSTART
IF V1 > 5 GOTO :EXIT 'Pound sign (#) must begin all macro commands
X + V1 (Only R variables can be brought into CNC words)
#V1 = SIN 30 '30 should be in parentheses
#V2 = R1 + SIN9(45.0) * V3) 'Too many right-side parentheses
```

Fortunately, syntax mistakes are quite easy to find and correct. Since the CNC control generates a specific alarm when it finds a syntax mistake, it is usually quite easy to locate the mistake *if you know which command is generating the alarm.* You must understand, however, that if you are running in the single-block mode, most controls will only stop at true CNC commands. Most will appear to skip calculation and logic commands. Consider this series of custom macro B commands from one of the programs shown earlier in this chapter.

```
O1021 (Circle milling parametric program)
(TEST MANDATORY ARGUMENTS)
IF [#1 EQ #0] GOTO 97 (Test argument A)
IF [#24 EQ #0] GOTO 97 (Test argument X)
IF [#25 EQ #0] GOTO 97 (Test argument Y)
IF [#26 EQ #0] GOTO 97 (Test argument Z)
IF [#7 EQ #0] GOTO 97 (Test argument D)
IF [#17 EQ #0] GOTO 97 (Test argument Q)
IF [#3 EQ #0] GOTO 97 (Test argument C)
IF [#9 EQ #0] GOTO 97 (Test argument F)
IF [#20 EQ #0] GOTO 97 (Test argument T)
IF [#13 EQ #0] GOTO 97 (Test argument M)
(TEST IF CLIMB OR CONVENTIONAL MILLING)
IF [#3 EQ 1] GOTO 5 (If climb milling, go to N5)
(XY CALCULATIONS FOR CONVENTIONAL MILLING)
#101 = #24 – #1 + #20/2 (X position at start of approach radius)
#102 = #24 + #1 – #20/2 (X position at end of escape radius)
#110 = 2 (G-code number for circular move)
GOTO 6
N5 (CALCULATIONS FOR CLIMB MILLING)
#101 = #24 + #1 – #20/2 (X position at start of approach radius)
#102 = #24 – #1 + #20/2 (X position at end of escape radius)
#110 = 3 (G-code number for circular move)
N6 (CALCULATIONS FOR BALANCE OF POSITIONS)
#103 = #25 + #13 – #1 (Y position at start and end of approach radius)
#104 = #1 – #20/2 (Radius of tool path center line during approach and escape)
#105 = #13 – #20/2 (Radius of tool path center line during full circle movement)
#106 = #25 + #13 – #20/2 (Y position at beginning of full circle movement)
(CALCULATIONS FOR NUMBER OF PASSES AND DEPTH OF EACH PASS
   FOR Z LOOP)
#111 = FUP[#7/#17] (Number of passes)
#112 = #7/#111 (Recalculated depth of each pass)
(INITIALIZE Z LOOP VARIABLES)
#121 = 1 (Counter)
#122 = #26 – #112 (Current Z position)
(MOVE TO STARTING POSITION)
G00 X#24 Y#25 (Move to hole center in XY)
Z[#26 + 0.1] (Rapid to just above work surface)
(BEGIN Z LOOP)
```

If you are in the single-block mode when this program is executed, the control will appear to skip all the calculation and logic commands and will actually make the rapid motion near the end of this sequence. The control will, of course, actually execute each calculation and logic command. It won't just stop for each one, which only makes sense. You would not want your operator having to press cycle start several times just to get to the first true CNC command.

Though the function of single block is as it should be (in our opinion), you may be fooled if you are running in single-block mode when an alarm is sounded. Consider this new series of commands that includes a syntax mistake:

```
O1021 (Circle milling parametric program)
(TEST MANDATORY ARGUMENTS)
IF [#1 EQ #0] GOTO 97 (Test argument A)
IF [#24 EQ #0] GOTO 97 (Test argument X)
IF [#25 EQ #0] GOTO 97 (Test argument Y)
IF [#26 EQ #0] GOTO 97 (Test argument Z)
IF [#7 EQ #0] GOTO 97 (Test argument D)
IF [#17 EQ #0] GOTO 97 (Test argument Q)
IF [#3 EQ #0] GOTO 97 (Test argument C)
IF [#9 EQ #0] GOTO 97 (Test argument F)
IF [#20 EQ #0] GOTO 97 (Test argument T)
IF [#13 EQ #0] GOTO 97 (Test argument M)
(TEST IF CLIMB OR CONVENTIONAL MILLING)
IF [#3 EQ 1] GOTO 5 (If climb milling, go to N5)
(XY CALCULATIONS FOR CONVENTIONAL MILLING)
#101 = #24 – #1 + #20/2 (X position at start of approach radius)
#102 = #24 + #1 – #20/2 (X position at end of escape radius)
#110 = 2 (G-code number for circular move)
GOTO 6
N5 (CALCULATIONS FOR CLIMB MILLING)
#101 = #24 + #1 – #20/2 (X position at start of approach radius)
#102 = #24 – #1 + #20/2 (X position at end of escape radius)
#110 = 3 (G-code number for circular move)
N6 (CALCULATIONS FOR BALANCE OF POSITIONS)
#103 = #25 + #13 – #1 (Y position at start and end of approach radius)
#104 = #1 – #20/2 (Radius of tool path center line during approach and escape)
#105 = #13 – #20/2 (Radius of tool path center line during full circle movement)
#106 = #25 + #13 – #20/2 (Y position at beginning of full circle movement)
(CALCULATIONS FOR NUMBER OF PASSES AND DEPTH OF EACH PASS
   FOR Z LOOP)
#111 = FUP[#7/#17] (Number of passes)
#112 = #7/#111 (Recalculated depth of each pass)
(INITIALIZE Z LOOP VARIABLES)
#121 = 1 (Counter)
```

```
#122 = #26 – #112] (Current Z position)
(MOVE TO STARTING POSITION)
G00 X#24 Y#25 (Move to hole center in XY)
Z[#26 + 0.1] (Rapid to just above work surface)
(BEGIN Z LOOP)
```

In single-block mode, an alarm will sound as soon as the operator presses cycle start at the beginning of this program. You may be tricked into thinking that the mistake is close to the beginning of the program, when in reality it can be anywhere in the series of calculation and logic commands. Knowing this, you should be able to find the mistake. If you cannot, there is a simple way to isolate the command causing the alarm. If you place an M00 (or any true CNC command) between each calculation and logic statement, single-block will cause the machine to stop after each parametric programming command. You will, of course eliminate these commands once you find the mistake.

Note that some controls also have a parameter that controls the function of single-block with calculation and logic commands. With the parameter set in its normal way, the single-block function behaves as we have described. If the parameter is changed, the control will halt after every command, including calculation and logic statements. However, it is usually easier to simply insert M00 commands in your program than it is to look up and change the single-block parameter.

Simple evaluation mistakes

Mistakes in this category are a little harder to find. Though no alarm will sound, the machine will behave in a peculiar manner.

Suppose, for example, you planned on using (custom macro B) common variable #101 as the *X* position for a hole location, yet by mistake, you give this motion command:

```
G00 X#110
```

Where will the machine go in *X*? It is impossible to say without knowing the value of #110.

Keep in mind that most controls give you the ability to view all of your variables. To diagnose this problem, you can simply check the value of #110.

Another example of a simple evaluation mistake is a misinterpretation of arithmetic priority. Again, the machine will not be going anywhere close to its intended position. (If you see a distance-to-go of 3,295.8337 in. [83,714.175 mm], it is likely that you have made this kind of mistake.)

A fatal mistake in your solution to the problem

While mistakes in the first two categories can eventually be corrected, mistakes in problem solving usually mean starting over. Although you were certain your solution to the problem was correct, it wasn't, and the program failed. If, for example, when developing the sphere milling program shown earlier, the whole approach to the arithmetic calculations was incorrect, the entire effort will be futile. By the time the mistake was discovered, you would have finished with the program and begun the verification process. A lot of time would be wasted.

The only protection against this kind of mistake is through preparation. You must be certain of your solutions before you begin programming. If in doubt, it helps to talk to others. By sharing what you intend to do with other programmers or operators, you can often pinpoint any holes in your thinking.

UNDERSTANDING ARGUMENT ASSIGNMENT NUMBER TWO

As mentioned during our discussion of variables in Chapter 2, there are two methods of argument assignment in custom macro B. To this point, however, all of our examples have used argument assignment number one. In this section, we will present argument assignment number two.

As you know, arguments placed in the call statement (the G65 command) are passed to the custom macro program. Each argument contains a letter address and a numerical value and is used to represent something about the application. You know that certain letter addresses (G, N, L, O, and P) cannot be used as arguments. We also avoid I, J, and K.

Once in the actual custom macro program, you can no longer reference arguments by their letter addresses. The letter addresses must be referenced in the custom macro program by local variables. You also know that the local variable numbers (#1 through #26) do not make much sense. You must reference a special chart to determine the local variable representations for all arguments used in the call statement. In this discussion, we explain *why* the local variable assignments in argument assignment number one don't make much sense. We also explain why we avoid using I, J, and K for most custom macro applications.

LOCAL VARIABLE ASSIGNMENTS FOR ARGUMENT ASSIGNMENT NUMBER TWO

Chart 6-1 shows how local variables are used to represent arguments in argument assignment number two. Notice that 10 sets of I, J, and K letter address arguments are used in conjunction with one set of A, B, and C arguments.

To actually make call statements with argument assignment number two, you specify a series of I, J, and K arguments in conjunction with A, B, and C (if required). Though it may not be entirely clear yet, here is an example call statement given in argument assignment number two.

```
N055 G65 P1000 A3.0 B4.5 C3.0 I2.235 J4.30 K3.23 I1.20 J2.23
K1.0 I0.30 J3.0 K1.0 ....
```

Chart 6-1. Local variables used to represent arguments.

A . . . #1	B . . . #2	C . . . #3
I(1) . . . #4	J(1) . . . #5	K(1) . . . #6
I(2) . . . #7	J(2) . . . #8	K(2) . . . #9
I(3) . . . #10	J(3) . . . #11	K(3) . . . #12
I(4) . . . #13	J(4) . . . #14	K(4) . . . #15
I(5) . . . #16	J(5) . . . #17	K(5) . . . #18
I(6) . . . #19	J(6) . . . #20	K(6) . . . #21
I(7) . . . #22	J(7) . . . #23	K(7) . . . #24
I(8) . . . #25	J(8) . . . #26	K(8) . . . #27
I(9) . . . #28	J(9) . . . #29	K(9) . . . #30
I(10) . . . #31	J(10) . . . #32	K(10) . . . #33

As with argument assignment number one, all arguments must include a decimal point in the numerical value.

When the custom macro program is executed, local variables #1, #2, and #3 will be used to represent A, B, and C (just as in argument assignment number one). The first I that the control comes across in this call statement (I2.235) will be stored in local variable #4. The first J will be stored in #5. The first K in #6. The second I (I1.2) will be stored in local variable #7. The second J in #8. The second K in #9. This continues through #33, representing the tenth occurrence of K.

How does the control know the difference? Actually, this is an automatic function, one that is rather transparent to the unsuspecting programmer. As the control assigns the local variables and happens to come across an I (the first one), it automatically places its value in local variable #4. If it comes across a second I, it places its value into #7. If it comes across a K (the first one), it places it into #5.

This is the reason we avoid using I, J, and K in argument assignment number one. If you avoid I, J, and K, you will never have the problem of the control misunderstanding which argument assignment type you are using.

There are times, however, when you may wish to use I, J, and K in argument assignment number one. If you are running out of letter addresses to choose from, for instance. Or perhaps the letter address name makes so much sense, you feel you should use it (using argument I as a flag to specify incremental mode, for example). In these cases, it is important to know how the control will behave with regard to local variable assignments.

First, if you use only one of the three (I, J, or K), you will never have a problem—if you feel that it just makes sense to use letter address I, J, or K, by all means, do so. But if you use I, J, and/or K together, you must be very careful with their placement order in the call statement. If you do not specify them in their correct alphabetical order, the control will not assign your local variables in the way you expect. Consider the following example call statement.

N065 G65 P1000 X3.0 Y2.0 Z0.30 F2.50 D0.25 I0.45 J5.0

In this example, everything will work according to plan. The *X* will be assigned as #24, *Y* as #25, *Z* as #26, D as #7, I as #4, and K as #5 (right from the table from argument assignment number one).

However, if by mistake you reverse the order of I and J, you're in for problems. Consider this new and incorrect call statement:

N065 G65 P1000 X3.0 Y2.0 Z0.30 F2.50 D0.25 J5.0 I0.45

In this case, notice that J comes before I and will be correctly assigned as #5. But when the control comes across the I, it will be taken as the second I in argument assignment number two (since it follows the first J). Instead of being assigned as #4, it will be assigned as #7, which will overwrite the value currently stored in #7 by the D argument. This, of course, will cause problems in the custom macro program.

WHEN TO USE ARGUMENT ASSIGNMENT NUMBER TWO

Owing to the complexity of setting up argument assignments, you should only consider using custom macro programs with argument assignment number two when the *number* of arguments needed in the call statement is *variable*. Consider the circular screw machine cam example in Figure 6-9.

If we establish that this is just one part in a family of cams, we know that variations will have to be accommodated. You need to develop a parametric program that will drive the tool around the cam, automatically generating the various rises and falls needed on the cam. This particular cam has five rises and falls. However, other workpieces in the family have a different number. Some have four, some have three, some have eight, and so on. With argument assignment number two, you can have up to 10 sets of arguments (rises and falls). Each set can contain up to three arguments (I, J, and K).

As you can see in Figure 6-9, only I and J (K is not required in this example) are required to describe each rise or fall. I is the incremental angle of each rise or fall, and J is the ending radius of the rise or fall. Also notice that the starting radius is specified with letter address A. The thickness of the workpiece (plus clearance) is specified with B. The incremental angle used for calculation is C.

Here is an example program that calls the cam custom macro program to machine an entire cam:

O2001 (Main program for cam)
N005 G54 (Set program zero)
N010 G65 P1002 A1.75 B–.03 C1.0 I12.0 J1.75 I35.0 J1.625 I175.0 J1.1 I25.0 J1.75 I113. J1.75 (Mill entire cam)
N015 M30 (End of program)

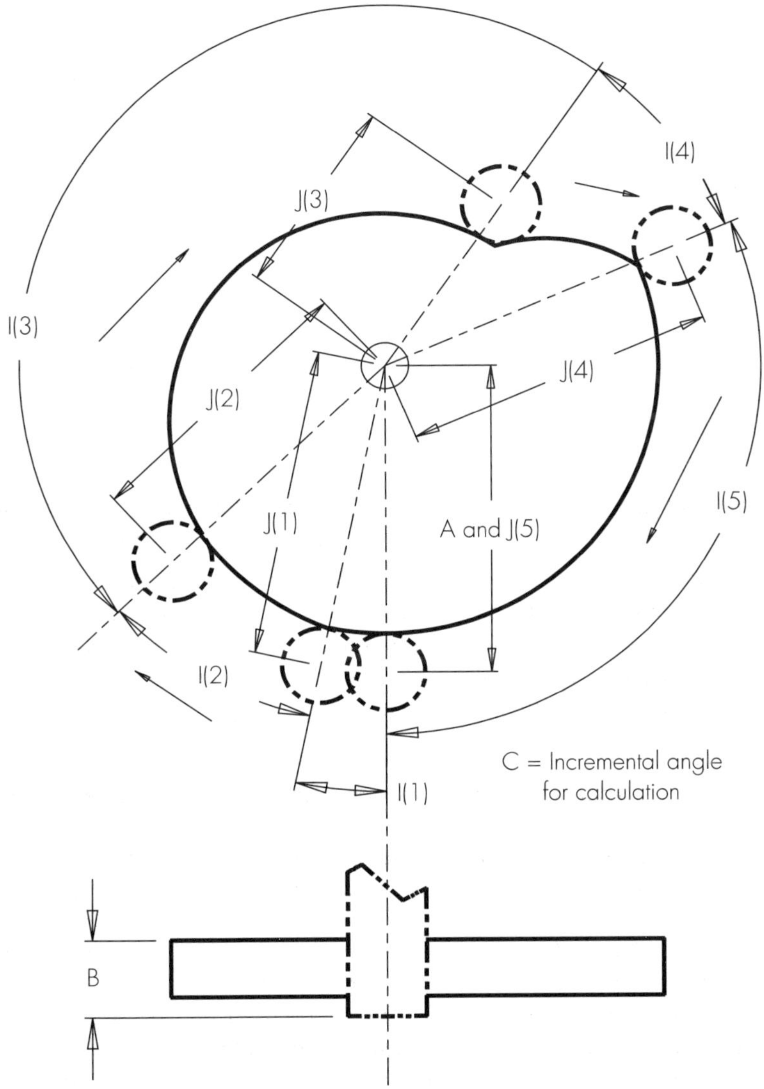

Figure 6-9. An application calling for argument assignment number two. Note that the number of rises and falls will vary from one cam to another.

How will the custom macro know when to stop? Whenever argument assignment number two is used, you must have your custom macro program track when the last set of arguments (I, J, and K) has passed. This can be done by monitoring for vacancy (#0). From the chart for argument assignment number two, notice that I(1) is placed in local variable #4. I(2) is placed in #7. I(3) is placed in #10, and so on. Note that each local variable for I is precisely three local variable numbers from the previous one. Though not simple to do, we can monitor the current value of each I (#4, #7, #10, #13, etc.) to determine the first time an I is vacant. The custom macro program can assume that when an I is vacant, there are no more rises or falls on the cam.

To do this, you actually create a loop to determine the number of rises and falls. Here is a portion of the loop:

```
O2001 (Program number)
#102 = 0 (Step value for I test)
N1 IF [ #[4 + #102] EQ #0] GOTO 99 (If current I is vacant, exit loop)
.
.
.
#102 = #102 + 3 (Step I tester by three)
GOTO 1 (Go back and test if finished)
N99 M99 (End of custom macro program)
```

With this loop, we can continue to test for the final occurrence of the set of arguments being passed.

Now look at this very powerful parametric program that contains a loop within a loop to machine the entire circular cam:

```
O1002 (Custom macro for cam)
#101 = 0 (Initialize current angle to zero)
#102 = 0 (Initialize stepper for last I test)
#104 = #1 (Initialize starting radius)
G90 S600 M03 (Start spindle)
G00 X0 Y – [#1 + .5] (Move to starting Y position at bottom)
G43 H01 Z.1 (Instate tool length compensation, move into Z position)
G01 Z#2 F25.0 (Fast feed to bottom in Z)
Y – [#1] F5.0 (Come flush to surface in Y)
N10 IF [ #[4 + #102] EQ #0] GOTO 99 (Start loop for ending I)
#103 = #101 + #[4 + #102] (Calculate ending angle for this rise or fall)
#105 = [#104 – #[5 + #102]]/#[4 + #102] * #3 (Calculate radius change per move)
```

```
IF [[#[5 + #102]] EQ #104] GOTO 15 (If start and end radius of rise or fall are
   equal, make simple circular move)
N1 #101 = #101 + #3 (Step current angle)
#104 = #104 – #105 (Step current radius)
IF [#101 GT #103] GOTO 50 (If finished with current rise or fall exit loop)
#110 = [SIN[#101] * #104] * [–1.] (Calculate current X position)
#111 = [COS[#101] * #104] * [–1.] (Calculate current Y position)
G01 X#110 Y#111 (Make current move)
GOTO 1 (Go back to beginning of loop)
N50 #104= #[5 + #102] (Calculate next beginning radius)
#102 = #102 + 3 (Step the current I and J)
#101 = #103 (Reset current angle)
GOTO 10 (Go to next rise or fall)
N15#101 = #103 (Begin to make simple radius)
#110 = [SIN[#101] * #104] * [–1.] (Calculate ending X)
#111 = [COS[#101] * #104] * [–1.] (Calculate ending Y)
G02 X#110 Y#111 R#104 (Make circular move to form dwell)
GOTO 50 (Go to next rise or fall)
N99 G00 Z.5 (Finished, get out in Z)
G91 G28 X0 Y0 Z0 (Return to zero return position)
M99 (End of custom macro)
```

While this custom macro will require study, it truly stresses the use of argument assignment number two. It is also one of the more powerful parametric programs you will ever use. With one loop inside another, notice how few commands are needed. Depending on the value of C (incremental angle between each movement), this custom macro program could generate many, many commands. If, for example, C is set to 0.001, this cam macro could generate as many as 360,000 motions (assuming there are no dwells on the cam).

PART TWO

INTRODUCTION TO PARAMETRIC PROGRAMMING FOR PROBING SYSTEMS

As we mentioned in Part One, one of the principal reasons for creating parametric programming was to provide a means of interfacing CNC programs with certain accessory devices that require a higher level of programming than G-code level manual programming. Touch probes are among the most sophisticated of these devices. Without parametric programming, in fact, it would be impossible to program probes with only the G-code level programming available from most CNC controls.

Probes are growing in popularity among CNC users worldwide, as are the number of feasible applications for probing systems. While early probing systems were easily damaged, short on accuracy, and rather slow, today's systems incorporate intelligent protection systems, are highly accurate, and quite efficient. Moreover, probing technology is still evolving, ever increasing the number of feasible applications. Given this continuous improvement trend, it is not unreasonable to expect that the day will come when all CNC machines will be designed with a standard probing system of one kind or another.

This trend notwithstanding, a great deal of confusion and controversy still surrounds the proper use of probing systems for CNC machine tool applications. Some companies swear by them while others feel they are wasteful. In Chapter 7 we try to dispel this confusion by introducing you to touch probes, showing you their applications, explaining how they work, describing how they are calibrated, and in general, providing the information needed to help you decide whether probing systems are right for your applications. While our discussion in Chapter 7 is for the most part positive in nature, it's only fair to play devil's advocate from time to time, showing you when probing systems may not *be the best solution to your CNC problems.*

WHAT MORE DO YOU HAVE TO KNOW?

With a firm understanding of G-code level manual programming and an understanding of parametric programming functions discussed thus far, it should be relatively easy to understand how probing systems are programmed. In fact, in our discussion in Chapter 5 on using an edge finder to help locate a program zero position, we discussed much of what you need to know about programming fully automatic probing systems. If you grasp that discussion, you can approach Part Two as simply an extension of what you already know.

Chapter 8 introduces all commands that have implications when probing. We even discuss certain G-code level CNC commands (commands that you likely have experience with), showing their special application with probing systems.

The diversity of probing systems cannot be stressed enough. Just as machine tool builders and control manufacturers differ in the way equipment must be programmed (M-codes, program formatting, available capabilities, etc.), so too do probe manufacturers with how their probing systems must be utilized. While the commands we show in Chapter 8 are representative of commands used with probing systems, you must be prepared for variations. In all cases, we bow to your probe manufacturer (and their documentation) to provide you with the specific commands needed for your own probing system.

DO YOU REALLY HAVE TO PROGRAM YOUR SYSTEM?

If you already have one (or more) probing system on your CNC machine tools, you likely know that your probe manufacturer (or machine tool builder) has provided a number of standard programs designed for a variety of applications. And depending upon your applications, these standard programs may be sufficient to handle your probing needs. There are certain probe types (discussed in Chapter 7) that, once programmed, never require additional programming, as long as they adequately satisfy the application needs.

In fact, for certain probing systems that have only one fixed application, the probe manufacturer (or machine tool builder) may not even use parametric programming to control the probe. With these probe types, programming is done at a lower level, the same level at which a control's canned cycle programs are written. Just as you cannot manipulate the inner workings of most controls' canned cycles, neither can you manipulate the probing programs for certain probing systems.

However, there is one kind of touch probe, the five-direction touch probe, that is highly diverse in its application. In fact, it is so diverse, some CNC users will find it necessary to provide new probing programs on a regular basis, possibly even for every setup or production run.

You can compare the need for special probing programs to the need for canned cycles that came with your machine tools (hole machining canned cycles, milling cycles, rough turning cycles, and threading cycles, for example). Some CNC users use canned cycles on a regular basis and find them entirely adequate. Others wish they had more. For example, few machining center controls come with a thread milling canned cycle. If you do a good deal of thread milling, you know from Part One that you can use parametric programming to develop your own user-created canned cycle for thread milling.

In similar fashion, some probe users find that the standard set of probing programs that come with their probe are completely sufficient. For them, utilizing their probing system will be a simple matter of using the previously written probing programs. There will be little need for custom programming. Other probe users need much more than what probe manufacturers provide. If your probing needs include functions your standard set of probing programs do not handle, additional probing programs will be needed. A host of highly talented consultants and contract programmers are available who specialize in writing custom probing programs. And, of course, probe manufacturers themselves also specialize in writing custom probing programs for their customers.

If you need only a few custom probing programs, and if you do not wish to write these programs yourself, you should seek the help of probe program writing specialists. On the other hand, if your probing needs are constantly changing, or if you do not agree with your probe manufacturer's standard probing programs, one alternative is to improve your own probing skills.

One way to look at probe programming for diverse applications is to compare it to normal CNC programming. Today, almost all CNC users have the ability to develop their own CNC programs. Very few depend on outside vendors for this relatively basic task. Conversely, very few probe users have the capability to develop their own probing programs at the present time. And for diverse applications, having a probing system and not knowing how to develop probing programs is like having a CNC machine and not being able to develop CNC programs. If you depend completely on consultants and contractors to develop your probing programs (or CNC programs), you will pay a premium for programs and never have complete control of your CNC environment.

Since they require the most in the way of custom programming, we will focus on five-direction touch probes in our discussions of how you actually develop probing programs in Chapter 9. Though you must be prepared for variations (and we will be pointing out throughout Part Two how probe programming varies from one manufacturer to another), if you understand Chapter 9, you will be well on your way to having the proficiency you need to write your own probing programs.

DIVIDE AND CONQUER

The task of developing probing programs for any probing system is a formidable one. When you contemplate all you must know, it can at first seem overwhelming. You will find, however, that the actual commands you use for probing are not at all difficult to understand or develop. As stated earlier, if you comprehend how you can use an edge finder as a kind of manual probing system (Chapter 5), you already know much of what you need to know to program a spindle probe.

You will see that much of writing probing programs is quite redundant. Just as you must follow the same format when writing CNC programs (program startup, tool changing, and program ending), so must you program with a strict format when writing spindle probe programs. You will see in Chapter 9 that several techniques are discussed to minimize redundant commands and keep you from having to remember them, thus eliminating the potential for mistakes. We call these divide and conquer *techniques.*

Chapter 7

Introduction to Probing

In this chapter we first introduce basic concepts of probing with touch probes. We then show the four types of touch probes addressed by this text and introduce the three general application categories for probing. Throughout Part Two, you will see that the actual act of probing remains much the same regardless of the probe type or application.

All touch probes are contact-sensitive, meaning they must be *triggered* by the *deflection* of a *stylus* within the probe. Most current probes are highly accurate and trigger with only a tiny amount of deflection. Though the mechanisms used to detect deflection vary from one probe maker to another, several of the manufacturers state that their probes will trigger within 20 *millionths* of an inch deflection (0.00002 in. [0.00051 mm]). In this sense, touch probes are highly sensitive contact switches.

THE CONCEPT OF PROBING

All parametric program-driven probing systems require that you understand three simple truths. First, that you have the ability to force a controlled collision between the probe stylus and a predetermined surface (workpiece, cutting tool, fixture, etc.). When this collision occurs, the probe will trigger and force the machine to halt its motion before any damage to the probing system occurs.

Second, that you have the ability to determine the machine's position very close to the instant in time that the probe stylus is triggered. Make special note of the words *very close*. There will always be some lag time between the instant that the probe is triggered to the instant the machine's position is attained. This lag time can lead to inaccurate probing and is one of the reasons why all probing systems must be *calibrated*.

Third, that you have the ability to utilize the machine's position after triggering in an appropriate manner, depending upon the application for the probe. If measuring a tool's length on a machining center, for example, the machine's position when a tool tip touches the touch probe stylus can be used to help determine the tool's length. This tool-length value can be automatically placed in the corresponding tool offset.

Does any of this sound a little familiar? During our discussion of system variables in Chapter 5, we detailed a technique that utilizes a simple edge finder to help the setup person locate the program zero location for a machining center application. We presented a corner pickup program to help the setup person find program zero. The techniques used in this example are amazingly similar to those used with any touch probe, and if you understand the presentation in Chapter 5, you already know much of what you need to for the programming of touch probes. The main difference is that the edge-finder example requires that the setup person *manually* touch the edge finder to each surface. With a true touch probe, this task is entirely automatic.

PARAMETRIC PROGRAMMING FUNCTIONS REQUIRED FOR PROBING

As stated in the introduction to Part Two, we assume you currently understand G-code level manual programming. From your manual programming experience, you know you have the ability to make the machine perform many functions. You can start the spindle at the desired speed, turn on the coolant, rapid a tool into position, and begin machining at the desired feedrate. Indeed, today's CNC machine tools allow you to control virtually *anything* you need to control.

For as powerful and functional as your manual programming commands are, you must understand that they are insufficient for handling even the simplest touch probe applications. With any probing application, you need additional programming capabilities (more than are allowed in most manual CNC programming languages).

All parametric programming functions in this section should be quite familiar from your exposure to them in Part One of this text. Here we mention only those that are of special interest to probing. The better you understand these functions, the easier it will be to begin writing probing programs.

Access to machine position after stylus contact

This capability is the most important function parametric programming allows that is not available through manual programming commands. Though the manner by which this position is used varies with each application, this capability is essential to all probing applications.

Variables

Many probing applications require multiple touches of the probe stylus to the surface(s) being probed. Comparisons or calculations involving the various machine positions after probing are usually required, meaning you must have a place to *store* machine positions for future use. Discussed in detail in Chapter 2, variables are used for this purpose.

Arithmetic

As mentioned, calculations will usually have to be made in probing programs. Just about anything that can be done on a scientific calculator can be done within a parametric program.

Access to offsets and other machine functions

Many probing applications require access to tool offsets. A machining center probing system designed to measure tool-length compensation values, for example, must have the ability to actually set a tool offset once the tool's length compensation value is determined. As you recall from Chapter 5, every kind of tool offset is accessible through custom macro B commands (tool offsets, fixture offsets, wear offsets, geometry offsets, workshift offsets, etc.).

Probing systems also require access to certain other machine functions. For example, probing must be done at a consistent feedrate. If an operator places the feedrate override switch at something other than 100%, it could result in inaccurate probing. For this reason, most versions of parametric programming allow the programmer to disable feedrate override (temporarily) during the probing cycle.

Logic and program flow control

Many probing applications require the probing program to make decisions and behave differently based on the result of the decisions made. For example, a programmer developing a probing program that measures the width of a slot will have to plan on several contingencies. If, for example, the measured slot is too narrow, it may be the programmer's approach to adjust an offset (automatically) and re-machine with the cutting tool. If the slot is too wide, possibly the workpiece is now scrap. In this case, an alarm could be sounded by the probing program to stop further machining. If the slot is within its tolerance band but not precisely at its mean value, perhaps an offset could be adjusted so the slot in the next workpiece is machined perfectly. The machine could then continue in the normal manner. (This is an example of one of the many times when logic commands discussed in Chapter 4 must be used within the probing program.)

THE PROBE ITSELF

In very simple terms, touch probes are nothing more than highly accurate contact switches. As stated earlier, all probes discussed in this text are touch probes, meaning they require but a tiny amount of deflection to trigger. *When* they trigger during a probing command (commonly within 20 millionths of an inch deflection), a signal is sent to the CNC control, which halts axis motion. This signal also tells the control to attain the machine's position as close as possible to the instant in time the probe triggered. This position (in all axes) is accessible within the probing program.

This understanding (that a touch probe simply tells the CNC control to halt motion and attain the machine's position), coupled with understanding the additional programming functions needed for probing and provided by parametric programming, should give you a sound base of knowledge for understanding *any* probing system, regardless of its complexity.

PROBE TYPES

Four types of touch probes are used with CNC machining centers and turning centers. Of primary concern at this point is understanding the *deflection directions* each allows.

THE FIVE-DIRECTION TOUCH PROBE

This type of probe can be mounted in the spindle of a machining center or in the turret of a turning center. As the name implies, it allows detection for *five deflection directions*. When mounted in the spindle of a machining center, deflection can be detected during *X*+, *X*–, *Y*+, *Y*–, and *Z*– movements. Figure 7-1 shows this type of probe.

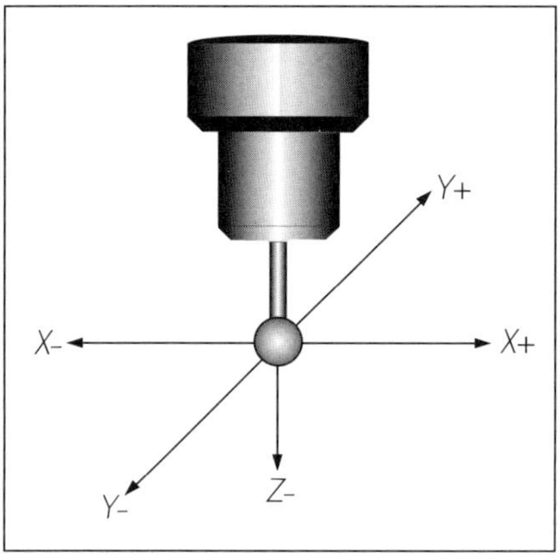

Figure 7-1. Directions of probing for a five-direction touch probe mounted in the spindle of a vertical machining center.

Five-direction touch probes used in machining centers are commonly called *spindle probes*. For a number of reasons, we place heavy emphasis throughout this text on the programming of five-direction spindle probes. First, they are the most popular form of spindle probe. Second, they commonly require the most in the way of custom programming from end users. It is not uncommon for a spindle probe user to require new programming for their probing system with each new setup. Third, they tend to be the most complicated probes with which to work. If you understand the programming of five-direction spindle probes, it will be easy for you to adapt what you know to other types of probes. Finally, of the probe types, spindle probes allow the most diversified applications (other probe types commonly have one specific purpose). Once they are initially programmed for that purpose (by the machine tool builder or the probe manufacturer), there will be little, if any, need for further programming.

The five-direction touch probe also can be mounted in the tool holder of the turret of a turning center. However, since most turning centers limit turret movement to only two directions, two directions of probing commonly go unused when the five-direction touch probe is used (unless the turning center has a third, or *Y*, axis). Figure 7-2 shows this kind of probe mounted in a tool holder of the turning center's turret.

Note that Figure 7-2 shows the most common type of stylus. For a two-axis turning center, when the probe is mounted parallel to the *Z* axis, the five-direction touch probe allows detection of deflection while moving in the *X*–, *X*+, and *Z*– directions. Depending on the application, however, it is possible to design the stylus to probe in different directions. For example, if the stylus for the probe in Figure 7-2 is bent 90 degrees, or if the touch probe is mounted in an *X* axis tool holder (parallel to the *X* axis), it will allow probing in *X*–, *Z*+, and *Z*–.

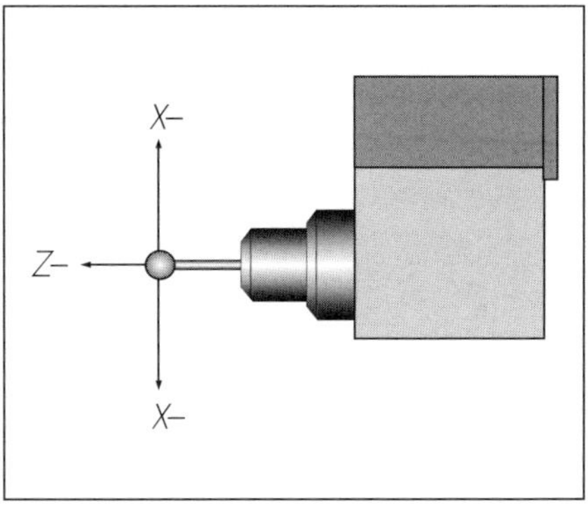

Figure 7-2. Directions of probing when a five-direction touch probe is mounted in the turret of a turning center parallel to the Z axis.

TWO-DIRECTION SPINDLE PROBES

The second touch probe type is a spindle probe that requires only two directions of probing. Shown in Figure 7-3, this kind of touch probe is mounted in the spindle of a machining center. Though it requires a special spindle orientation function, this probe is less expensive and eliminates droop, one of the factors contributing to inaccurate probing (more on droop later). Note that this spindle orientation is similar to that specified with many automatic tool changers used for aligning the tool changer arm with the tool holder (specified by M19 on most machines).

At least four spindle orientation positions must be programmable if this kind of probe is to be used (*X*+, *X*–, *Y*+, and *Y*–). With most, the actual angle of spindle orientation can be specified in one-degree increments (more on how later). Since the spindle rotates to its proper orientation prior to probing, only one direction of probing is needed to probe any number of angular positions in the *XY* plane (*X*+, *X*–, *Y*+, *Y*–, or any angular combination). The second direction of probing for this probe type is *Z*–. Note that with only these two directions of probing, this touch probe type can perform any task possible with a five-direction touch probe used in the spindle of a machining center.

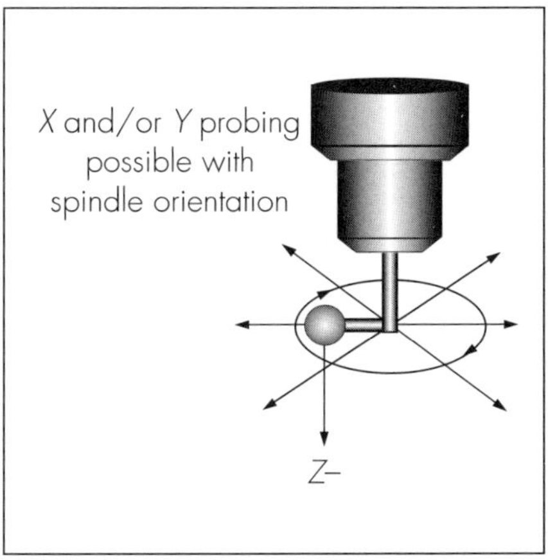

Figure 7-3. Directions of probing for a two-direction spindle probe. Note that X and Y probing requires precise spindle orientation.

TOOL-LENGTH MEASURING PROBES

There are actually two variations for machining center tool-length measuring probes. The first, and most common use of this form of probe, is as a tool for measuring offset values that are used with *tool-length compensation*. If used solely to find tool-length values, only one direction of probing is required (*Z*–). Figure 7-4 shows this kind of probe.

With the probe placed somewhere in the machine's work envelope (possibly mounted right to the machine table), a tool tip is brought into contact with the probe stylus.

The second variation of this kind of probe is its possible use as a tool to measure *cutter-radius compensation* offset values. When used for this purpose, an additional direction of probing is necessary (*X*+, *X*–, *Y*+, or *Y*–). Of course, cutter-radius compensation is only required for milling cutters, and only when the milling cutter will be milling on its periphery (as when contour milling). For this reason, probing for a cutter's radius is only required for certain milling cutters. Figure 7-5 shows this kind of probe.

To come up with an accurate cutter-radius compensation offset value, most probe manufacturers recommend that the spindle be started prior to measuring the cutter's radius; to avoid actually machining the probe stylus, manufacturers recommend starting the spindle in reverse (M04 for right-hand milling cutters).

Note that Figures 7-4 and 7-5 depict a kind of tool-length measuring probe that must be mounted to the table top. This can be inconvenient, since it can often interfere with the mounting of the workholding fixture. It is especially inconvenient with horizontal machining centers. For this reason, most current tool-length measuring probes are not physically mounted to the table top. Instead, they are mounted on a *swing arm* that can be programmed to enter the work area and retract from it automatically (commonly with two M codes).

TOOL TOUCH-OFF PROBES

This turning center probing system is used primarily during setup to determine the program zero assigning values for each tool held in the machine's turret. Most current-model turning center controls use some form of tool offset with which to assign program zero (one control builder calls the set of program zero assigning offsets *geometry offsets*). When used with this kind of program zero setting, the tool touch-off probe will measure the needed offset values and automatically store them into the corresponding offsets as part of the probing routine.

Figure 7-6 shows an overall view of a turning center's work area including a swing-arm-mounted tool touch-off probe, and Figure 7-7 shows a close-up of the probe itself and its deflection directions.

This kind of probe allows all single-point cutting tools held in turning centers to be easily touched. Notice that each cutting tool will require two touches, one to

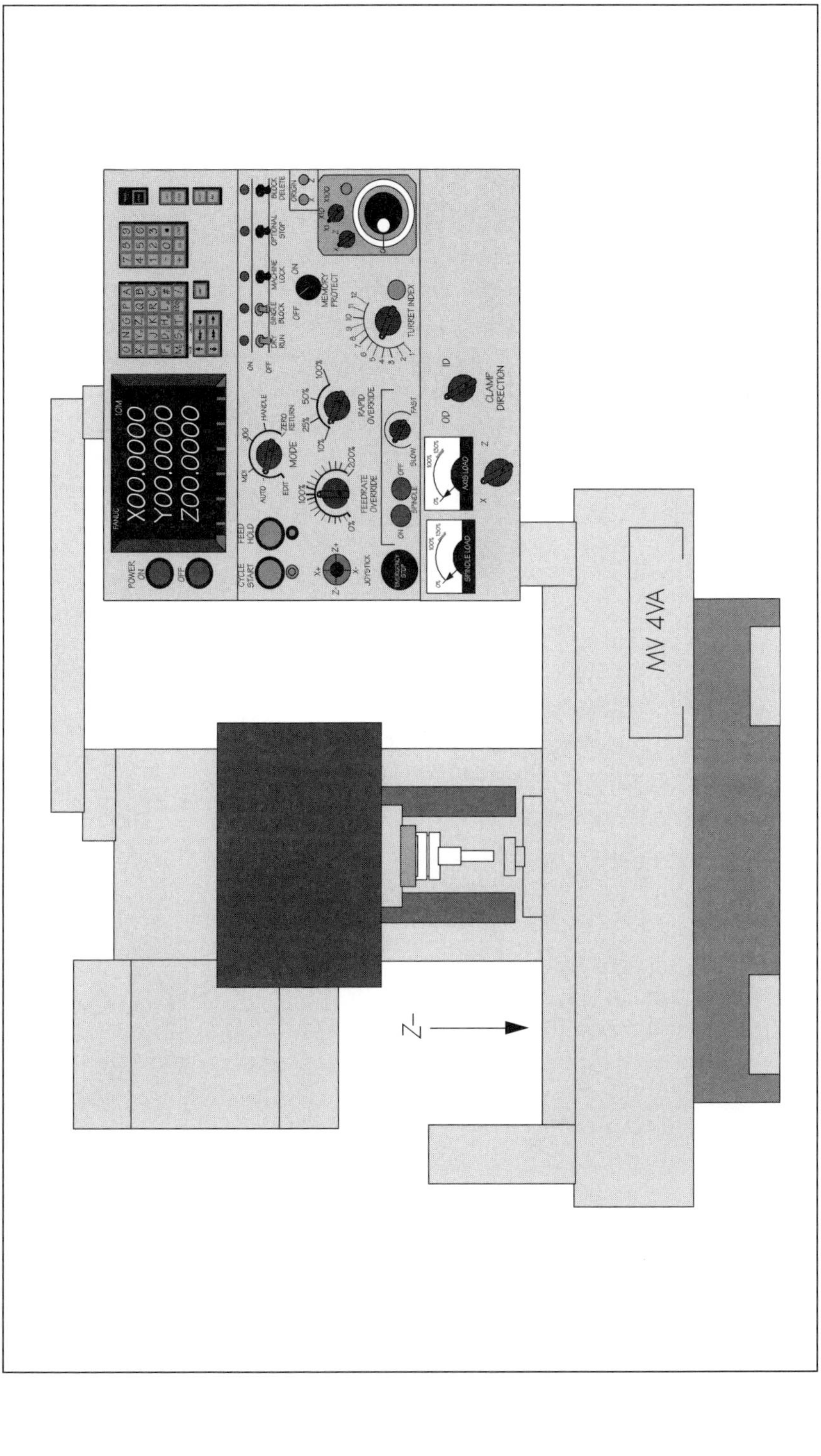

Figure 7-4. Direction of motion for a tool-length measuring probe.

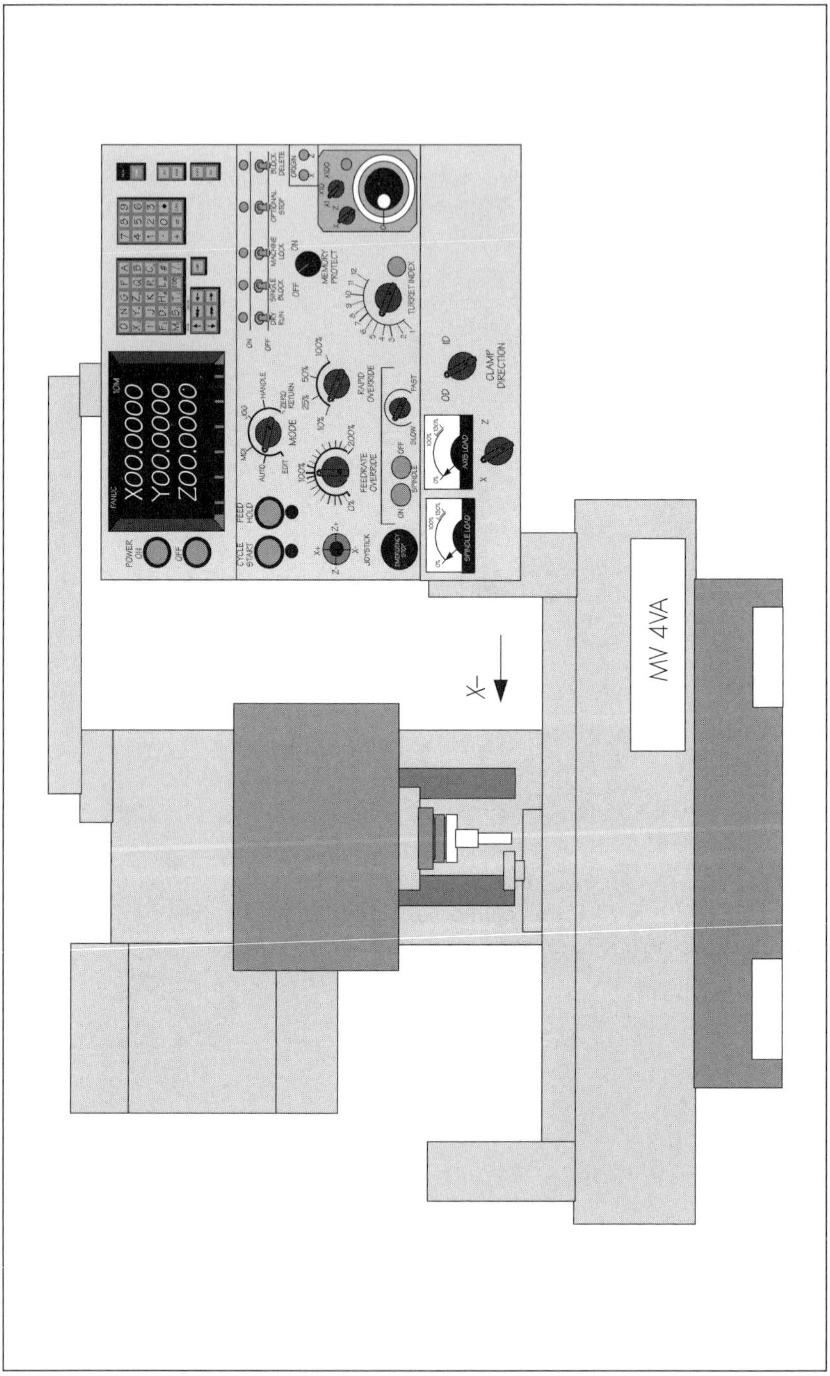

Figure 7-5. Tool-length measuring probe having the additional capability of measuring a cutter's radius.

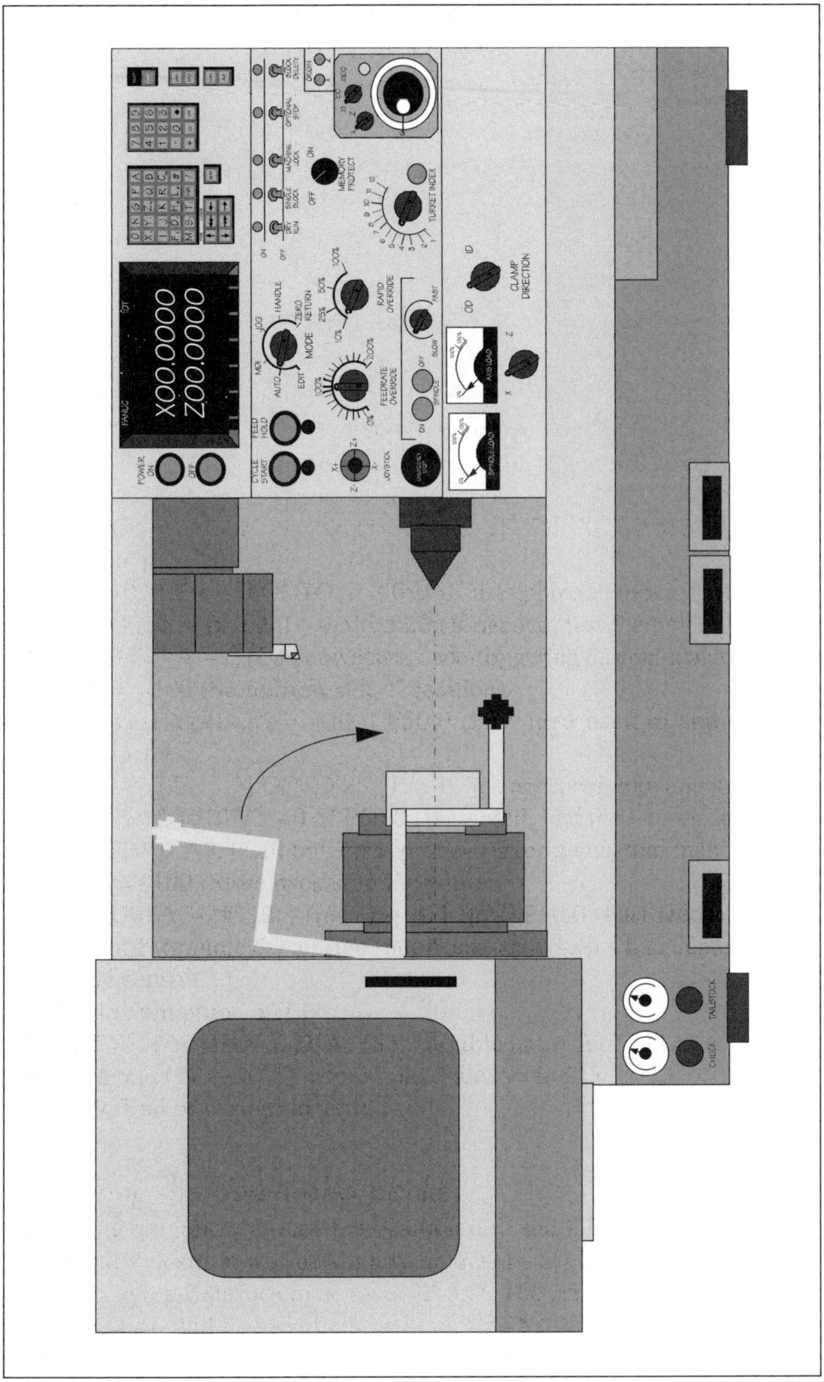

Figure 7-6. Overall view of turning center work area, including the swing-arm-mounted tool touch-off probe.

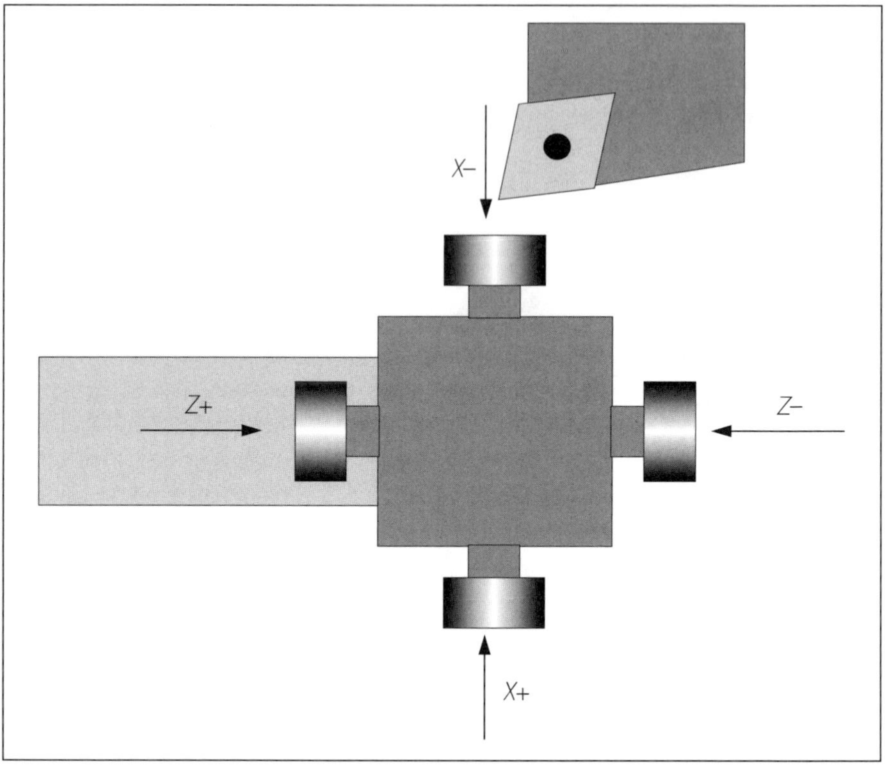

Figure 7-7. Deflection directions for tool touch-off probe.

determine the *X*-axis program zero setting value and one to determine the *Z*-axis program zero setting value. A normal outside diameter turning tool, for example, must be probed in the *X*– direction as well as in the *Z*– direction.

One type of tool presents a special problem for the tool touch-off probing system. The tool touch-off probe cannot determine the *X*-axis program zero setting value for center-cutting tools, like drills, taps, reamers, and counterboring tools (it can still probe the *Z*-axis program zero setting value). Fortunately, most turning center turrets are designed in such a way that the *X*-axis program zero setting value will remain the same for all center-cutting tools. Once it is measured one time (manually), the resulting value can be used for all center-cutting tools.

As is the case with tool-length measuring probes, machine tool builders differ in the way they mount tool touch-off probes to the machine tool. Almost all incorporate some form of swing-arm to bring the probe into the machine's work envelope and allow access to cutting tools. Some machines incorporate manually activated swing-arms, meaning that the operator or setup person must be available to activate the swing-arm. This means, of course, that completely unattended operation is not possible, which may be one of the primary goals of the tool touch-off probing system. Most turning centers incorporate automatic swing-arms. Two M-codes are commonly used: one to bring the arm into the work envelope and one to put it away.

A secondary feature of tool touch-off probes is that of automatically compensating for the minor deviations in cutting-tool position after tool maintenance is performed. As you know, when a turning center cutting tool machines workpieces, it will eventually begin to wear. While it may not be ready for replacement, the surfaces it machines will not be perfectly to size. An operator will likely make several wear offset changes during an insert's life. When the insert is exhausted and in need of replacement, it can be difficult for the operator to remember just how much total offset was required during the insert's life. This problem, combined with the fact that it may be difficult (if not impossible) for the operator to place the new insert into its tool holder in the exact same position as the last insert, makes it feasible to use tool touch-off probes after each tool maintenance. In effect, the tool touch-off probe simply remeasures and sets the cutting tool's current program zero setting values (commonly in geometry offsets) and clears the related wear offsets. This can be done whenever a cutting tool (or insert) is changed during the production run.

APPLICATIONS FOR PROBING

Following the pattern we set in Part One during our introduction to parametric programming, we will illustrate parametric probing by presenting real-world applications. When you see its potential, you will likely want to learn as much as you can about preparing your own probing programs. Though emphasis in this discussion is on feasible applications, we will venture to the periphery from time to time and point out possible times when probes may be *misapplied*. For the purpose of organization, we divide all applications for probing into three categories; setup help, in-process gaging, and utilities.

SETUP HELP

Some of the best applications for probing are related to providing help during the setup process. This is especially true when the CNC setup person must perform certain setup tasks *on line* (while the machine is down between production runs). Though it is not our intention to discuss setup time reduction principles in detail, there are actually only three ways to reduce setup time. The first is to eliminate on-line tasks. This can often be accomplished through planning and ingenuity.

The second way is to move on-line tasks off line. In essence, this means getting ready for the next setup during the current production run. If certain tasks related to the next setup, like tool assembly and measurement, are performed during the current production run, setup time can be reduced. However, this technique assumes you have the personnel and time to perform setup tasks off line.

If you cannot eliminate the task, and if you cannot move the task off line, your only alternative for minimizing setup time is to *facilitate on-line tasks*. Frankly, probing systems are best applied for setup help in this area. *You should only consider probing systems to facilitate setup tasks when you have exhausted the first two ways to reduce setup time.* (There are some exceptions to this statement. Some setup applications require probing, regardless of how much time is available during the previous production run.)

In many companies, for example, and especially smaller firms, a CNC operator is expected to perform all CNC-related functions, including processing, programming, setup, and program verification, as well as running the machine during production. If there is no CNC support staff (setup person, tool setter, programmer, etc.), it is likely that this operator will be very busy, and have little time to do anything off line to get ready for the next setup. This is also true in companies that have very short cycle times. With short cycles, the operator will be very busy loading and unloading workpieces, performing tool maintenance, and sizing with tool offsets. This operator can do nothing to prepare for the next setup during the production run. And in companies that have very low quantities *in conjunction with short cycle times*, absolutely nothing can be done off line, regardless of how many people work in the CNC environment. In this kind of CNC environment, most of the tasks related to making setups must be done on line. And when setup tasks must be done on line, probing systems can dramatically facilitate certain on-line tasks. Moreover, since the probing system can usually perform its task automatically, the operator is commonly freed to do other things while the probing system works.

This is the major reason why so much controversy surrounds the application of a probing system for the purpose of providing help during setups. People working for a company that has lengthy production runs, is adequately staffed, and has relatively long cycle times commonly feel that applying probes for setup help is *wasteful*. People working for a company that has short production runs, with only one person doing many CNC-related tasks, and that has short cycle times commonly feel that applying probes for setup help is *essential*.

Other factors like the diversity of work, the amount of repeat business, and even workpiece size, shape, and weight also play a large role in determining the wisdom of employing a probing system to help with setup-related tasks. Generally speaking, the more diverse your work, the more difficult (and expensive) it is to design and build *qualified* workholding tooling (tooling that can be mounted over and over without deviations in placement) that can eliminate the setup-related needs for probing. Similarly, the fewer repeat jobs your company machines, the less you will want to invest in special tooling of any kind. And the larger, heavier, and more awkward your workpieces, the more difficult it is for your CNC operators to correctly load them. *A probing system can easily compensate for inaccurate and unqualified workholding tooling*. In essence, probing systems make it possible to get by with lower quality (and less expensive) workholding devices for those times when expensive workholding tooling cannot be justified.

While it is the intent of this text to teach you how to program probing systems, probing systems themselves must be used wisely. As we have been stressing, certain setup and cycle-time compromises may be made when probing systems are applied. When it comes to applying them to setup-related tasks, be on the lookout for alternatives to probing if:

1. *Cycle times are high.* During lengthy cycles your people (and maybe even your CNC operators) can be performing many of the next job's setup tasks during the current production run.
2. *Production quantities are medium to large.* During long production runs, much work can be done in preparation for the next setup.
3. *You have an adequate staff in your CNC environment.* If your CNC operators do not have time to perform setup tasks off line, possibly others can (setup people, tool setters, programmers, etc.)
4. *You have workpieces in a close family.* Possibly qualified workholding tooling can be designed that will eliminate the need for probe functions during setup.
5. *You have a large percentage of repeat business.* The more repeat business you do, the more times you will be duplicating setups. With this kind of work it should be relatively easy to justify qualified workholding tooling that eliminates the need to use probes during setup.
6. *Your workpieces are small and easy to handle.* Again, relatively inexpensive qualified workholding tooling can be built for this kind of work.

Help measuring the program zero location

All CNC machine tools, including machining centers and turning centers, allow floating zero point programming. As you know, this important feature permits the programmer to determine the position from which all coordinates in the program are taken. If the program zero point is chosen wisely, coordinates going into the program commonly match print dimensions. As you also know, the program zero assignment is not magic. The CNC control must *know* the location of the program zero point before correct motions can occur. This requires that the setup person, operator, or programmer specify to the CNC control, by one means or another, the location of the program zero point.

The most popular machining center program zero designator is *fixture offsets* (also called *work coordinate system shifting*). For turning centers, the most popular program zero designator is *geometry offsets*. We use these methods of program zero assignment throughout this text to illustrate specific techniques for program zero assignment with probing systems.

The specification of the program zero location is an extremely important setup-related task. The difficulties related to this task range from extremely efficient, easy, and fail-safe to time-consuming, difficult, and error-prone. For example, when qualified workholding tools are used on machining centers (a keyed fixture, for example), the setup person can be confident that whenever a setup requiring this fixture is made, the location of program zero will remain in the same location (relative to the machine's reference position) as it was the last time the setup was made.

Additionally, by knowing some basic and consistent distances on the machine tool (distance from spindle center line at the reference position to each table slot in *X* and *Y* and the distance from the spindle nose to the table top in *Z*), as well as the distances from the placement keys on the fixture to workpiece location surfaces, a programmer can *calculate* the location of program zero before the setup is even made. The program zero assignment values can even be part of the CNC program since fixture offsets and geometry offsets can be entered by programmed command, meaning no on-line time will be taken for measuring program zero.

Of course, what we have just described is the ideal situation (though many companies work very hard to achieve this). Many companies simply cannot justify the costs of using qualified workholding tooling for all of their setups. For those setups that cannot be qualified, the location of program zero must be determined *after* the workholding setup is made, meaning that the measuring and entering of the program zero location will be an on-line task. Probing systems can dramatically facilitate this on-line task.

Machining center program zero measuring help. When workholding setups cannot be qualified, determining the program zero point location on machining centers involves taking measurements on the machine during setup. Without a spindle probe to help, very tedious, time-consuming, and error-prone manual measuring methods must be used. Cumbersome tools like edge finders and dial indicators must be used in conjunction with manual use of the machine's axis position displays (used as a kind of digital readout during program zero measuring). If your setup people are using these crude techniques to assign program zero, a great deal of setup time is likely being wasted and the potential for measuring and entry mistakes is high.

For example, square- or rectangular-shaped workpieces require that edge finders be used to locate the program zero point in *X* and *Y*. A third measurement is required for the *Z* axis program zero point. With round workpieces, a dial indicator is commonly used to align the spindle center line with the center of the round workpiece for *X* and *Y* program zero location. Again, another measurement is required to determine the *Z*– axis program zero location. Depending on the skill level of the operator and the device being used, measuring a single program zero point location can take from 5 to 15 minutes of setup time. A spindle probe—such as shown in Figure 7-8—to perform the same functions will take about 30 seconds. And remember, the program zero entry is part of the probing program. There will be no chance for errors in calculation or entry mistakes if the spindle probe is used. The more setups you make, the more time the spindle probe can save. Figure 7-9 shows a spindle probe being used to determine the program zero point location for a rectangular-shaped workpiece, while Figure 7-10 shows the same for a round workpiece.

Figure 7-8. Five-direction spindle probe used to measure program zero location on machining centers. (Courtesy Marposs Corporation)

In essence, everything an operator would do manually with an edge finder or dial indicator is done automatically with the spindle probe. With the rectangular-shaped workpiece, for example, the program zero point location is the lower left-hand corner of the workpiece in *X* and *Y* and the top surface of the workpiece in *Z* (*X*– surface, *Y*– surface, and *Z*+ surface). The spindle probe can first be brought into a position within about 0.50 in. (12.7 mm) of this program zero corner in all axes (as shown in Figure 7-8). The probing program might first move plus in *Y* and minus in *Z*, to a position making it possible to probe the left surface of the workpiece in *X*. The probing command would then be given in the *X*+ direction. Once the stylus touches the surface in *X*, the probe will trigger, halting motion and causing the control to store the current machine position. The parametric program can then calculate the distance from the machine's *reference position* (also called the *zero return position*) to the *X* surface being touched, taking into consideration the stylus size and any overshoot experienced when probing in the *X*+ direction (overshoot will be discussed

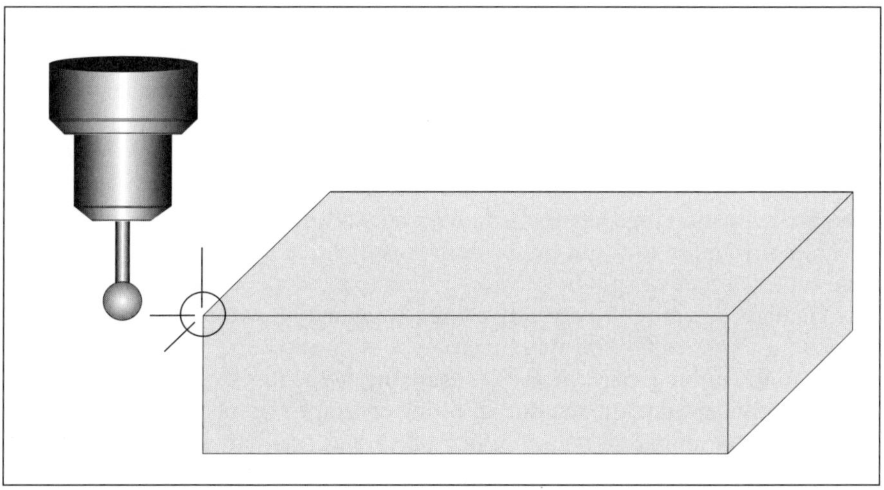

Figure 7-9. Initial position of a spindle probe used to measure the program zero point location for a rectangular workpiece.

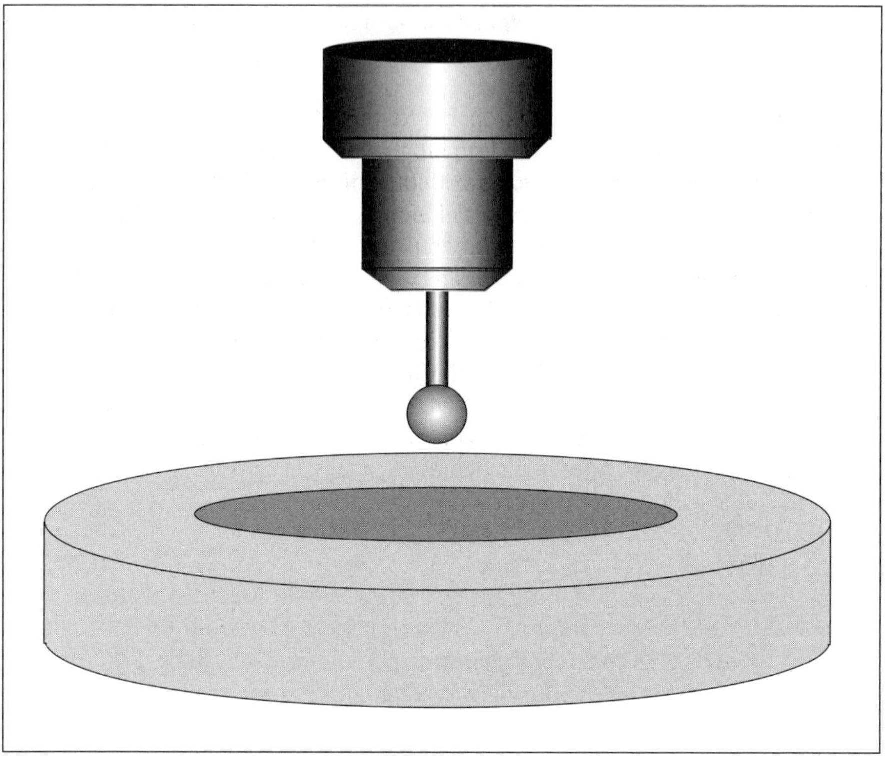

Figure 7-10. Initial position of a spindle probe used to measure the program zero-point location for a round workpiece.

in great detail later). The *X*-axis program zero designating value can then be stored into the corresponding fixture offset. The same steps can then be repeated for the *Y* and *Z* program zero surfaces.

Though we have stressed the use of spindle probes to measure the program zero point location for rectangular and round workpieces, keep in mind that any program zero point location can be determined with a spindle probe (center of a slot, center of a dowel pin hole, special surface locators on fixtures, etc.). Truly, any methods you may be currently using to manually measure program zero can be done automatically with spindle probes.

Turning center program zero measuring help. Just as spindle probes can dramatically facilitate the measuring of the program zero point location for machining center setups, so do tool touch-off probes facilitate program zero point measuring on turning centers. In fact, the need for tool touch-off probes in turning center operations is amplified (as compared to machining center spindle probes) for two reasons.

First, *every* cutting tool used in turning centers requires its own program zero point designation (as compared to machining centers that require only one set of program zero assigning values for each program zero point location, regardless of how many tools are used in the program). A turning center program using 10 tools will require 10 sets of program zero assigning values. If the program zero point measurements are done manually during setup (on line), much time can be wasted by the setup person taking measurements. And once measured, two values for each tool must be entered into offsets, meaning the same points made earlier regarding mistakes in calculations and entry still apply.

Second, it is much more difficult to perfectly *qualify* turning center workholding setups, especially when the tooling holding the workpiece must be machined during setup (as is the case with soft jaws). Even if workholding tools are perfectly qualified, turning centers present additional problems since the precise location of cutting tools must also be known to truly qualify the setup. This means it may be almost impossible for a programmer to perfectly calculate the program zero assignment values *up front*, as is commonly possible with machining center qualified workholding tooling.

Manual measurement of program zero assigning values can be very tedious on turning centers. It usually requires actually machining a workpiece (manually) during the measurement. For example, one common method for determining the *X*-axis program zero assigning value for a turning tool involves skim cutting a small amount of material from a diameter. Without moving the tool in the *X* axis, the setup person stops the spindle and measures the diameter just machined and enters it into the *X*-axis position display. The machine is then sent to its *X*-axis reference position. At the reference position, the *X*-axis position display will show the program zero assignment value in *X*. The same techniques must be used to measure the *Z*-axis program zero assignment value.

Experienced setup people may be able to accomplish these program zero measurements in less than 5 minutes per tool. Novices will take longer. And of course, there is always the potential for errors in measurement and offset entry. A tool touch-off probe will take about 20 to 30 seconds per tool, including the turret index. And the potential for measurement and entry mistakes is eliminated.

Though we are focusing on probing applications aimed at helping during setup, bear in mind that these same problems recur during any tool maintenance performed *during the production run*. It may be impossible for operators to perfectly track the amount of offset adjustment used for the previous tool insert. Additionally, they may have problems perfectly replacing the worn insert with a new one. If either of these two problems is mishandled, the next workpiece to be machined will likely be scrapped. For these two reasons, *the same tasks related to measuring program zero assigning values during setup are commonly required during tool maintenance*. Tool touch-off probes (Figure 7-11) not only facilitate program zero assignment, they also facilitate tool maintenance. We deal with facilitating tool maintenance in greater depth later, when we discuss the utility applications for probing.

Figure 7-11. Turning center tool touch-off probe used to measure program zero position (geometry offset value) during setup and during tool maintenance. (Courtesy Marposs Corporation)

Remember that each tool must touch the tool touch-off probe twice, once for *X* and once for *Z*. For a turning tool, the tool is first brought to a position above the *X*– deflection detection stylus. The probe command is then given to drive the tool into the stylus. The instant the probe senses contact, it stores the current machine position and halts motion. The probing program, having access to the distance between the spindle center line and the *X*- deflection detection stylus (through a special variable we discuss later), can calculate the distance between the machine's reference position and the spindle center line. This value is automatically placed in the *X*-axis value of the corresponding (geometry) offset. Similar techniques are then used to determine the *Z*-axis program zero assignment value.

You may have noted that the program zero point measurement just described ignores one important point. This measurement is simply a *static measurement* that does not take into consideration the dynamic forces of a machining operation. When the turning tool actually begins machining, it is likely that tool pressure will deflect the workpiece from the cutting tool, causing an imperfection in the workpiece. Most current tool touch-off probes have ways to compensate for this deviation.

Measuring tool-length and cutter radius

As stated, nothing beats a tool-length and radius measuring probe when tool measurements must be made on line during setup. Manually measuring tool-length compensation values on the machine requires that the CNC setup person use the axis position displays as a form of digital readout. If measuring the actual length of each tool (the most common method of utilizing tool-length compensation offsets) on a vertical machining center, for example, the setup person must first bring the spindle nose into contact with a smooth flat surface that is parallel to the table top. Many setup people use a gage block for this purpose. With the spindle nose flush with the gage block, they reset the *Z*-axis position display to zero. Next they retract the spindle nose far enough in the *Z* axis to load a tool into the spindle. They then cautiously bring the tip of the tool to the same surface that the spindle nose had touched (the gage block). At this time, the *Z* axis position display will be showing the tool's length, and this value will be entered into the tool's corresponding tool-length compensation offset. From this point, the setup person will load each tool to be measured, bring its tip to the gage block, and enter the tool's length into the proper tool-length compensation offset. While some controls have special features aimed at simplifying the task of tool-length measurement and offset entry, many do not, which opens the door to wasted time and offset entry mistakes.

Figure 7-12. Tool-length measuring probe. (Courtesy Marposs Corporation)

An experienced setup person will take from 2 to 5 minutes per tool. A novice will take longer. A tool-length measuring probe (Figure 7-12) can perform tool-length measurements in less than 30 seconds per tool. And most tool-length measuring programs run unattended for the time it takes to measure all tool-length compensation values, freeing the operator to perform other tasks during the tool-length measuring operation.

Figure 7-13. Tool-length measuring probe that has the additional ability to measure a cutter's radius. (Courtesy Marposs Corporation)

Many tool-length measuring probes have the additional ability to determine the radius of milling cutters (Figure 7-13). For those that do allow cutter-radius compensation value measuring, the command specifying which tools must be measured must have the added capability to designate which tools require cutter-radius compensation offset measurement. Tool-length measuring systems vary with regard to how the command is given. Here is one example in custom macro B format using argument assignment number two (introduced at the end of Chapter 6):

```
G65 P9050 I1. J7. I8. J8. K1. I10. J14. I17. J21
```

As we stated in Chapter 6, up to 10 sets of I, J, and K can be specified in the G65 command. For our example, each I represents the first station number in a series of tool stations that contains a tool to be measured. Each J represents the last tool station in each series. If a K is specified in a series, the parametric program will assume this series of tools contains milling cutters that require the additional measurement of the cutter's radius.

In our specific example call, notice that in the second set of I and J, both values are eight. In this case, only one tool is to be measured. The K word in set number two (K1.) specifies whether this tool group requires cutter-radius compensation offset value measuring. If K is left out of the command (as it is in all other I/J sets of our example call statement), the control only performs tool-length measurement. If K is set to one (1.), the control will additionally perform cutter-radius compensation offset value measurement on all tools in the group. Note that since the tool diameter is *not* specified with our example command, the probing program must allow for the largest possible cutter when measuring, which could lead to time-consuming measurements. Most companies using this kind of probe are unconcerned with the time wasted, however, since the operator can be performing other setup-related tasks during tool-radius measurements.

Compensating for difficult workholding setups

Any machinist is well aware that some workpieces are more difficult to locate and hold than others. One factor that contributes to the level of setup difficulty is the degree of consistency from one workpiece in the production run to the next. For machining center applications, square or rectangular workpieces can be easily held by simple table vises. In similar fashion, round workpieces can be easily held by a three-jaw chuck. For turning center applications, round workpieces (possibly from round bar stock) can be easily held in a three-jaw chuck. As long as there is a high degree of consistency from one workpiece to another, an operator can easily load each workpiece and be confident that all machining operations will be performed correctly for every workpiece.

However, some workpieces are much more challenging to locate and hold in the workholding setup. Fixtures designers will agree that any size and shape variations among workpiece blanks will increase the challenge of designing and fabricating fixtures that will work for all workpieces in the production run. And if workpiece positioning within the setup varies from one cycle to the next, most CNC machine tools by themselves can do nothing about it. Always remember that without probing systems, a CNC cycle will run in exactly the same manner, regardless of the raw material inconsistencies. Even a small deviation in raw material can result in scrapped workpieces. Larger deviations can result in damaged cutting tools, damaged fixtures, and damaged machine tools, not to mention an increase in the potential for operator injury. Sand castings are notorious for deviations that wreak havoc with CNC machine tools, and generally speaking, the larger the casting, the more potential for rough stock variations.

Traditionally, inconsistencies among rough stock for CNC operations are handled within the workholding devices themselves. Many casting-holding fixtures incorporate adjustable locators and targeting mechanisms to allow the setup person or operator to actually move the location of the workpiece during the production run (commonly during the loading of *every* workpiece). Adjustments to workholding tools can be tedious, time-consuming, and error-prone, especially if they must be done while the machine is down between cycles.

With a probing system, there will be no need to adjust workpiece position. Instead of moving the workpiece to match the CNC program commands, the positioning movements of the CNC program will be changed (by the probing system) to match the location of the current workpiece.

Figure 7-14 shows an example of this problem. In this case, notice that there is a window pattern of holes machined around the cored cavity. The holes must be machined relative to the cavity, since a cover plate is mounted over this cavity that must seal properly (with a gasket). For the sake of illustration, let's suppose this workpiece is rather large, and the variation from one workpiece to the next is substantial.

Perhaps production quantities for this workpiece are low, meaning the design and fabrication of elaborate fixturing that adequately targets the workpiece from the cored cavity is not feasible or practical. Or maybe this workpiece is so large, that targeting from the cored cavity is not possible. Of course, any cavity variations in the casting procedure will result in the holes not being machined in the correct relationship to the cavity.

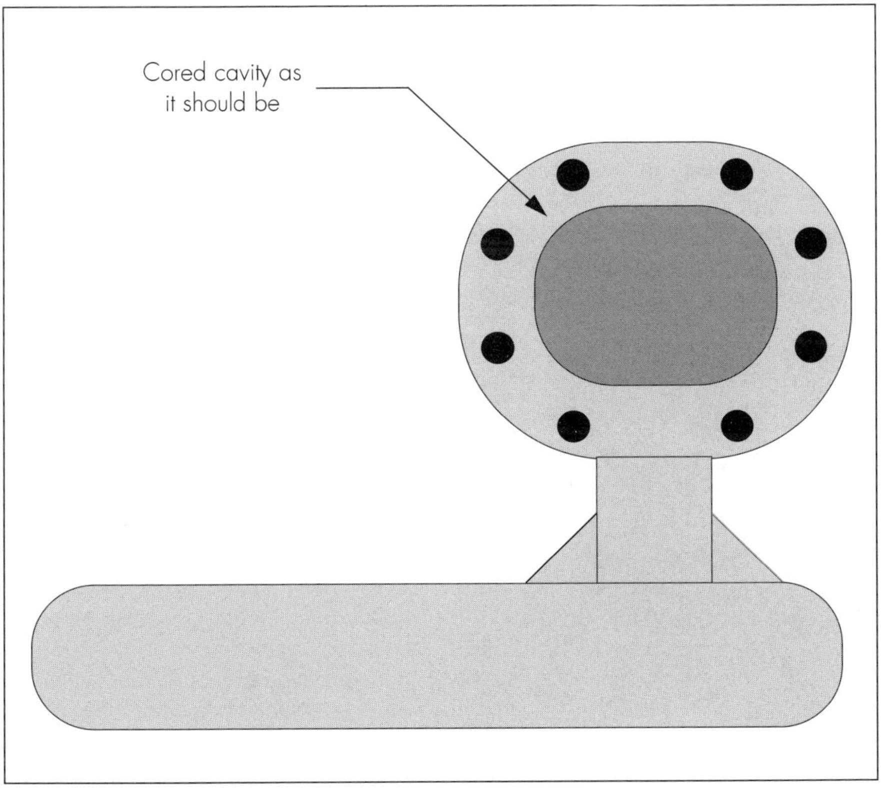

Figure 7-14. Casting with a cored cavity that varies from one workpiece to the next. This variation will cause problems with the placement of the holes.

A spindle probe could easily find the *X* and *Y* location of the cavity by probing any one of the four corners. Once the corner location is determined, a fixture offset could be set to locate that corner as the current program zero point location for machining the holes.

Variations in size and shape typify cases in which probing systems are of great value. Based on the specific criteria for the setups you make, you may have other issues to deal with. For example, you may have such large workpieces that it is difficult to perfectly locate them square in the setup. A probing system can be used to adjust for any misalignment. Figure 7-15 shows an example.

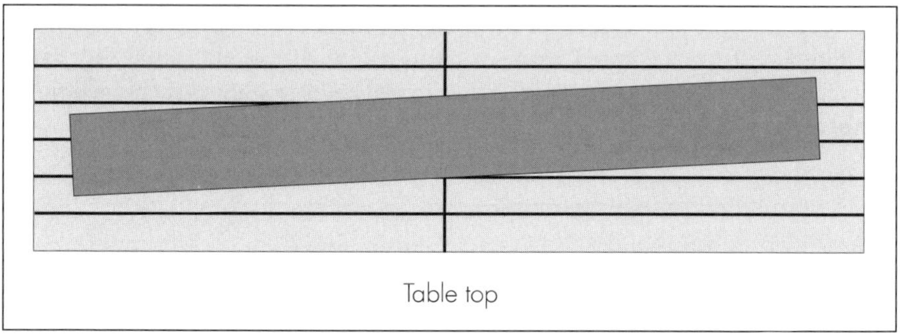

Figure 7-15. An angular misalignment problem with a very large workpiece.

Figure 7-16. Spindle probes can be used to help align workholding fixture. (Courtesy Marposs Corporation)

Though we intentionally exaggerate the problem in Figure 7-15, any small angular misalignment will, of course, cause machining operations to be out of sync with the outside surfaces of this workpiece. A spindle probe could easily touch the workpiece in two places along the *X* axis to determine the amount of angular misalignment (see Figure 7-16). The positioning movements in the program could then be rotated to allow for the misalignment (with the CNC feature called *coordinate system rotation*).

Inconsistencies in machining operations performed before the CNC cycle will cause the same problems as inconsistencies in raw material size and shape. A cutoff saw, for example, may be sawing round bar stock for a turning center chucking application. Any length deviation among the slugs being sawed can result in disaster during the CNC turning center operation.

In this case, a five-direction touch probe could be brought into contact with the very end of the workpiece to determine how much rough stock must be machined. The CNC program could automatically make just enough passes to efficiently machine this stock.

Cycle time warning. Be sure you exhaust all possibilities before you incorporate a probing system to compensate for problems that occur in every cycle. With these applications, the probing system will be required for every workpiece being run, and the probing time required to locate workpieces will be added to *every* cycle. In essence, workpiece loading time (which is actually part of cycle time) is being increased. If the need for probing is eliminated through ensuring raw material consistency or designing better workholding tooling, a great deal of production time may be saved, especially for large production quantities.

IN-PROCESS GAGING

Once workpieces are machined by a CNC machine tool, the CNC user must ensure that they have been machined correctly. By one means or another, inspection must be done to confirm the quality of machined parts. With the high degree of emphasis that has been placed on quality control in recent years, most manufacturing companies have even implemented statistical process control (SPC) to ensure the quality of the workpieces they machine. It is not uncommon, for example, for companies to check at least the most critical attributes of *every* workpiece machined. Our second category of application for probing systems, which relates to this kind of quality control, is *in-process gaging*.

A great deal of misconception, confusion, and controversy has surrounded the application of probing devices for in-process gaging. In many of my CNC courses, for example, I ask students if they are using probes for in-process gaging. Almost all probe users will say they are, but when we examine their specific applications, we commonly find that they are using their probe with another category of probing applications. In this discussion, we explain why there are misconceptions, attempt to clear up the confusion, and discuss the controversy surrounding the use of probing systems for in-process gaging. Then we cite specific applications.

Almost all in-process gaging applications for probing require five-direction touch probes or two-direction spindle probes (Figure 7-17). For this reason, many five-direction touch probes, including spindle probes, have been commonly referred to as in-process gaging probes, regardless of their true application. As we have seen, five-direction touch probes have many applications that facilitate setup. This notwithstanding, many CNC people, including some probe manufacturers, refer to five-direction touch probes as in-process gaging probes. This somewhat improper naming helps explain the misconception about the true task of in-process gaging.

Figure 7-17. Spindle probe used for in-process gaging. (Courtesy Marposs Corporation)

For CNC probing applications, we define in-process gaging as *the act of measuring one or more workpiece dimension on the CNC machine tool, after all machining operations are complete, yet during the CNC cycle, while they are still held in the workholding setup, for the purpose of quality control.* The key words in this definition are *measuring one or more workpiece dimensions.* If you are not measuring workpieces with your probing system for the purpose of quality control, you are not using your probe for in-process gaging in the true sense of the term.

The issues surrounding in-process gaging with touch probes

Some CNC users use touch probes for in-process gaging on a regular basis and swear by their use. Others claim that in-process gaging is just not feasible. As you may have guessed, the criteria that determine the wisdom of applying five-direction touch probes for in-process gaging are related to the end user's own *application*, and several factors are involved.

Cycle time required. One of the primary factors that detracts from the feasibility of performing in-process gaging with touch probes is that gaging time is *on line* (done during the CNC cycle). The higher the production quantities, the more time will be taken during the production run for measuring workpieces, assuming every workpiece is inspected. Many companies perform sampling inspections to minimize the production time required.

Generally speaking, it takes a current probing system about 2 seconds to probe one surface. Each workpiece attribute being measured requires a different number of touches. To find the center location in *XY* for a hole will take at least four touches (about 8 seconds). To measure a slot width will take two touches (about 4 seconds). To measure a pocket depth will take two touches, which translates into about 4 seconds. Even though current probes are relatively fast (2 seconds per touch is about as good as it gets at the present time), measuring time can add up fast if several workpiece attributes must be measured during the cycle. It is not uncommon for the measuring time to equal or even exceed machining time. Where cycle time reduction is a key objective, any on-line task that can be moved *off line* (like workpiece measuring) will effectively reduce cycle time. In many cases, a *postprocess gaging system* may be more feasible, since it will move the task of workpiece measuring off line.

There are those applications, however, where cycle time is not the most important issue. In these cases, in-process gaging may be a very practicable way of measuring workpieces. In unattended operation, for example, the most basic criteria may be running production without requiring a person to monitor the CNC cycle (such as during the night when there is no one in the company). Companies performing unattended work are usually more than willing to compromise cycle time to achieve the larger goal of unattended operation. In this case, not only will workpieces be machined automatically and unattended, but the workpiece inspection will be automatic and unattended as well.

Another instance when cycle time may be happily compromised to gain the benefits of in-process gaging is when workpieces are so large and bulky that they cannot be easily measured in any other way. Proponents of in-process gaging also will be quick to point out that if dimensions are found to be out of tolerance for these large workpieces, at least the workpiece is still being held in the setup. Chances are good that it can be repaired with a minimum of trouble and effort.

Accuracy of in-process gaging probes. Current probes are extremely accurate. As mentioned earlier, many will trigger within 20 millionths of an inch (0.00051 mm) deflection. However, the triggering of the probe is only the first step in the probing *cycle*. After triggering, the machine's position at the time of contact must be attained and stored. There will always be some lag time between the time the probe triggers and the time the machine's position can be attained. Additional factors like machine size and weight, workholding device weight, workpiece weight, probing feedrate, thermal deviation during warmup, and even the accuracy of the automatic tool changer placing the spindle probe into a machining center spindle can affect the accuracy and consistency of probing.

Probing technology is constantly improving—there are now calibration methods with current probing systems to minimize accuracy-related problems. However, since probes are made by probe manufacturers and CNC machine tools are made machine tool builders (two different entities), it can be difficult to get either to *guarantee* anything related to the accuracy of the overall probing system as it will be applied to your particular CNC machine tool. The best *real-world* accuracy claims (again claims, not guarantees) of current probing systems are about 0.0002 in. (0.0051 mm). While this may sound quite good, most tool designers would agree that *the accuracy of your measuring tool should be within 10% of the tolerance band you expect the measuring tool to measure*.

When using a probing system with an overall accuracy of 0.0002 in. (0.00051 mm), the best overall tolerance you should expect the probing system to measure is 0.002 in. (0.051 mm) (10% of 0.002 is 0.0002). However, 0.002 in. is not at all considered by most manufacturing people to be a very close tolerance. If you would like to measure a 0.0005 in. (0.0127 mm) overall tolerance, the accuracy of your measuring device (the probing system) should be *50 millionths of an inch* (0.00127 mm). Remember the 10% rule-of-thumb as you decide whether to apply touch probes to in-process gaging applications. Also remember: if a probing system is just barely able to meet your accuracy needs when it is new, the system may soon fail as the machine starts to show signs of wear.

Workpiece condition after machining. Any inspector would agree that before a workpiece dimension can be measured, it should be clean and free of burrs. It also should be in a relieved state (not under the stress of clamping). This, of course, is not usually the case in the CNC machine tool's working area when in-process gaging is done. It is likely that chips and coolant have built up around the workpiece to be measured, making for a poor measuring environment. For this reason, most in-process gaging systems incorporate some kind of workpiece cleaning system (possibly a high-pressure air blowing system) to clean chips and coolant from the workpiece.

In the case of machining centers, at least in-process gaging probes can be kept in the tool-changer magazine, out of the work area and free from chips and coolant during the machining portion of the cycle. With many turning centers, however, the in-process gaging probe is exposed to a very grimy environment during machining.

Another workpiece condition-related problem that could result in false measurements is the clamping stress a workpiece experiences during the machining cycle. After machining, if the workpiece remains clamped during measurements, it remains under a great deal of stress. If the workpiece is the least bit flimsy, it will spring when clamps are removed. Measurements taken while the workpiece is clamped will not be correct after the workpiece is released.

Using machine tools to measure. Some CNC people simply feel that if a machine tool did not machine a workpiece correctly, any measuring the machine tool does will be equally incorrect. The saying "the blind leading the blind" might aptly apply. However, during the rather violent dynamics of machining, it is possible that cutting tools deflect slightly due to tool pressure. Workpieces may bow slightly for the same reason. Even the bed of the machine will shift a small amount during powerful machining operations. Given the stable and static condition of the machine during measuring, it is possible for imperfections in the workpieces to be determined, even by the CNC machine tool that machined the workpiece. As long as the accuracy of the machine's in-process gaging system falls within 10% of the tolerances being held, using the machine tool to measure is entirely valid and feasible.

Workpiece attributes that can be measured

Bear in mind that probe manufacturers and machine tool builders are constantly striving to curtail or eliminate the limitations that have fueled the controversy discussed to this point. As time goes on and each limitation is overcome, applications for in-process gaging will become more and more commonplace. If you are not already doing so, it is likely that at some future date you will be working with touch probes for in-process gaging. Here we discuss the various workpiece attributes that can be measured with in-process gaging probes.

Dimensional tolerances. With a little ingenuity, in-process gaging probes can be made to check almost any dimension given on a workpiece print. In the following paragraphs we offer the techniques used for checking a few common types of workpiece attributes. While we show only a few simple examples, this discussion is intended to show the kind of ingenuity that goes into solving the problems related to developing in-process gaging probing programs.

Determing hole center position and diameter. In machining center applications where sidewalls of a hole are smooth (as with drilled, reamed, or bored—not tapped—holes), the center position and diameter of the hole can be determined with four touches of a ball stylus. With the probe stylus beginning near the center of the hole, the probing parametric program can first direct the stylus to touch the *X*– side of the hole and retain the surface value in a variable. The probe can then touch the *X*+ side and retain the surface value. The probing parametric program can then easily calculate the center position of the hole in the *X* axis. The process will then be repeated in the *Y* axis to determine the *Y* axis center position of the hole. Since the center of the stylus is perfectly aligned with the center of the hole in *X* during the *Y* axis center measurement, the probing program can easily calculate the hole diameter after the *Y* axis center is determined, as long as the stylus radius is known.

Slot or pocket depth. Measuring slot or pocket depth along the *Z* axis requires only two touches of the probe stylus. First, the top surface of the slot or pocket is touched and the surface position is retained in a variable. Next the bottom surface is touched and the surface position retained. Simple subtraction will render the pocket depth.

Slot or pocket width. In similar fashion, slot or pocket width can be measured with two touches of the probe. First, one sidewall is touched and the surface location is attained (along *X* or *Y*) and its value retained in a variable. Then the opposing sidewall is touched and the surface is attained. Again, simple arithmetic will render the slot or pocket width.

Groove width. Five-direction touch probes placed in the turret of a turning center can measure critical dimensions of turned workpieces. Groove width, for example, can be determined with two touches of the probe stylus. With the probe stylus near the center of the groove, first one side of the groove is touched and the corresponding *Z* surface position is placed in a variable. Then the opposing side of the groove is touched and the *Z*– surface position attained. Simple arithmetic will render the groove width.

Surface relationship (geometric) tolerances. Dimensional tolerances are not the only dimensions that can be measured by in-process gaging probes. Surface relationship tolerances (sometimes called *geometric* tolerances), like parallelism, concentricity, tapered surfaces, and angular relationships can be measured as well. In fact, some of the best applications for in-process gaging are related to surface relationship tolerance measuring, since they may be quite difficult to measure after the workpiece is removed from the machine.

What about the results of in-process gaging?

As we have seen, the actual techniques to take measurements with an in-process gaging probe are simple to understand which makes the system relatively easy to program. However, this is where the simplicity commonly ends for in-process gaging applications. The simplest in-process gaging systems merely report the results of measurements to the SPC system for quality tracking purposes. As you know from Chapter 5, most versions of parametric programming provide the capability to send data out through the communications (RS-232C) port. The SPC system computer simply retrieves this data and no change in the CNC process is made. You can think of this kind of system as any automatic gaging fixture that simply records SPC data. However, most CNC in-process gaging system users want more. They don't want just a passive recording system; they want the additional capability of making dynamic changes in the way the CNC machine behaves, based on the results of in-process gaging.

Consider the complexity of an in-process gaging program that must measure several dimensions. Each measurement will have its own criteria for what the gaging program should do based on probe findings. If, for example, a pocket depth is being measured, one of at least three conditions will result from the measurement. First, the slot may be to size and pass inspection, yet it may not be exactly in the middle of its tolerance band. In this case, it may be necessary to adjust a tool-length compensation offset value to make the slot in the next workpiece the proper size. Second, the slot depth may be too shallow. In this case, since the workpiece is not yet scrap, the in-process gaging program could immediately adjust the tool-length compensation value, recall the tool, and remachine the slot. Third, the probe may indicate that the slot is too deep. In that case, of course, the workpiece is scrap, and an alarm may be sounded to halt the production run.

Can you think of other possibilities? They are virtually unlimited. What if the slot is not even in the workpiece (maybe the milling cutter was damaged during machining)? What if the slot is precisely in the middle of its tolerance band? What if your company wants to incorporate more SPC logic, making adjustments to offsets only after a specified *trend* is completed? The "what ifs?" are endless and the range of possibilities is virtually unlimited—even for seemingly simple measurements. Given the likelihood of several measurements per workpiece, it is not difficult to see how in-process gaging program applications can quickly become quite complicated.

UTILITIES

The third probe application category is *utilities*. Probes can be used to accomplish a host of tasks you may never have previously considered. In many cases, they can actually perform monitoring functions a CNC operator normally handles, and in some cases they can facilitate unattended operation. In the following paragraphs we offer a few specific examples.

Automating trial machining

Many companies perform trial machining operations to ensure that the first workpiece in a production run is machined correctly. Trial machining commonly involves:

- Adjusting an offset in such a way that the corresponding tool will leave excess stock.
- Machining with the tool leaving excess stock.
- Halting the cycle to measure the workpiece.
- Making an adjustment (most often to an offset) based upon the measurement.
- Re-running the tool.

Sound familiar? These are common techniques many CNC users use on a regular basis when tolerances are quite critical. Most traditional trial machining techniques require a great deal of operator intervention. They also require a high level of skill from the person doing the trial machining. In many cases, a probing system can automate the process and eliminate the potential for operator error in measurement and offset entry.

Workpiece identification

In many flexible manufacturing systems, several unrelated workpieces may be in the system. Depending on the sophistication of the flexible manufacturing system itself, it may or may not be possible for the CNC machine tool to know which workpiece is to be machined next. In these systems, probes can be used as a kind of *workpiece identifier* (Figure 7-18). By knowing the identifying attributes of workpieces in the system, the probe programmer can direct the probe to touch one or more workpiece features to determine which workpiece is currently in the machine's work area. From the findings of the probe, the correct CNC machining program can be activated by the workpiece-identifying probing program.

Figure 7-18. Spindle probe used for workpiece identification.

Turning center users who machine workpieces in a close family may have similar problems with similar solutions. If, for instance, all workpieces of a given diameter are run consecutively to minimize setup time, at the completion of a given production run, a five-direction touch probe mounted in the turret can probe the length of the *first* workpiece in the next production run. Based on the workpiece length, the correct CNC machining program can be executed.

To check for chip clearance

Certain machining operations require that chips formed by prior operations be cleared from the machining area before continuing with subsequent operations. For totally unmanned operation, most companies will incorporate some form of chip clearing device. One common style is an air blowing system. By one means or another, the air blowing system will clear the chips from the drilled holes (hopefully). However, most chip clearing devices work with somewhat limited success. Tapping, for example, requires that chips be cleared from drilled holes before tapping can be done. In unattended applications, a probe can be brought into the area where chips may accumulate to determine if it is safe to proceed. If the probe does not contact chips along its probing command, it can be assumed that no chips block the way. If it is determined that chips are going to interfere with the next machining operation, possibly the chip blowing system (commonly a high-pressure air blast), could be commanded in an attempt to clear the chips. Or possibly an alarm would be sounded to call an operator to the machine.

To check for tool breakage

A probe can be used to check machining operations for tools that are prone to breakage. In many cases, if tools do break, other operations will be affected. If, for example, a tap drill breaks during the drilling of the first of 10 holes, the subsequent tap will be damaged when it is called upon to machine. A five-direction spindle probe can be used to ensure that the last drilled hole has been properly drilled prior to allowing the tapping operation.

For machining center applications, a tool-length measuring probe could then be used to confirm that the drill had truly broken. The tap drill could be (automatically) remeasured and its current length compared to its original length. If the tool is currently much shorter than it was when originally measured, it can be assumed that the tap drill is broken.

To help with tool maintenance

As mentioned during our discussion of setup help, tool-length measuring probes for machining centers and tool touch-off probes for turning centers enable the operator to quickly remeasure toolsetting offset values whenever tool maintenance is required. Especially helpful on turning centers (since tool measuring must commonly be done on line), these devices can dramatically streamline the on-line tasks related to tool *maintenance.*

Bear in mind that the same cutting-tool tasks needed during setup must be repeated as tool maintenance is required. When a machining center tool is dull and must be replaced, its new length must be measured. If done on line, the tool-length measuring probe can minimize the amount of time it will take.

HOW TOUCH PROBES WORK

Throughout this chapter, we have attempted to stimulate your interest in what touch probes can help you do. You should now have a fairly sound basic knowledge of the touch probe types and their applications. However, we have been speaking in rather general terms, and not much has been said about the physical functions of touch probes or how they are actually programmed. We now introduce the actual workings of touch probes, showing you much of what you need to know to begin programming them.

We first describe the basic design of touch probes, including the probe components, the types of probe signals, and special considerations for battery-powered probes. While probe manufacturers do differ in their specific design techniques, basic variations are few. We go on to discuss what actually happens as the probe comes into contact with a surface and define two important probing factors called *overshoot* and *droop*. A discussion follows on how probes must be *calibrated* to compensate for overshoot and droop, and, finally, we show the two most popular *probing cycles*.

TOUCH PROBE COMPONENTS

While you do not have to be a probe designer to effectively use a probing system, it helps to understand your probe's design if for no other reason than to learn how to care for it properly. Figure 7-19 shows the basic components of one popular probe type, the five-direction touch probe as it is equipped for placement in the spindle of a machining center (also called a spindle probe).

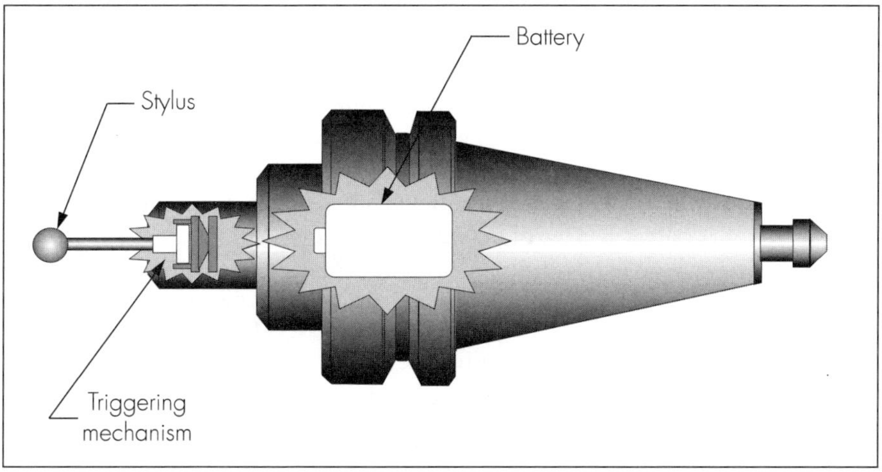

Figure 7-19. The major components of a five-direction touch probe used as a machining center spindle probe.

As shown in the figure, most current five-direction spindle probes incorporate an *integral shank,* meaning they have built-in adapters that accommodate the machine tool with as few separate components as possible. For spindle probes used in machining centers, the probe shank is the same configuration as that for any other tool held in the spindle (commonly V-flange tooling). This facilitates accurate probing, since there can be no deviation between the probe shank and the machine's tool holder. Five-direction touch probes used in turning centers commonly have a *straight shank* that adapts to the internal tool holder of the machine's turret.

The most important and delicate probe component is the triggering mechanism. Though the various probe manufacturers differ in the actual designs of their mechanisms, all probes are designed to detect the most minute amount of stylus deflection. The third major component is the *stylus*, which is the element that is actually brought into contact with surfaces being probed.

Stylus types

Spindle probe manufacturers know that the stylus is commonly exposed to extreme conditions. When a mishap occurs, the stylus is usually the first probe component to be damaged. For this reason (and since the probe stylus shank must be extremely rigid to ensure that the stylus will not bend during probing) many spindle probe manufacturers recommend that the stylus shank be made of ceramic. A ceramic stylus shank will not bend, and if there is a mishap which causes the stylus to crash into a surface, the shank will shatter before any damage can be done to the delicate probing mechanism. (Other crash protection functions for spindle probes will be discussed later in this chapter.)

The actual stylus shank shape can also vary. While most applications for spindle probes require a straight shank, there are applications for which the stylus shape must vary. For example, some turning center applications for five-direction touch probes require that the stylus be at a 90 degree angle to the probe itself. This type of stylus is commonly referred to as a *bent stylus*.

We emphasize that the contact surface of any touch probe is called a stylus. The contact surface of a tool-length measuring probe, for example, is called a stylus, though it in no way resembles the stylus of a five-direction touch probe. Similarly, each contact surface of a turning center's tool touch-off probe is called a stylus.

The actual contact surface shape and material also vary with application. Spindle probes commonly have a ball-shaped contact surface. This shape is easy to deal with during probing as are the calculations needed within the probing program. Some spindle probe applications require a cylindrical stylus. Tool touch-off probes and tool-length measuring probes generally have a smooth flat surface on each stylus contact surface.

The stylus contact surface material also varies with application. Though many are made of steel, more and more probe manufacturers recommend ruby or some other highly stable nonmetallic contact surface material. These materials preclude electricity generation (from two dissimilar metals) or discharge during stylus contact with metallic surfaces (from static electricity buildup). An electrical discharge can interfere with the triggering of some probe designs.

PROBE SIGNAL TRANSMISSION TYPES

As stated earlier, when a probe triggers, it sends a signal to the CNC control, telling the control to halt motion and to attain the machine's current positions (in all axes) as soon after triggering as possible. The method by which this signal is sent varies with the application and the probe manufacturer's design. Here we show the four most popular signal transmission types.

Hard-wired transmission

The most positive form of signal transmission is the hard-wired type. With this method, a permanent, consistent power source is always available to power the probe. Additionally, nothing short of a loose connector (or general power outage) will stand in the way of a successful signal transmission. While this is the most positive form of signal transmission, unfortunately, hard wiring can only be applied to probing systems when the probe itself remains relatively stationary throughout its life on the CNC machine tool. Probe types that can be easily hard-wired include tool-length measuring probes on machining centers and tool touch-off probes on turning centers.

Not all probe types can be easily hard-wired. A spindle probe, for example, must be stored in the tool changer magazine and called into the spindle only when needed. Since the probe must move with the automatic tool changer magazine, it would be difficult, if not impossible, to hard-wire. In similar manner, a five-direction touch probe held in a turning center turret would be difficult to hard-wire, given that the turret must be free to rotate in both directions as needed. If a probe cannot be hard-wired, a *battery* must be included within the probe to provide the power needed for signal transmission. We discuss specific considerations of battery powered probes a little later.

Inductive transmission

Using technology identical to a highly accurate form of CNC axis-positioning feedback system called *inductisyn scales*, inductive probe signal transmission incorporates an inductive (magnetic) transmitter to send the signal and an inductive sensor to receive the signal. While this is a highly positive signal transmission type (second only to hard-wired type), its major limitation is that the inductive sensor must be placed extremely close to the inductive transmitter for successful transmissions to occur.

If used with a five-direction touch probe mounted in the turret of a turning center, for example, the inductive sensor may be hard-mounted to the machine side of the turret, in a location from which wires can be easily connected to the

CNC control. This sensor must be placed close to the turret's working station so that when the turret is indexed to the station containing the five-direction touch probe, the inductive transmitter on the probe will swing into a position very close to the sensor. Figure 7-20 shows an example. Similar configurations could be designed for machining center applications for spindle probes. As the spindle probe is placed into the machine's spindle, the inductive transmitter on the probe shank is brought into proximity with the inductive sensor mounted and hard-wired to the spindle nose.

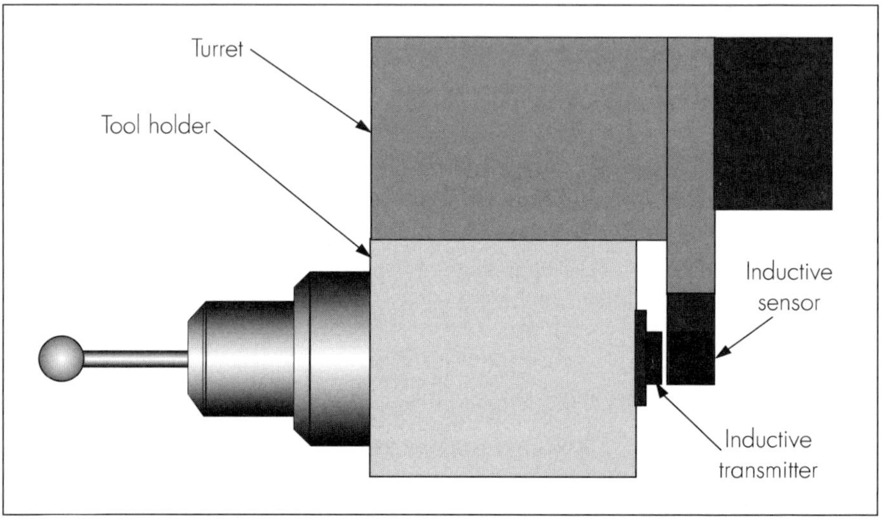

Figure 7-20. Inductive-type signal transmission used with a five-direction touch probe held in the turret of a turning center.

Infrared transmission

A signal transmission type that is growing in popularity employs the same technology found in most remote channel selectors for televisions, VCRs, and stereo systems. In the same basic fashion as inductive-type signal transmissions, an infrared transmitter is mounted within the probe. It must be *aimed* at the infrared receiver, just as you must point your television channel remote selector at the infrared receiver of the television. Like the inductive receiver, the infrared receiver is hard-wired to the CNC control.

The advantage of infrared over inductive transmission is the increased effective range provided (distance) between the sensor and the transmitter. With infrared, transmission can be accomplished along a *line of sight*. This makes it possible to place the probe sensor at a significant distance from the probe itself, which is most helpful with machining center spindle probe applications.

The major limitation of this kind of transmission detection is anything that presents a barrier to line of sight. The design of the probing system must be such that no object can obstruct the line of sight from the infrared transmitter to the infrared receiver. Some probe manufacturers equip their probing systems with two or more sets of transmitters and receivers to overcome this potential problem. Additionally, the infrared transmitter and receiver must be kept very clean to ensure accurate transmissions. Given the harsh and grimy working environment of a CNC machine tool, it is likely that the infrared devices will need cleaning on a regular basis.

Be aware that there may be other devices in the manufacturing environment that use infrared technology, meaning interference can cause problems when probing: thus, care must be taken when purchasing probing systems incorporating infrared. Automatic transfer vehicles (ATVs) used to move workpieces through the shop, for example, commonly use infrared technology.

Radio transmission

A final method of signal transmission is RF, or radio frequency. A radio transmitter mounted in the probe sends the transmission to a radio receiver mounted somewhere on or close to the machine tool. This form of transmission overcomes the line-of-sight problem associated with infrared. However, radio frequency interference caused by other machinery in the shop like welders, electrical discharge machines, and any other equipment with heavy electricity usage may present radio interference problems.

BATTERY CONSIDERATIONS

As stated, any transmission type except the hard-wired variety requires a battery in the probe to provide the energy needed to transmit signals. When batteries are used for this purpose, the probe manufacturer must incorporate safety measures to guard against battery failure.

First, we have been talking about the probe signal transmission as if the only time a signal is transmitted is when the probe is triggered. In reality, once activated, the probe is *constantly* communicating with the CNC control through transmissions. When a battery-driven probe is activated, the CNC control will expect to receive a *confirmation signal* soon after activation. This confirmation signal sent by the probe tells the CNC control that the probe is activated and ready (the battery is not dead). If for any reason the signal is not received, the machine will go into an alarm state, alerting the operator that something is wrong with the probe. This confirmation signal is *continuous*, meaning that if the battery fails at any time during probe usage, the battery failure alarm will be sounded.

A second battery-related consideration is *shelf life*. In many companies, a probe is only required during setups. It will sit idle for long periods of time during production runs. The longer the production run, the longer the probe will sit idle.

For this reason, and to reduce the amount of battery maintenance required, many probe manufacturers recommend using *lithium* batteries for their battery-driven probes.

OTHER PROBE PROTECTION SYSTEMS

In addition to the continuous confirmation signal used to confirm that the probe's battery is not dead, more and more probe manufacturers are designing other protection systems to keep the probe from being damaged in case of a mishap. It's important for the probe user to remember that these protection systems are designed to protect only the sensitive and delicate inner workings of the triggering system and stylus. If the shank of the probe is crashed into a workpiece or fixture, there will be no protection.

The most common probe protection system is built into the stylus deflection travel portion of the probing sequence. If the stylus deflects to an extreme degree (more than it should through normal use), it will send a kind of *overtravel signal* to the CNC control to halt motion, regardless of the kind of motion command currently instated, including the rapid mode. Most probe manufacturers claim that even if the probe is currently turned off when this kind of overtravel is detected, the probe will automatically self-activate and send an immediate signal to the machine's control to halt motion. This makes the probe stylus almost crash-proof, since the machine will always stop before damage to the stylus can occur.

UNDERSTANDING OVERSHOOT AND DROOP

As you know, when a touch-probe triggers, it sends a signal to the CNC control telling it to attain and store the machine's position as close to the instant of triggering as possible. Figure 7-21 illustrates what really happens during stylus contact with a probed surface.

While the illustration in Figure 7-21 exaggerates the problem to help you visualize what happens during stylus contact and probe triggering, there will always be some lag time from the instant the stylus makes contact to the time when the machine's position is attained and stored. This is true with *all* forms of touch probes, including spindle probes, tool touch-off probes and tool-length measuring probes. We call the amount of motion the probe makes during this period of lag time *overshoot*. This overshoot-causing time lag is more a CNC control problem than it is a probe system problem—and CNC controls vary dramatically with regard to how quickly the machine's position can be attained and stored during stylus contact.

With older controls (those more than 10 years old), it is likely that the machine must come to a *complete stop* before the machine's position can be attained after probe triggering. For this reason, overshoot with early probing systems is excessive. *The greater the overshoot, the more potential there will be for inconsistencies in overshoot, and the more difficult it will be to perform consistent, accurate*

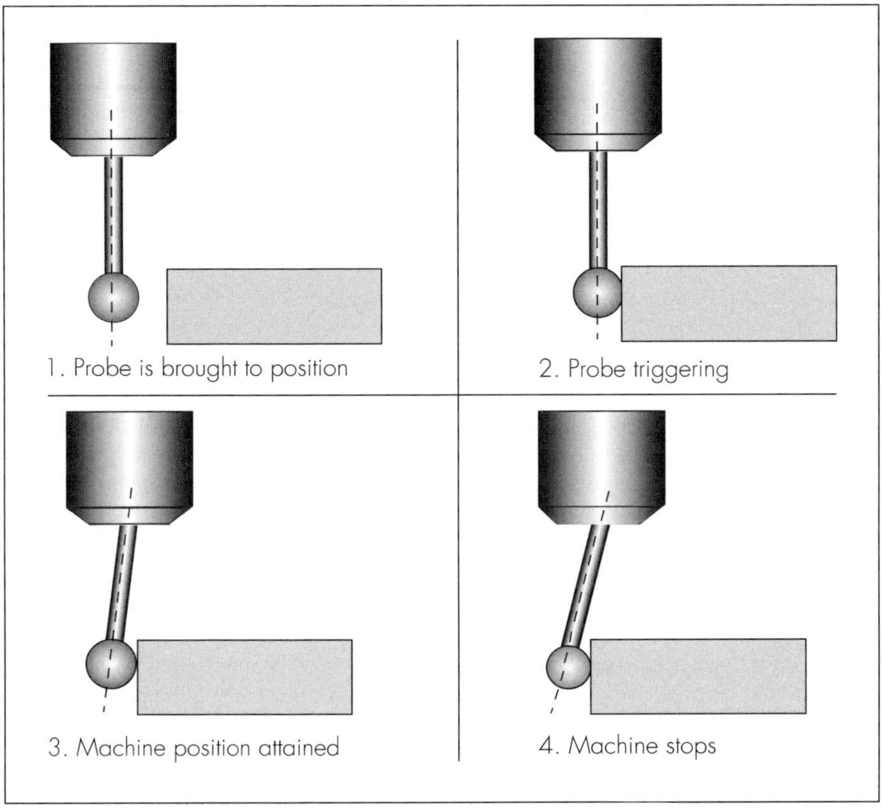

Figure 7-21. Sequence of events that occur during stylus contact with a probed surface.

probing. As you evaluate new CNC equipment for purchase, and probing systems for that equipment, you should look into this very important CNC control function. Generally speaking, digital-axis drive systems allow shorter lag times than analog axis drives.

Just because you are purchasing a new machine, you cannot assume your CNC control has short lag time for probing purposes. Even some current controls take about 10 milliseconds (10 thousandths of a second) from the instant the probe triggers to the time machine's position can be attained. At a probing feedrate of 30 in. (762 mm) per minute (considered a little slow by current standards), the overshoot during 10 milliseconds will be 0.005 in. (0.127 mm). At a more desirable feedrate of 80 in. (2032 mm) per minute (F80.0), the overshoot during 10 milliseconds will be 0.013 in. (0.330 mm). The greater the degree of overshoot, the greater the potential for *variation* in overshoot. And variations in overshoot will lead to inconsistent, inaccurate probing.

The best current lag time claims from one popular CNC control manufacturer is 2 *microseconds*. That's 2 millionths of a second. With this extremely short lag time, it is truly as if the CNC control can take a snapshot of the machine's position the very instant the probe is triggered. Even at the extremely excessive feedrate of 1,000 in. (25,400 mm) per minute (ridiculously fast for current probing systems) the machine can only move 0.000033 in. (33 millionths of an inch [0.00084 mm]) during 2 microseconds. With more realistic probing feedrates (up to 150 in. [3810 mm] per minute), CNC controls with only 2 microseconds of lag time virtually *eliminate* problems caused by overshoot.

The larger the overshoot, the more other machine-related factors can affect the accuracy of probing. Machine size and weight, probing feedrate, workpiece size and weight, and even the machine's thermal deviation during warmup can cause overshoot values to vary during the production run. As stated, variations in overshoot lead to inaccurate probing, and this is especially troublesome with older controls that must wait until the machine comes to a complete stop before attaining and storing the machine's position. About the only way to minimize the problems caused by varying overshoot amount is to slow down the probing feedrate, which in turn increases probing time. Fortunately, as long as overshoot remains consistent, it can be easily compensated for through probe *calibration*.

Understanding droop

While overshoot must be considered in every form of probing system, *droop* affects only spindle probes. Droop is the deviation in *X* and *Y* from the spindle center line to the center line of the stylus ball. Figure 7-22 illustrates droop.

Droop results from a variety of causes. If the stylus center line is not perfectly aligned with the spindle center line, droop results. If the stylus is slightly bent, droop results. If the probe or stylus shank is not perfectly aligned with the spindle center line, droop results. Whatever the reason for droop, if not taken into account, it will have a negative impact on the accuracy of probing.

Like overshoot, droop can be compensated for through proper probe calibration. However, be aware that *droop must remain consistent* after calibration and from one probing session to the next. Probe manufacturers have two ways to ensure the consistency of droop. First, if the spindle probe is a five-direction touch probe, *the spindle must be positioned to its orient position* (commonly with the spindle orientation command, M19) prior to calibration and prior to any probing session. As long as the probe is consistently loaded, spindle orientation will ensure that any droop compensated during calibration will remain consistent whenever probing is done.

There is one more concern that spindle probe users must be wary of. Remember that some machining centers allow V-flange tooling to be loaded into the automatic tool changer magazine and spindle in two orientations (180 degrees apart). When a spindle probe is used with this kind of machine, the probe user must confirm that it is *always* loaded into the spindle and magazine *in the same manner*, during calibration and during every subsequent probing session.

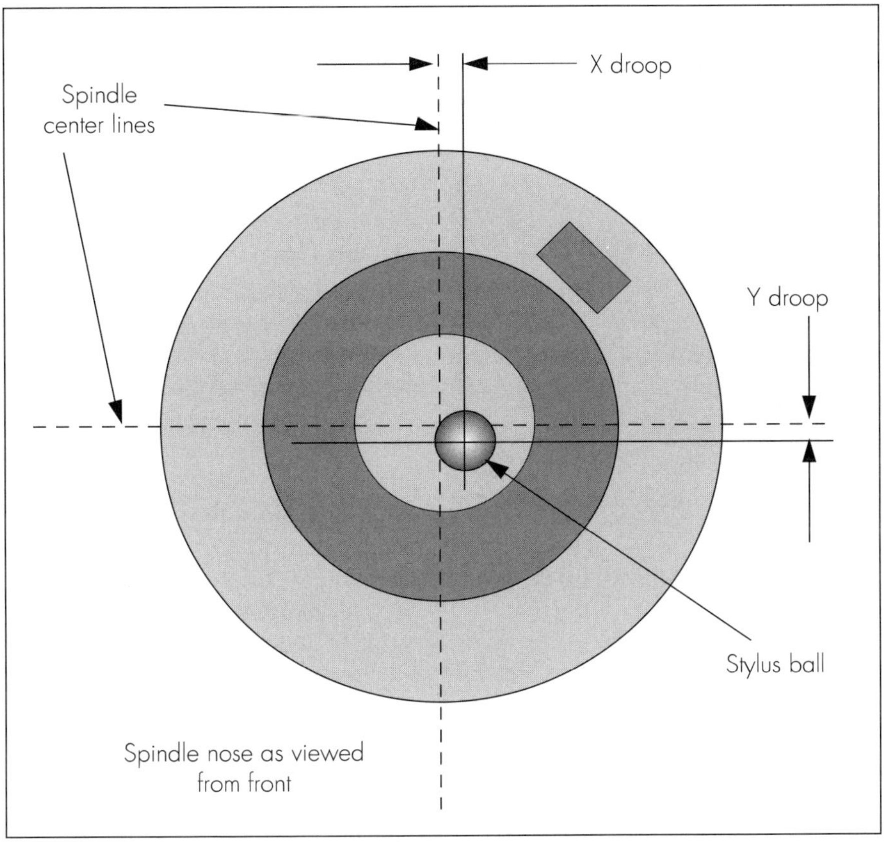

Figure 7-22. Drawing illustrates droop—the deviation in X and Y from the spindle center line to the stylus center line.

Second, not all spindle probes are five-direction touch probes. Some only require probing in two directions (*Z*– and along the *XY* plane). When *X* or *Y* probing is required, the spindle is programmed to orient to the correct angular position to allow the stylus to contact the surface to be probed. By rotating the stylus into the proper probing position, droop will remain consistent for every stylus touch and can be easily compensated for through calibration.

How calibration compensates for overshoot and droop

All probe manufacturers have their own special probe calibration methods, and it is not the intention of this text to have you stray from their methods. Rather it is our intention to help you understand the methods they may be using. You must be prepared for variations. Calibration usually involves using a special fixture as well one or more pre-written parametric programs. The fixture and programs are supplied by the probe manufacturer. For this reason, we discuss the topic in only general terms. While in Chapter 9 we do actually describe one method of probe calibration for each of the probe types, our intention is more to show you *why* probes must be calibrated than *how* they are actually calibrated.

Figure 7-23. Calibration fixture used for the calibration of spindle probes. (Courtesy Marposs Corporation)

In all cases, calibration is necessary to determine values that consider overshoot. These values may be the actual overshoot amounts (one for each direction of probing), or they may be system constants representing measured distances from the machine's reference position to probe stylus surfaces. With five-direction spindle probes (Figure 7-23), calibration is also done for the purpose of determining droop amounts. Most five-direction spindle probe manufacturers utilize just one value to incorporate both overshoot and droop for a given probing direction.

To be able to write your own probing programs, you must understand how your own probing system is calibrated. Though the calibration functions described in Chapter 9 should help you understand probe calibration, we must bow to your probe manufacturer to explain the specific techniques needed to calibrate your probing system. With an understanding of our discussion, you should be able to adapt what you learn to your own probe's recommended calibration methods.

All forms of probe calibration will render at least one calibration value that must be used when actually probing. These calibration values must be permanently stored and available for use by probing programs. As you know from part one, permanent common variables in custom macro B are nonvolatile, and most probe manufacturers use them for calibration value storage when custom macro B is the parametric programming language. In this sense, permanent common variables used for calibration values are a form of *system constant*. With Okuma's user task 2, a special set of system variables are set aside for probe calibration values, as is the case with Fadal's macro.

As stated, calibration is most commonly done to determine overshoot and droop values, and the actual method of calibration will vary dramatically from one probe type to another. However, *the method of calibration will always affect the method by which probing must be done*. Stated another way, how probing programs must be written requires a thorough understanding of how calibration is done. You must have a very good understanding of your own probe's calibration methods before you will be able to write probing programs.

With spindle probes, for example, you will at least have to know which permanent common variables are used to store the various calibration values. For tool-length measuring probes, you must also know what the probing system is measuring, the tool's length, or the distance from the tool tip at the machine's reference position to the program zero point. For tool touch-off probes, you must know how geometry offsets are being used.

Though you must be prepared for variations in the way your own probing system functions, we will use the calibration methods given in this section for all example probing programs presented in this text (unless otherwise specified). Armed with a firm understanding of our methods of probe calibration, and with an understanding of one way to handle your own specific probing problems (shown through example probing programs later in this text), you should be able to adapt to your own probing system with relative ease.

The probing cycle must reflect expected accuracy and the overshoot amount

As stated, calibration is performed to compensate for overshoot and droop. Similarly, the actual techniques used when probing must consider *how much* overshoot is experienced in the probing system. We refer to how the probe attains a given surface location as a *probing cycle*. With a spindle probe, for example, the probing cycle includes everything that happens from the positioning of the probe stylus to an approach position until the stylus is retracted from the surface position after the surface position is determined.

Remember, the greater the overshoot amount, the more potential there will be for variations in overshoot, and these variations lead to inaccurate probing. Generally speaking, probe manufacturers recommend using one of three probing cycles, depending on the accuracy you expect and the typical amount of overshoot experienced in the overall probing system. Of course, the probing cycle used during the probe's calibration must be used for actually probing surfaces during the probe's use. This means that the choice of which probing cycle to use is an important one. The appropriate probing cycle is commonly recommended by the probe manufacturer.

The one-touch probing cycle. As the name implies, this cycle requires only one stylus contact, meaning probing time can be kept to a minimum. The approach position is kept as close as possible to the probe surface to minimize the travel distance prior to probe triggering. Contact is done at a relatively fast feedrate, again, to minimize probing time. For our example programs, we use a feedrate of 30.0 in. (762 mm) per minute. Using this cycle, probing time can be kept to less than 2 seconds per cycle.

This probing cycle is recommended if you are working with a probing system that has a lag time of only 2 microseconds (digital-axis drives), since the problem of overshoot is all but eliminated. Even when you expect very accurate probing, the one-touch probing cycle can be used, even with very fast probing feedrates. With probing systems having longer lag times of over, say, 5 milliseconds, the one-touch probing cycle should be used only if probing accuracy is not critical. Even if lag time is longer than 2 milliseconds, for example, the one-touch probing cycle is commonly the cycle of choice with tool-length measuring probes and spindle probes used for measuring program zero locations.

The two-touch probing cycle. With this cycle, the first probe touch will be done at a very fast feedrate (80.0 in. [2032 mm] per minute). The probe (or cutting tool) will be retracted slightly, just more than the overshoot amount (about 0.010 in. [0.254 mm]). The second probe touch will be done at a very slow feedrate (3.0 in. [76.2 mm] per minute). For probing systems with longer lag times (more than 2 milliseconds), this probing cycle will render relatively fast and accurate probing results.

Taking the average of several touches. Early probing systems require that the machine axes come to a complete stop before the machine's position can be attained. With these older probing systems, overshoot can be excessive, as is variation in overshoot. For this reason, about the only way to render consistent probing is to average surface location over several touches. As you can surmise, this method will require much probing time and should only be used as a last resort.

Chapter 8

CNC Commands Used with Probing

This book is written on the assumption that you already have a basic understanding of G-code level manual programming and that you can write programs for the CNC machine types with which you are working. In this chapter, we introduce all CNC commands that play a role (even a minor one) in the use of a probing system. While several of the commands in this chapter may be new to you, you may be surprised at how many you already know that have slightly different implications when probing.

We begin by introducing the standard manual programming commands used for probing (most are G-codes and M-codes). We then show the additional probing commands available through parametric programming. As mentioned in the last chapter, we show only the custom macro B commands to minimize the length of the text. With a firm grasp of the material in Part One, you should be able to convert this information to any version of parametric programming. Generally speaking, most custom macro B commands offered in this chapter are related to *system variables*. We end this discussion by summarizing all probing commands, which should provide you with an excellent quick and easy reference to probing commands. Finally, we introduce specific probe calibration programs and discuss the importance of understanding your own calibration programs so that you can write your own probing programs.

NORMAL G-CODE LEVEL COMMANDS

Here we introduce those commands that are considered part of standard manual programming. Though you will find these commands documented in your control manufacturer's programming manual, our focus here is on their specific implications for probing. Note that we begin with those commands that are common to all types of probing systems and continue with a discussion of those that are specific to one form of probing system.

SKIP SIGNAL COMMAND (G31)

The *skip signal command*, specified by G31 on most controls, is the heart of the probing cycle. It is the command that tells the CNC control to keep the machine's axes moving until probe triggering occurs. When the control receives the triggering signal, it will halt axis motion and cause the balance of the motion command to be skipped. Hence the name *skip signal.*

You can think of the G31 command as being very similar to a G01 straight-line motion command. When G31 is commanded, for example, the motion generated will be in the form of a perfectly straight line (just like G01). Additionally, the machine will move at the programmed feedrate (programmed with the F word). And just as with G01, the current programming mode (absolute or incremental) affects the programmed departure. In incremental mode, the motion commanded is the departure direction and distance. In absolute mode, the motion's end point is specified relative to the program zero point.

Aside from halting axis motion and skipping the balance of the motion command at probe triggering, G31 differs from G01 in one important way. When probe triggering occurs, it will cause the machine's current position, relative to the current program zero point of the current coordinate system (the machine's current absolute position) to be automatically stored. For three-axis machining centers, three system variables are involved (#5061 for *X*, #5062 for *Y*, and #5063 for *Z*). For two-axis turning centers, two system variables are involved (#5061 for *X* and #5062 for *Z*). We discuss these system variables in greater detail later in this chapter.

Remember that the current feedrate mode (feed-per-minute and feed-per-revolution) also applies to G31, just it does for G01. Since almost all machining centers are programmed exclusively in the feed-per-minute mode (inches or millimeters per minute), and since this feedrate mode is almost always initialized at powerup, there is no need for a special command to invoke the feed-per-minute mode when programming probes for machining center applications.

On the other hand, turning centers have *two feedrate modes*, and are most often programmed in the *feed-per-revolution* mode (commonly selected with a G99 command). Additionally, the feed-per-revolution mode is initialized at powerup, meaning the turning center will assume the feed-per-revolution mode if no feedrate mode is selected. For all turning center probing applications, feedrate *must* be specified in the inches-per-minute mode, since probing is done with the spindle stopped. You must select the feed-per-minute mode (with a G98 on most turning center controls) prior to probing. You also must remember to reselect the feed-per-revolution mode (G99) when finished with the probing cycle to ensure that subsequent machining operations will be done in the correct feedrate mode.

If you have a machine that incorporates *digital-axis drives* (as opposed to analog drives), you may have a different probing command. With some Fanuc controls, G31.1, called the *high-speed skip-signal* command, tells the control to take advantage of the very fast *sweep time* of the control and minimize the lag time prior to attaining the machine's position. You must confirm which of these two skip-cutting commands your control uses by checking your probe manufacturer's documentation.

Examples of the skip-signal command

Figure 8-1 shows an example of the skip-signal command for the probe types addressed by this text. Notice that, in all cases, the probe stylus or cutting tool is relatively close to the probing surface. This distance must reflect your confidence in the stylus/tool position. As when approaching any unqualified surface (like a cast surface) with a cutting tool in a normal CNC program, approach distance must be increased when you are unsure of the surface location.

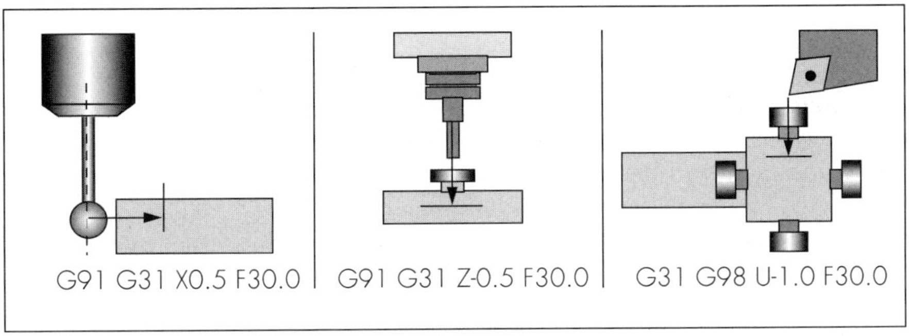

Figure 8-1. Examples of skip-signal command with three types of touch probes.

It is often convenient to use the incremental mode to specify departure direction and distance. This is especially helpful for those times when the probe is manually positioned prior to activating the probing program, since you will have no idea of the probe stylus or surface location in the absolute mode. For almost all machining center controls, incremental mode is specified with G91. Though there are some turning center controls that also use G91 to specify incremental mode, most turning center controls use U and W to specify incremental departures. The letter U specifies an incremental *X* departure while W specifies an incremental *Z* departure. Note that with most controls, incremental *X* motion means a diameter change. In the inch mode, for example, U1.0 will cause 0.5 in. (12.7 mm) of actual motion in the positive *X* direction.

And finally, notice that the departure distance for each example ensures that stylus contact will occur long before the departure can be completed. Though this may take some getting used to (it may feel as if you are intentionally crashing the stylus into each surface), this is how stylus triggering is done. As mentioned earlier, to confirm that overshoot remains consistent, the programmed feedrate in the skip-cutting command used in probing programs must match the feedrate used for calibration.

Do *not* forget the G31 word in your probing command. Since the departure distance specified within this command will intentionally drive the stylus/tool well into the surface being probed, forgetting the G31 could be disastrous, especially for older probing systems. As mentione . in the previous chapter, most newer probes have a kind of overtravel protection that will sound an alarm if excessive deflection is detected (possibly even if the probe is turned off). However, damage will occur to probes without this kind of protection. At the very least, the probe stylus will bend or break.

RAPID AND STRAIGHT-LINE MOTION COMMANDS

The G00 in normal CNC programs is commonly used to minimize positioning time as tools approach and retract to and from machined surfaces. While G00 is still used with probing systems, you must use it carefully. We recommend minimizing its use until you have proven your probing program, especially when using a spindle probe for setup help-related applications when an operator must perform tasks cautiously. Allowing rapid motion during these cautionary times is dangerous.

Instead of using G00 for probe positioning movements, you can use a G01 straight-line cutting command with a feedrate that is faster than normal, say 50 in. (1270 mm) per minute (F50.0). An additional advantage of using this technique with probing programs that require manual intervention is that the operator can control the rate of *all* positioning movements with *feedrate override*.

In cases where probing is done on every workpiece, or when cycle time is critical (as is often the case with in-process gaging applications), or when you are absolutely certain that your probing programs are correct, the rapid-motion command can be used for positioning movements.

OFFSET SETTING BY PROGRAMMED COMMAND

As you know from our discussion of probe applications in Chapter 7, in many instances, a programmer must have access to offsets when probing. A tool-length measuring probe, for example, must set the tool-length compensation offset for each tool after measurement. While there is a series of *system variables* that also allow you access to tool offsets, and we normally recommend using them instead of G10 (since system variables allow you to write *and read* offset values), there are times when the G10 may be more convenient to use than offset-related system variables. For example, one G10 command allows the entry of all values (*X*, *Y*, and *Z*) for a given fixture offset. Using system variables to accomplish the same task requires three separate commands.

The actual use of G10 varies from machining centers to turning centers, and among control manufacturers and even among control *models* within a given control manufacturer's product line. While we offer a few examples of G10 to help you understand how it can help when probing, you must read your control's programming manual to learn more about how the offset setting by programmed command is used for your control. Note that G10 is *nonmodal*. It must be included in every offset setting command.

Machining center examples of G10

For most controls, an *L word* in the G10 command tells the control what kind of offset is being set. For one popular control model, L1 specifies that simple tool offsets used for tool-length and cutter-radius compensation are being set. (If the L word is omitted, by the way, L1 is assumed.) On the other hand, L2 specifies that fixture offsets are being set. A *P word* in the G10 command specifies the offset number of the offset being set.

For tool offsets, an *R word* specifies the amount of offset. For fixture offsets, the corresponding axis designations are used (*X*, *Y*, and *Z*). Keep in mind that the amount of offset specified in the G10 command is influenced by the departure mode (incremental or absolute). If in the incremental mode, the current value in the offset is changed (incrementally) by the value specified in the G10 command. If in the absolute mode, the current value of the offset is *overwritten* by the value in the G10 command. Here are a few examples that should help you understand the G10 command as it is used for machining centers.

N1 G90 G10 L1 P1 R6.3366 (Set offset number one to 6.3366 in. [160.95 mm])
N2 G91 G10 L1 P31 R0.005 (Increase offset number 31 by 0.005 in. [0.127 mm])
N3 G90 G10 L2 P1 X – 12.2454 Y – 10.2345 Z – 9.2139 (Set fixture offset number one to values given in *X*, *Y*, and *Z*)

Turning center examples of G10

An *L word* in the G10 command also specifies the type of offset being set with turning center controls. With Fanuc controls, the choices will be for setting wear offsets, geometry offsets, and work shift offsets. For one popular control model, L1 specifies wear offsets and will be assumed if the L word is omitted, L2 specifies geometry offsets, and L3 specifies work shift offsets. A P word still tells the control the offset number to be set. The axis name (*X* or *Z*) is used to specify the actual value for wear and geometry offsets. An R word is used to specify the tool radius and a T word is used to specify the tool type for tool-nose radius compensation. The points just made about incremental/absolute still apply (remember, many turning centers use U and W to specify incremental values).

N1 G10 L2 P1 X – 12.2544 Z – 6.4422 (Set geometry offset number one)
N2 G10 L1 P1 U – 0.0021 (Decrease the value of wear offset number one by 0.0021 in. [0.0533 mm])
N3 G10 L3 Z – 2.0298 (Set the work shift value in *Z*)

A final note about G10: All examples shown above use hard and fixed values for each offset entry. However, any actual value of any word used in the G10 (L, P, R, *X*, *Y*, etc.) can, of course, be specified with a *variable*. This allows the probe programmer to adjust the value of an offset based on the findings of the probe. The actual offset value being set can (and probably will) change every time the probe is used.

INCH AND METRIC MODES

Just as with any normal CNC program, G20 and G21 allow the probing programmer to select the input unit for probing. And, generally speaking, the same input unit used for machining should be used for probing. If you machine a workpiece in the inch mode, use the inch mode for probing. If you use the metric mode for machining, use the metric mode for probing.

You must understand, however, that there is an advantage in accuracy level to using the metric mode—for both machining and probing. The advantage has to do with the *least input increment* of the input unit you are allowed to program. For most CNC controls, the least input unit in the inch mode is 0.0001 in. In the metric mode, the least input increment is 0.001 mm. The measure 0.001 mm is less than half the size of 0.0001 in. (one thousandth of a millimeter is actually 0.000039 in.). Figure 8-2 shows the relationship (to scale) of inch versus metric least input increment.

As you can see, the metric mode allows a much finer resolution than the inch mode. This has important implications when probing, since machine positions *attained* during probing in the metric mode will reflect the improved resolution. If you have ever worked with a graphic-type computer application program (draw-

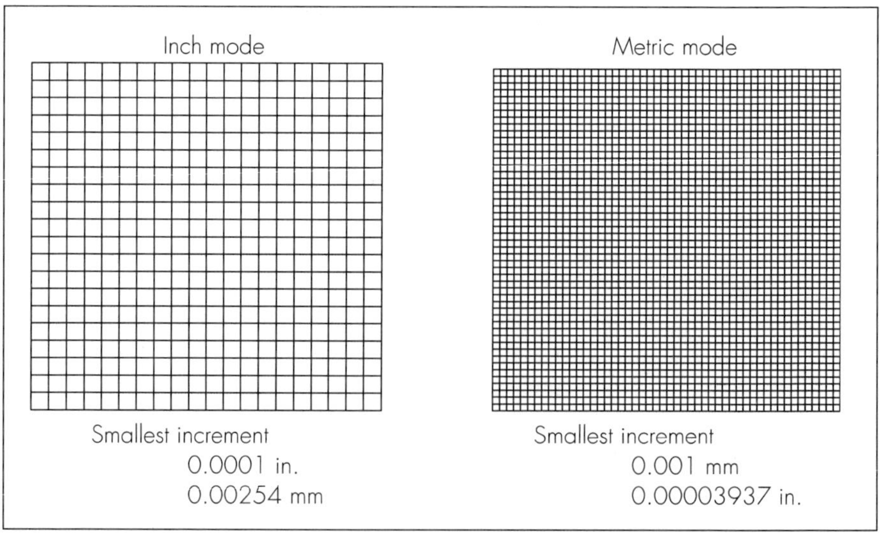

Figure 8-2. Note the finer resolution allowed by the metric mode with most CNC machines.

ing program, CAD system, paint program, etc.) that has a *snap-to-grid* function, you should be able to easily understand this resolution-related advantage of the metric mode.

With a drawing program, the snap-to-grid function will confirm that the cursor position selected by your computer's mouse will always fall on a grid. This is great for making sure that objects in your drawing are kept in perfect alignment. The larger the grid increment value, the easier it will be to see the snapping effect this function has on objects placed in your drawing.

When probing on a CNC machine tool, the machine's position is attained soon after triggering and the position values are stored in a series of system variables, one for each axis. The preciseness of each stored position is directly affected by the input unit you are using. In the inch mode, it will be as if your computer paint program's *snap-to-grid* function is set to a rather large value, since the CNC control must round the position value to the closest 0.0001 in. grid position. In the metric mode, it will be as if the drawing program's *snap-to-grid* function is set to less than half of what it is in the inch mode, since the control will round the position value to the closest 0.001 mm grid position (0.001 mm is less than half of 0.0001 in.).

If you have been working exclusively in the inch mode, it may be somewhat difficult to make the adjustment to working in the metric mode. For this reason, and since most people in the U.S. understand the inch mode as well as or better than metric mode, all examples given in this text are shown in the inch mode first, followed in parentheses by the metric conversion. However, if you expect to perform highly accurate probing (especially required when performing in-process gaging), metric should be your input unit mode of choice.

TOOL-LENGTH COMPENSATION CANCEL COMMAND

The value of system variable #5063 (the machine's *Z*-axis position after probing) is affected by tool-length compensation. That is, the value of #5063 will differ based on whether tool-length compensation is instated or not. If tool-length compensation is instated, the value of #5063 will reflect the stylus tip position. If not, #5063 will reflect the spindle nose position. For this reason, it is often necessary to ensure that tool-length compensation is canceled (with G49) prior to probing on machining centers. This is commonly necessary with spindle probes (especially with manually-activated spindle probing programs) as well as when using tool-length measuring probes to determine each tool's tool-length compensation value.

COORDINATE MANIPULATION COMMANDS

Though we minimize the use of coordinate manipulation commands for the sake of clarity, you must know that certain of these commands such as mirror image, axis exchange, axis rotation, and scaling can be used to minimize the number of commands required in a probing program. For example, the same series of spindle probing commands used to find the location of the lower left-hand corner of a rectangular workpiece in *X* and *Y* can be manipulated with mirror image and axis exchange to find any other corner of the rectangular workpiece. This will shorten the probing program to about one quarter the size of the program if coordinate manipulation commands were not used.

Since probe manufacturers are highly concerned with the length of their probing programs (as are many probe users), they tend to heavily favor the use of these coordinate manipulation commands to shorten their probing programs. However, it can be difficult to understand what is happening in those programs that use these functions, especially for beginning probe programmers. For the sake of clarity, we urge you to avoid use of these functions unless program storage capacity is a highly important concern. Even then, wait until you have a few applications under your belt before you start using them.

CUSTOM MACRO B CALLING COMMANDS

These commands (G65, G66, and G67) were well covered in Part One. We simply wish to restate the importance of understanding parametric programming for writing probing programs. The more you know about parametric programming, the easier it will be to write probing programs.

As with family-of-parts and user-created canned cycles, there are two common ways to get data into the parametric program. Generally speaking, if your probing program is to be totally automatic (no operator intervention), it must be called by your machining (main) program. In this case, you will want to treat your probing custom macro program like a user-created canned cycle, activating it with a G65 command that includes a series of arguments. On the other hand, if you will be expecting the operator or setup person to perform some manual functions prior to probing (machine positioning, probe loading, etc.), you will treat your probing program like a family-of-parts application, and include all arguments at the beginning of your probing program. Common variables are normally used for this purpose.

M-CODES USED WITH PROBING

All probing systems use M-codes to control functions related to the probe. For example, the probe must be activated by one means or another. M-codes are commonly used for this purpose. Some spindle probes, for example, use M13 to activate the probe and M14 to turn it off. Others require that the spindle be started for probe activation (with M03). Tool-length measuring probes commonly have one M-code to swing the probe into position (which also turns on the probe) and another to get it out of the working area. In similar fashion, turning center tool touch-off probes incorporate M-codes to swing the probe into and out of the probing position.

Unfortunately, we cannot be specific about the use of the actual M-code numbers used for probes. As is the case with standard CNC programming, machine tool builders differ dramatically with regard to which M-code numbers they use to control specific machine functions. M-codes used with probing systems vary in the same manner. While we have attempted to make each probing example as realistic as possible, and we use specific M-codes in our examples, you must be prepared for variations. You will need to confirm the M-codes available with your own probe by checking in your probing system's documentation. Following are the M-codes we use throughout this text, along with their meanings. Again, importantly, these are *not* universal. You will have to make substitutions based on your own probe's related M-codes. But at the very least this information should let you know what to look for.

M-codes related to tool-length measuring probes

M04 and M05—Spindle on counterclockwise and spindle stop. Many tool-length measuring probes have the additional ability to measure a milling cutter's radius. These probes require that the spindle be turned on (in reverse) prior to measuring a tool's radius. The reverse spindle direction ensures that the cutting tool will not actually machine the probe stylus. Once finished, of course the spindle must be stopped (with M05).

M13—Swing probe into position and activate probe.
M14—Swing probe out of position and turn off.

M-codes related to turning center tool touch-off probes

M13—Swing probe into position and activate.
M14—Swing probe out of position and turn off.

M-codes related to spindle probes

M13—Activate probe. (Note that some spindle probes require that the spindle be turned on with M03 in order to activate. Of course, the spindle must be turned off and oriented prior to probing.)

M14—Turn probe off. (Notice that some spindle probes incorporate a time delay to turn off the probe. After a given length of non-use time, they automatically turn off.)

M19—Spindle orient (used to ensure that droop remains consistent).

M03—Spindle on clockwise (used to activate some spindle probes).

M15—Spindle off.

COMMANDS USED SPECIFICALLY WITH TURNING CENTER TOOL TOUCH-OFF PROBES

Tool touch-off probes have just a few more commands to help with their use.

Program zero assignment command

As you know, when the probe triggers, the control stores the machine's current position and halts motion. The position stored will be relative to the *current program zero point*. It is sometimes helpful to modify the location of program zero in the probing program to set a logical point of reference. The G50 command (G92 on some turning center controls) allows the programmer to specify the distance from the desired program zero point to the current machine position in *X* and *Z*. For example, the command

```
G50 X10. Z10.
```

will set the program zero point 10 in. (254 mm) from the machine's current position in both axes (*X* is a diameter specification). There is more discussion on this command as we show specific applications that require its use.

Feedrate mode selection

As mentioned earlier, feedrate during any probing operation must be done in the *per-minute* mode. With many turning center controls, G98 is used to select this mode. And before ending the probing program, you should place the machine back in the per-revolution mode (with G99 on most controls) for subsequent machining operations.

COMMANDS USED SPECIFICALLY WITH SPINDLE PROBES

There are a few more points to be made about commands related to spindle probes.

Spindle orientation

As mentioned earlier, a spindle orientation must be done (with M19) with five-direction spindle probes to ensure that droop will remain consistent for all probing. Remember that some machine spindles have a neutral setting to allow easy (manual) rotation of the spindle. For M19 to function, the spindle cannot be in neutral; it must be in an actual spindle range. If the machine is in neutral when M19 is commanded, an alarm will sound, halting the program's activation. For this reason, we recommend including an S word (like S2000 shown below) with the M19 command to ensure that a spindle range is selected. To ensure a minimum of spindle vibration due to harmonics, most probe manufacturers recommend using the high spindle range for the orientation.

```
S2000 M19 (Select spindle range and orient spindle)
```

Fixture offset commands

Commands G54 through G59 in many controls select the desired fixture offset, which in turn assigns the desired program zero location. Since most spindle probe applications require a point of reference for motion commands (program zero point), it can be quite helpful to assign a fixture offset within your probing program prior to actually probing. For example, it is sometimes helpful to use the machine's current position as the program zero point.

As you know, fixture offsets can be designated from the machine's reference position to the specified program zero point. Additionally, three system variables (#5021 for *X*, #5022 for *Y*, and #5023 for *Z*) constantly render the distance from the machine's reference position to the current machine position. Knowing this, and knowing how the G10 command (offset setting by program command) works, you should agree that the commands

```
G90 G10 L2 P1 X#5021 Y#5022 Z#5023
G54
```

will store the current distances in *X*, *Y*, and *Z* from the machine's reference position to the machine's position in fixture offset number one and then invoke this coordinate system. This effectively sets the machine's current position as the program zero point. There is more on the specific implications of assigning program zero points with fixture offsets later in the section on actual spindle probing applications that require this ability.

Program zero assignment command

Like G50 on turning centers, G92 allows the machining center programmer to specify the distance from the program zero point to the machine's current position. Though this can be helpful, we recommend using fixture offsets for this purpose whenever possible, since for many controls, G92 is not completely compatible with fixture offsets.

Spindle orientation for two-direction spindle probes

All spindle probe example programs given in this text assume your are using a five-direction touch probe. You now know, however, that there is another spindle probe type that can be used for spindle probe applications. This probe type has only two directions of probing (*Z* minus and along the *XY* plane). When performing any *X* or *Y* probing, the spindle must be oriented to a precise angular location. This of course, requires that the machine tool allow a more elaborate form of spindle orientation than the simple M19 discussed earlier.

Machine tool builders differ with regard to how this precise spindle orientation must be programmed. If only four orientation directions are needed (*X*+, *X*–, *Y*+, and *Y*–), possibly four M-codes will be used, one for each orientation direction.

More likely, the machine tool builder will allow angular positioning in one-degree increments (360 positions). In this case, a special word (like C) may be used to specify the angle of orientation. The word C45, for example, might orient the spindle to a 45-degree angular position. We must bow to your probe manufacturer or machine tool builder to show you how to command this kind of spindle orientation.

More on deactivating the probe

Some spindle probes do not require a special command to deactivate the probe. Instead, a time delay is used to deactivate the probe. After a given period of non-use (10 minutes let's say), the probe will be *automatically* turned off.

CONCLUSION TO CNC PROBING COMMANDS

Though we have not given every possible probing-related implication of CNC programming commands, you should be starting to understand the subtle programming differences required for probe programming. Let's continue this discussion by studying the custom macro B system variables also required for probing applications. Many of these system variables should be familiar from Part One.

CUSTOM MACRO B SYSTEM VARIABLES FOR PROBING

As stated in Part One, some system variable numbers change from one control model to the next (especially those related to offsets). Additionally, there will be other system variable number differences from machining centers to turning centers. You must confirm each system variable number we show by comparing them to the ones shown in the custom macro B section of your control's programming manual. Though you may have to do a little digging, you can be confident that the functions we show are available on all current controls equipped with custom macro B. To minimize confusion and to maintain consistency, note that all probing example programs shown in this text use the system variable numbers as we show them in this discussion.

ACCESS TO TOOL OFFSETS (#2000 SERIES)

We have emphasized several times that probing requires complete access to tool offsets. Recall from Part One of this text that you have this access to any offset stored within the CNC control. While the G10 command allows you to (only) write offset values, this series of system variables allows you the additional capability to *read from* the offset table as well, making the value of any offset value available to your probing programs. Refer to Chapter 5 for more information about offset-related system variables.

ALARM GENERATION (#3000)

Custom macro B provides a means by which an alarm can be sounded from within the execution of a CNC program. This is especially helpful for probing programmers, since there may often be times when the probing program can detect some error or mistake which could cause problems. The alarm-generating command is helpful for instance, during in-process gaging: if a dimension is detected to be out of tolerance and you wish to stop the cycle and alert the operator, the #3000 command, when executed, will cause the machine to go into an alarm state, similar to any other program storage alarm. The following program tests the width of a groove which is designed to be between 0.249 and 0.251 in. (6.324 and 6.375 mm) wide. If the groove is too narrow or too wide, a specific alarm is generated. The value of the groove's width as been measured and stored in common variable #101 prior to this command:

```
.
.
.
IF [#101 GT 0.251] GOTO 97
IF [#101 LT 0.249] GOTO 98
.
.
.
GOTO 99 (Skip alarm generation commands)
N97 #3000 = 110 (GROOVE TOO WIDE)
N98 #3000 = 111 (GROOVE TOO NARROW)
N99 M30 (End of probing program)
```

If either of the IF statements is true, the control will branch to a #3000 alarm generation command.

SINGLE-BLOCK SUPPRESSION (#3003)

This system variable allows you to temporarily disable the single-block function. Even if single block is turned on, the control will proceed to the next programmed command. This can sometimes be helpful if you wish to give your probing commands the appearance of a kind of canned cycle. The command

```
#3003 = 1 (Disable single block)
```

will disable the single-block function.

From this point, you can perform those probing motions you wish. When finished probing, you must of course relinquish control of single block and return the machine to its normal function relative to single block. As shown in Chapter 5, the command to do this is:

```
#3003 = 0 (Enable single block)
```

Feed hold and feedrate override suppression (#3004)

You can also disable the feed hold and feedrate override function with custom macro B commands. Here are the related commands:

#3004 = 1 (Disable feed hold)
#3004 = 2 (Disable feedrate override)
#3004 = 3 (Disable both feed hold and feedrate override)
#3004 = 0 (Enable feed hold and feedrate override [the initialized state])

As we have pointed out several times, it is important to confirm that the feedrate used for probing (and calibration) remains consistent to ensure that overshoot remains consistent. For this reason, you will give the command #3004 = 2 prior to probing and the command #3004 = 0 just after probing.

STOP WITH MESSAGE (#3006)

This command is similar in format to the #3000 alarm-generating command. As you now know, #3000, if read, will put the control into alarm state and display an alarm message on the screen. To continue with the program, the operator must reset the alarm.

There are times when your probing programs will require manual intervention. In these cases, you will need to stop the machine to give the operator a message, but you will want the operator to be able to continue with the program by simply pressing cycle start. Most programmers normally use the simple M00 (program stop) command for this purpose. If you wish to stop the machine so the setup person can position the probe stylus, you could give this command.

N050 M00 (MOVE STYLUS ABOVE SURFACE)

However, if the operator is not monitoring the program mode while running the program, the message in parentheses will *not* be seen. The operator will have to switch display screen modes to see the message. The programmer can make the message appear (similar to an alarm message) with the use of system variable #3006. But with #3006, the operator will also be able to continue with the cycle by simply pressing cycle start. Here is an example.

#3006 = 1 (MOVE STYLUS ABOVE SURFACE)

See Chapter 5 for more about the stop-with-message system variable.

ACCESS TO MACHINE'S POSITION (#5000 SERIES)

This series of system variables gives you access to the machine's current position in several ways. However, only three of these ways are important when probing.

Since these system variables are well covered in Chapter 5, only additional points relative to probing will be discussed here.

Current absolute position (#5001, #5002, and #5003)

There may be times during a probing cycle when it is helpful to attain the machine's current absolute position relative to the program zero point. During an in-process gaging program, for example, you may wish to utilize the machine's current absolute position as the reference position for probing. This would allow you to work in the absolute mode when making probing commands.

Current position relative to the machine's reference position (#5021, #5022, and #5023)

It is also useful to attain the current machine position relative to the machine's reference point. One instance when this technique can be helpful is when you wish to assign the machine's reference position as the program zero point. The turning center command to perform this function is

```
G50 X#5021 Z#5022
```

Absolute machine position at the time of probe triggering (#5061, #5062, and #5063)

At least one of these system variables will be used *every* time probing is done. They render the machine's position as close to the time of triggering as possible (in *X*, *Y*, and *Z*, respectively). Again, remember that these values are relative to the currently-instated program zero point, meaning that you may have to change the position of program zero prior to probing in order for these values to be as you want them.

ACCESS TO FIXTURE OFFSETS (#5200 SERIES)

The last series of system variables related to probing allow access to fixture offsets. As with other offset-related system variables, you must check to confirm their actual numbers for your particular control model. The list of actual system variable numbers related to fixture offsets is given in Chapter 5.

For any probing application that must assign a program zero point with fixture offsets, it is a good idea to temporarily store each fixture offset value to avoid losing important data during probing. You can, for example, use a common variable in which you will place each important fixture offset value that must be changed at the beginning of your probing program. You can then manipulate the fixture offset in any required manner to accomplish your probing task. When finished with your probing function, you can reset the fixture offset values back to their original values with thc temporary #100 series common variables. Here is an example using fixture offset number one.

```
.
.
.
#101 = #5221 (Store current value of fixture offset number one X )
#102 = #5222 (Store current value of fixture offset number one Y )
#103 = #5223 (Store current value of fixture offset number one Z )
#5221 = 0 (Set fixture offset number one X as needed)
#5222 = 0 (Set fixture offset number one Y as needed)
#5223 = 0 (Set fixture offset number one Z as needed)
G54 (Invoke fixture offset number one)
(Perform needed probing task using fixture offset number one)
.
.
.
#5221 = #101 (Reset X )
#5222 = #102 (Reset Y )
#5223 = #103 (Reset Z )
G54 (Invoke fixture offset number one with original values)
M30 (End of program)
```

Remember that simply changing the values in a given fixture offset does not actually change the program zero position. You must additionally *invoke* the fixture offset to ensure the most recently-entered values are used.

QUICK REFERENCE CNC COMMANDS USED WITH PROBING WITH CUSTOM MACRO B

Figures 8-3 and 8-4 offer a quick reference guide to all commands used with probing when using custom macro B as the parametric programming language.

Custom Macro B Quick Reference

Calling Words	Examples
M98 - Subroutine call	M98 P1000 (Call subprogram O1000)
G65 - Custom macro call	G65 P1000 A0.3 B5.0 C2.3 (Call custom macro O1000 with arguments)
G66 - Custom macro modal call	G65 P1000 A0.3 B5.0 C2.3 L5 (Execute custom macro O1000 five times)
G66.1 - Custom macro modal call	G66 P1000 X3.4 Y1.2 Z3.2 (Set arguments for custom macro O1000)
G67 - Modal call cancel	G66.1 P1000 U1.3 V2.2 W1.1 (Modal call to custom macro O1000)
M99 - End of custom macro	G67 (Cancel either modal call)
P word - Program number being called	
L word - Number of executions	Note: See manual to learn the differences between G66 and G66.1

Custom Macro Arguments (assignment number one)

A, B, C, D, E, F, H, I, J, K, M, Q, R, S, T, U, V, W, X, Y, Z (Avoid I, J, and K if possible. See manual for reasons.)

Custom Macro Arguments (assignment number two)

A, B, C, I, J, K, I, J, K, I, J, K, I, J, K, I, J, K, I, J, K, I, J, K. I, J, K, I, J, K, I, J, K (A, B, C and ten sets of I, J, K)

Local Variable Assignments (assignment 1)

A: #1	H: #11	R: #18	X: #24
B: #2	I: #4	S: #19	Y: #25
C: #3	J: #5	T: #20	Z: #26
D: #7	K: #6	U: #21	
E: #8	M: #13	V: #22	
F: #9	Q: #17	W: #23	

Local Variable Assignments (assignment 2)

A: #1	I(2): #7	I(4): #13	I(6) #19	I(8) #25	I(10) #31
B: #2	J(2): #8	J(4): #14	J(6): #20	J(8): #26	J(10): #32
C: #3	K(2): #9	K(4): #15	K(6): #21	K(8): #27	K(10): #33
I(1): #4	I(3): #10	I(5): #16	I(7): #22	I(9): #28	
J(1): #5	J(3): #11	J(5): #17	J(7): #23	J(9): #29	
K(1): #6	K(3): #12	K(5): #18	K(7): #24	K(9): #30	

Common Variables	Option A	Option B	Option C	Option D
Common Variables	#100 - #149	#100 - #199	#100 - #199	#100 - #199
Permanent Common Variables	#500 - #549	#500 - #599	#500 - #699	#500 - #999

System Variables (See reverse for more detailed information)

- #1000 - #1035 ... Input signals
- #1100 - #1135 ... Output signals
- #2000 - #2999 ... Tool offsets
- #3000 ... Alarm generation
- #3001 ... Milliseconds timer
- #3002 ... Hour timer
- #3003 ... Single block suppression & M-code wait
- #3004 ... Feed hold, feedrate override, and exact stop check suppression
- #3006 ... Stop with message
- #3007 ... Mirror image on/off
- #4001 - #4120 ... Modal information
- #5001 - #5015 ... Current absolute position
- #5021 - #5035 ... Current machine position
- #5061 - #5075 ... Current absolute position at G31
- #5201 - #5335 ... Fixture offsets

Note: For OM controls, #2501 - 2623 are related to fixture offsets

Arithmetic

Priority:
1) Brackets
2) Functions
3) Multiply
4) Diviide
5) Add
6) Subtract

Function	Symbol	Example
Equality	=	#100 = 3
Addition	+	#100 = 3 + 6
Subtraction	-	#100 = 9 - 2
Multiplication	*	#100 = 3 * 4
Division	/	#100 = 9 / 3
Absolute Value	**ABS**	#100 = ABS [2-5]
Rounding	**ROUND**	#100 = ROUND [4.4]
Round Down	**FIX**	#100 = FIX [4.8]
Round Up	**FUP**	#100 = FUP [3.2]

Function	Symbol	Example
Sine	**SIN**	#100 = SIN [45]
Cosine	**COS**	#100 = COS [45]
Tangent	**TAN**	#100 = TAN [45]
Arc Tangent	**ATAN**	#100 = ATAN [3] / [9]
Square Root	**SQRT**	#100 = SQRT [9]
Newer	**Models**	ONLY!
Arc Cosine	**ACOS**	#100 = ACOS [.23]
Logrisms	**LN**	#100 = LN [2.3]
Eponent	**EXP**	#100 = EXP [3]
Decimal adding	**ADP**	#100 = ADP [#3]

Logic Statements

Statement Label:	**N Word**
Unconditional Branching:	**GOTO**
Conditional Branching:	**IF**
Looping:	**WHILE**

Logical operators:

Equal To:	**EQ**
Not Equal To:	**NE**
Less Than:	**LT**
Greater Than:	**GT**
Less Than Or Equal To:	**LE**
Greater Than Or Equal To:	**GE**

Examples

```
.
.
IF [#3 NE 0] GOTO 5
M09
GOTO 6
N5 M08
N6...
.
```

```
.
.
#101 = 1
N1 IF [#101 GT 10] GOTO 1
.
.
.
#101 = #101 + 1
GOTO 1
N1 ...
.
```

```
.
.
#101 = 1
WHILE [#101 LE 10] DO 1
.
.
.
#101 = #101 + 1
END 1
.
```

Figure 8-3. Page one of quick-reference guide for custom macro B and probing-related commands, describing call words, arguments, variables, arithmetic, and logic semantics.

Custom Macro B Quick Reference

System Variables

#1000 - #1035: Input signals for device interface
- #1000 corresponds to input signal UI000
- #1001 corresponds to input signal UI001
- .
- #1031 corresponds to input signal UI031

#1100 - #1135: Output signals for device interface
- #1100 corresponds to input signal UO000
- #1101 corresponds to input signal UO001
- .
- #1131 corresponds to input signal UO031

#2000-#2999: Tool offsets
Machining Centers
(For machining centers having one value per offset)
- #2001: Offset 1
- #2002: Offset 2
- #2003: Offset 3
- .
- #2199: Offset 199

#2000-#2999: Tool offsets **Turning Centers**

	Wear Offsets					Geometry Offsets			
#	X	Z	R	T	#	X	Z	R	T
1	#2001	#2101	#2201	#2301	1	#2401	#2501	#2601	#2701
2	#2002	#2102	#2202	#2302	2	#2402	#2502	#2602	#2702
3	#2003	#2103	#2203	#2303	3	#2403	#2503	#2603	#2703
.	.	.	.	.	.	.	.	.	.
32	#2032	#2132	#2232	#2332	32	#2432	#2532	#2632	#2732

#3000: Alarm generation	Example: #3000 = 100 (TOOL TOO BIG)	26 characters in message, alarm no. 100-255
#3006: Stop w/ message	Example: #3006 = 100 (CLEAR CHIPS)	26 characters in message, alarm no. 100-255

#3001: Millisecond timer	#3001 = 0 (Resets timer)	**#3002**: Hour timer	#3002 = 0 (Resets timer)

#3003: Suppress single block	#3003 = 0 (Enable single block)	#3003 = 1 (Disable single block)
#3003: M-code await	#3003 = 0 (Await M-codes)	#3003 = 2 (Don't await M-codes)
#3004: Suppress feed hold	#3003 = 0 (Enable feed hold)	#3004 = 1 (Disable feed hold)
#3004: Suppress feed override	#3003 = 0 (Enable feed override)	#3004 = 2 (Disable feedrate override

#4000-#4130: Current states (**machining centers**)

#4001: G00...	#4004: G22...	#4007: G40...	#4010: G98...	#4013: G95...
#4002: G17...	#4005: G94...	#4008: G43...	#4011: G50...	#4014: G54...
#4003: G90...	#4006: G20...	#4009: G80...	#4012: G66...	#4015: G61...
				#4016: G68

#4000-#4130: (**turning centers**)

#4001: G00...	#4004: G68...	#4007: G40...
#4002: G96...	#4005: G98...	#4008: G25...
#4003: ABS...	#4006: G20...	

#5201-#5335 Fixture offsets (not 0 series)

Offset	X	Y	Z
Common	#5201	#5202	#5203
G54	#5221	#5222	#5223
G55	#5241	#5242	#5243
G56	#5261	#5262	#5263
G57	#5281	#5282	#5283
G58	#5301	#5302	#5303
G59	#5321	#5322	#5323

#2501-#2523 Fixture offsets (For 0 series)

Offset	X	Y	Z
G54	#2501	#2502	#2503
G55	#2521	#2522	#2523
G56	#2541	#2542	#2543
G57	#2561	#2562	#2563
G58	#2581	#2582	#2583
G59	#2601	#2602	#2603

Machining & turning
- #4102: B
- #4107: D
- #4108: E
- #4109: F
- #4111: H
- #4113: M
- #4114: N
- #4115: O
- #4119: S
- #4120: T
- #4130: Additional fixture offset number

Probing Related Commands

G-codes
- G00 & G01: Motion commands
- G10: Offset setting
- G20 & G21: Inch & metric modes
- G31: Skip cutting
- G49: Cancel tool-length compensation
- G50: Turning center program zero
- G53: Motion relative to zero return
- G54-G59: Fixture offsets
- G65: Custom macro call
- G66: Modal custom macro call
- G66.1: Modal custom macro call
- G67: Cancel modal call
- G90: Absolute mode
- G91: Incremental mode
- G92: Machining center program zero
- G98: Turning center feed per minute
- G99: Turning center feed per revolution

M-codes
- M00: Program stop
- M01: Optional stop
- M03: Spindle on CW
- M04: Spindle on CCW
- M05: Spindle stop
- M13: Activate probe
- M14: Deactivate probe
- M19: Spindle orient
- M98: Subprogram call
- M99: Subprogram return

Special program numbers
- O9000: T-code calling
- O9001-O9009: M-code calling
- O9010-O9020: G-code calling

External output commands
- POPEN: Open RS-232C port
- PCLOS: Close RS-232C port
- BPRNT: Print (no decimal point)
- DPRNT: Print (with decimal point)

Calibration values
- #500: Stylus radius
- #501: X+ overshoot & droop
- #502: X- overshoot & droop
- #503: Y+ overshoot & droop
- #504: Y- overshoot & droop
- #505: Probe length

Naming #500 series variables
- SETVN 500 [RADIUS]
- SETVN 501 [X+ O&D]
- SETVN 502 [X- O&D]
- SETVN 503 [Y+ O&D]
- SETVN 504 [Y- O&D]
- SETVN 505 [LENGTH]

Figure 8-4. Page two of quick-reference guide.

Chapter 9

Spindle Probe Programming

At this point you should have a firm understanding of the various programming functions available to probe programmers in the custom macro B programming language. Programming commands in other versions of parametric programming will be quite similar to those used in custom macro B. If you understand the commands shown in Chapter 8, you should have few problems converting what you know to your own version of parametric programming.

While you have not actually written any probing programs (yet), you have been exposed to the *raw tools* available to probe programmers. Where you go from here really depends on your own specific probing needs; it is therefore difficult for this text to be highly specific for all readers. To make this text as valuable as possible, we focus the balance of our discussions on the most common and current spindle probing applications. We show how to approach a variety of specific applications and program for them. It's possible that we may show an application identical to one you must undertake. More likely, though, you will use our examples as a template from which to create your own probing programs or build on the techniques we present.

As mentioned in the introduction to Part Two, our approach to helping you understand how to program probes is with *a divide-and-conquer* method. By breaking down (dividing) your programming task into smaller, easier-to-handle pieces, you can make the overall task much simpler. While the programs we present as examples may not perfectly parallel your applications, their use to describe the divide-and-conquer method of probe programming is intended to dramatically simplify the programming of any probing system

As we have tried to make clear, different probe types require different approaches to programming. A spindle probe is handled much differently than a tool-length measuring probe or a tool touch-off probe. The amount of end-user programming required also varies with probe type. Generally speaking, once a tool-length measuring or turning center tool touch-off probe user is satisfied with the method by which his or her probes are functioning, there will be little, if any, need for further programming. In fact, many tool-length measuring and turning center tool touch-off probe users are completely satisfied with the pre-written parametric programs their probe manufacturers supply and will never have a need to actually write probing programs.

Spindle probes, on the other hand, commonly require ongoing programming based on changing applications. In fact, the spindle probe's usage may change with every setup and with each production run. It is quite likely that the probe manufacturer's pre-written programs, while they are very helpful, will not completely satisfy a probe user's needs. For this reason, we highly emphasize spindle probes during the balance of this text. And the form of spindle probe we use for specific examples is the most common, the five-direction touch probe.

We begin by showing one method to calibrate a spindle probe. As we know from Chapter 7, it is mandatory that the probe programmers understand the calibration method used with their spindle probing systems. Though variations do indeed exist among systems, the calibration values rendered by our calibration method will be used for all example programs in this chapter.

We also discuss specific programming methods for spindle probes based on the three application categories (help with setup, in-process gaging, and utilities). Unfortunately, it is difficult to present the best and most efficient probing techniques at the same time we teach *how* to write probing programs. To begin by showing the most complex techniques would be foolish. For this reason, we use a building-blocks approach throughout this chapter and constantly build on previously-presented techniques. This means that you may find later that there are more efficient ways to handle the applications we show first.

CALIBRATION FOR FIVE-DIRECTION SPINDLE PROBES

We cannot stress enough the importance of understanding the specific calibration methods of your probing system as they relate to writing your own probing programs. Unfortunately, because of the highly specific nature of probing applications, this text cannot possibly address all the differences in program techniques, and you must be prepared for variations. The general method we use for our sample calibration is quite common, but the specific techniques will change based on your probe manufacturer's way of doing things. We bow to them (and their documentation) to explain specific methods.

Among the most important things you must know are

- Probing cycle used (one touch or two touch);
- Feedrates used for stylus contact;
- Where the calibration values are stored (in #500 series permanent common variables with custom macro B) and the meanings of each value;
- Any other system constants used by the probing system, such as the stylus radius (commonly stored in permanent common variables).

In the sample calibration programs, we use the one-touch probing cycle (we'll say lag time; our probing system is only 2 microseconds). The feedrate will be 30.0 in. (762 mm) per minute (though some current probing systems go much faster). Overshoot and droop values will be stored in permanent common variables as follows:

#500 — Stylus radius
#501 — *X*+ direction overshoot and droop value
#502 — *X*– direction overshoot and droop value
#503 — *Y*+ direction overshoot and droop value
#504 — *Y*– direction overshoot and droop value
#505 — Probe length after triggering (from spindle nose to stylus tip).

Stylus radius. The stylus radius (#500) *must* be considered when locating surfaces in *X* and *Y*, and depending on your specific application, there may be times when the stylus radius must change. For example, you may need to use a very small stylus for an application that has very small workpiece attributes and a larger stylus for the bulk of your probing applications. Interference and obstructions on the device being probed (workpiece or fixture) dictate the maximum possible stylus radius. The probing system must, of course, be recalibrated whenever a stylus is changed, even if you are replacing the stylus with one that is identical (possibly after stylus damage).

Fortunately, the vast majority of spindle probe users use only one stylus configuration for all probing operations. In that case, the value of #500 can remain a hard-and-fixed value, set in the calibration program itself. However, if you must change stylus radius from one application to another, you will want to make it as easy as possible to change the value of #500. An easy way to do this is to include a stop-with-message command (with system variable #3006) in the calibration program, telling the operator to manually set the value of #500 to the stylus radius being used before continuing with the calibration program. Note that our calibration program assumes the use of a 0.10-in. (2.54-mm)-diameter stylus (0.050-in. [1.27-mm] radius) and automatically sets #500 accordingly.

XY overshoot and droop values. In our calibration program, #501 through #504 are rendered as simple amounts of overshoot and droop, meaning there is no direction (plus or minus) implied in any of these values. We are simply determining the *amount* of overshoot and droop when probing in each direction. Unless you are working with a bent stylus (bent by more than the total of overshoot and droop), each of these values will be positive. Most probe manufacturers calibrate in this manner. However, this must be confirmed before you can write probing programs for your probing system.

Probe length. This is the length of the probe *after deflection*, meaning you do not need to know the actual *value* of *Z*– direction overshoot when probing in the *Z* axis; it is incorporated in #505.

This value (#505) may not be needed for all applications. If only probing in *XY*, for example, there may be no need for the probe length. Even when probing in the *Z* axis, it's possible that you will not need to consider the value of #505. If your probing application requires that you simply compare *Z*-axis position to measure the depth of a pocket, for example, you may be simply subtracting one surface position from the other, and the probe's length will not be important. (We go into greater detail on whether the probe's length is important when we address specific probing applications.)

Remember that the value of system variable #5063 (the machine's absolute position in *Z* after probing) is affected by whether or not *tool-length compensation* is instated. If it is instated (using the probe's length as the tool-length compensation value), the value of #5063 after probing will be relative to the very tip of the probe. If tool-length compensation is not instated (in the cancel state), the value of #5063 after probing will be relative to the spindle nose.

Our sample calibration method requires two separate programs, and both involve some manual intervention. Keep in mind that our intention is as much to help you understand *what happens during calibration* as it is to come up with calibration values. Our methods may be somewhat simplistic compared to those actually used by spindle probe manufacturers.

One of our programs determines the probe's length after triggering. Again, this length is the distance from the extreme *tip* of the probe (not the stylus center line) to the spindle nose and can be used when probing *Z*-axis surfaces. Since it already takes into consideration the overshoot value when probing in the *Z* direction, there will be no need to consider actual *Z*-overshoot *amount* when probing.

The second calibration program determines overshoot and droop values for *X*+, *X*–, *Y*+, and *Y*– probing directions (one value for each direction). Note again that the *XY* calibration program assumes the stylus radius to be 0.050 in. (1.27 mm) (a 0.10-in. [2.54-mm]-diameter probe stylus).

The results of the calibration programs are stored into permanent common variables as discussed above. All spindle probe programs shown in this text assume these permanent common variable settings.

CALIBRATION PROGRAM FOR DETERMINING PROBE LENGTH

This program requires the use of a flat block mounted to the table top. For a vertical machining center, a simple gage block will work. For horizontal machining centers, care must be taken to ensure that the block is square with the table.

```
O9050 (Probe-length measuring program)
N1 #101 = #5323 (Store current Z value of fixture offset number six for resetting later)
N2 #3006 = 101 (TOUCH SPINDLE NOSE TO BLOCK)
N3 #5323 = #5023 (Set fixture offset number six's Z value to the machine position)
N4 G59 (Invoke fixture offset number six)
N5 G91 G01 Z15.0 F50.0 (Move away far enough to load spindle probe)
```

```
N6 #3006 = 102 (LOAD PRB & CTR 0.5 ABV BLK)
N7 S2000 M19 (Orient spindle)
N8 M13 (Activate probe)
N9 #3004 = 2 (Disable feedrate override)
N10 G91 G31 Z – 0.75 F30.0 (Drive probe stylus into block)
N11 #505 = #5063 (Store probe length after triggering into permanent common
   variable #505)
N12 #3004 = 0 (Enable feedrate override)
N13 G91 G01 Z0.5 F50.0 (Move stylus away from block)
N14 #5323 = #101 (Reset fixture offset number six's Z value)
N15 M14 (Turn probe off)
N16 M30 (End of program)
```

Since this is the first complete probing program we have shown, it may be a little difficult to understand. For this reason we now give a line-by-line explanation of the program. Note first that we are using the one-touch probing cycle during calibration. This cycle must be used whenever probing.

Line N1 stores the current *Z* value in fixture offset number six to be reset later. Though this is only necessary if you are concerned with maintaining the fixture offset values for use after calibration, it is a good technique to use. With it, calibration will not corrupt the fixture offset values.

Line N2, like an M00 program stop, will halt the machine. But as you know from Chapter 5 (and unlike M00), the message TOUCH SPINDLE NOSE TO BLOCK will be given on the display screen. At this point, the operator will place the machine in a manual mode (most likely the handwheel mode) and cautiously bring the nose of the spindle to touch the block. Then he or she will place the mode switch back to automatic and reactivate the cycle (with cycle start).

Since the block face is the reference point for probe-length measuring, line N3 sets fixture offset number six's *Z* value to the machine's current *Z* position (relative to the machine's *Z*-axis reference position). This means the distance from the machine's reference position to the block surface (in system variable #5023) will be placed in the *Z* register of fixture offset number six. In essence, this sets the block surface to the *Z*-axis program zero point.

Line N4 invokes fixture offset number six, which actually sets the block surface to the program zero point in *Z*. Line N5 moves the spindle far enough away to load the probe. Line N6 stops again so the operator can load the probe into the spindle and manually position it so that the stylus is about 0.50 in. (12.7 mm) above the block. This message is abbreviated since only 26 characters are allowed in the stop-with-message command. Line N7 orients the spindle (though not altogether necessary for *Z*-axis calibration) and line N8 activates the probe.

Line N9 temporarily disables feedrate override. Line N10 drives the probe stylus into contact with the block. Remember that the feedrate in this command must be used in all probing programs when probing in the *Z* axis.

Line N11 stores the current machine position (relative to the block surface) in permanent common variable #505. Remember that system variable #5063 is relative to the program zero point of the current coordinate system. This is why we went to the trouble of setting the block surface as the program zero point. The value in permanent common variable #505 (the probe's length after overshoot) will be used when probing in the *Z* axis. Line N12 enables feedrate override.

Line N13 moves the stylus away from the block. Line N14 resets the value of fixture offset number six's *Z* to its original value, leaving the fixture offset as we found it. Line N14 turns off the probe and line N16 ends the calibration program.

Note that since this program is manually activated for calibration, it is a main program, and must be ended with an M30 or M02 (as opposed to M99).

CALIBRATION PROGRAM TO DETERMINE *XY* OVERSHOOT AND DROOP AMOUNTS

Our *XY* calibration program requires the use of the special calibration fixture shown in Figure 9-1. As stated, this kind of fixture is required when calibrating any five-direction spindle probe.

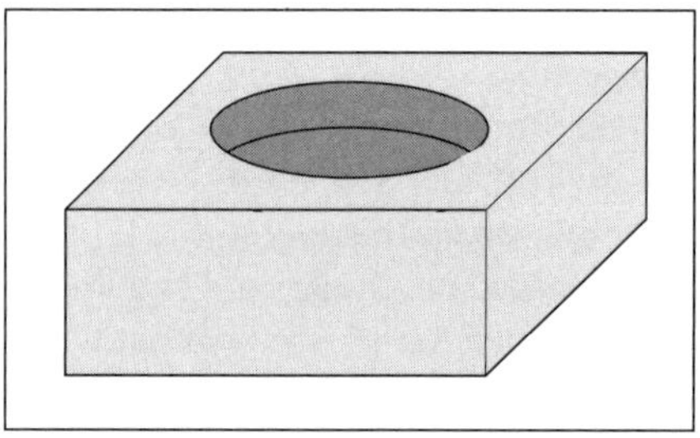

Figure 9-1. Calibration fixture used to determine X and Y overshoot and droop values for five-direction spindle probes.

This calibration fixture is perfectly square in *X* and *Y* (our fixture is precisely 2.0 in. [50.8 mm] square) and a round hole is machined perfectly in the center of the square. To properly determine overshoot and droop values, our method requires that the person doing the calibration first use a dial indicator mounted in the spindle to perfectly (and manually) align the spindle center line with the center of the hole in the calibration fixture. The calibration program will assume the machine is at this position when it is activated. The probe is then loaded and brought to a position in *Z* about 0.50 in. (12.7 mm) above the top surface of the fixture (while the spindle center line is still perfectly centered on the fixture). This calibration program is then run.

```
O9051 (Calibration program for determining XY overshoot and droop values)
N1 #100 = 2.0 (Size of square fixture)
N2 #500 = 0.05 (Set stylus radius in permanent common variable)
N3 #101 = #5321 (Store X value of fixture offset number six for resetting later)
N4 #102 = #5322 (Store Y value of fixture offset number six for resetting later)
N5 #5321 = #5021 (Set current X position to program zero in X )
N6 #5322 = #5022 (Set current Y position to program zero in Y )
```

```
N7 G59 M13 (Invoke fixture offset six, activate spindle probe)
N8 S2000 M19 (Orient spindle to ensure consistency in droop)
N9 G91 G01 X – [#100/2 + #500 + 0.5] F50.0 (Fast feed to clear left side of square)
N10 G01 Z – 0.65 (Fast feed below Z surface of square)
N11 #3004 = 2 (Disable feedrate override)
N12 G91 G31 X0.75 F30.0 (Drive stylus into left side of square)
N13 #501 = #500 + #100/2 – ABS[#5061] (Store overshoot and droop for X+ probing)
N14 #3004 = 0 (Enable feedrate override)
N15 G01 X – 0.25 F50.0 (Fast feed away from square in X )
N16 Z0.65 (Fast feed above square in Z )
N17 G90 X0 (Fast feed back to center in X )
N18 G91 G01 X[#100/2 + #500 + 0.5] F50.0 (Fast feed clear of right side of square)
N19 G01 Z – 0.65 (Fast feed below Z surface of square)
N20 #3004 = 2 (Disable feedrate override)
N21 G91 G31 X – 0.75 F30.0 (Drive stylus into right side of square)
N22 #502 = #500 + #100/2 – ABS[#5061] (Store overshoot and droop for X– probing)
N23 #3004 = 0 (Enable feedrate override)
N24 G01 X0.25 F50.0 (Fast feed away from square in X )
N25 Z0.65 (Fast feed above square in Z )
N26 G90 X0 (Fast feed back to center in X )
N27 G91 G01 Y – [#100/2 + #500 + 0.5] F50.0 (Fast feed clear of Y– side of square)
N28 G01 Z – 0.65 (Fast feed below Z surface of square)
N29 #3004 = 2 (Disable feedrate override)
N30 G91 G31 Y0.75 F30.0 (Drive stylus into lower side of square)
N31 #503 = #500 + #100/2 – ABS[#5062] (Store overshoot and droop for Y+ probing)
N32 #3004 = 0 (Enable feedrate override)
N33 G01 Y – 0.25 F50.0 (Fast feed away from square in Y )
N34 Z0.65 (Fast feed above square in Z )
N35 G90 Y0 (Fast feed back to center in Y )
N36 G91 G01 Y[#100/2 + #500 + 0.5] F50.0 (Fast feed clear of Y+ side of square)
N37 G01 Z – 0.65 (Fast feed below Z surface of square)
N38 #3004 = 2 (Disable feedrate override)
N39 G91 G31 Y – 0.75 F30.0 (Drive stylus into Y+ side of square)
N40 #504 = #500 + #100/2 – ABS[#5062] (Store overshoot and droop for Y– probing)
N41 #3004 = 0 (Enable feedrate override)
N42 G01 Y0.25 F50.0 (Fast feed away from square in Y )
```

```
N43 Z0.65 (Fast feed above square in Z )
N44 G90 Y0 (Fast feed back to center in Y )
N45 #5321 = #101 (Reset X value of fixture offset number six)
N46 #5422 = #102 (Reset Y value of fixture offset number six)
N47 M14 (Turn probe off)
N48 M30
```

While this calibration program may at first seem somewhat complicated, note that it is performing essentially the same actions four times (once for *X*+, once for *X*–, once for *Y*+, and once for *Y*–). Again, notice that we are using the one-touch probing cycle. This cycle must be used for future probing. We begin by specifying the square's size for easy manipulation (line N1) and storing the stylus radius in line N2. We then store the current *X* and *Y* value of fixture offset number six for resetting later in lines N3 and N4. Since the center of the square is first manually indicated with the dial indicator to ensure that the spindle center line is true with the center of the square, we next set the machine's current *X* and *Y* position as the program zero point in lines N5 and N6. Line N7 selects the coordinate system and turns on the probe. Line N8 ensures that droop will remain consistent by orienting the spindle.

Line N9 begins the redundant task of finding the overshoot and droop values. We begin by finding the value needed when probing in the *X*+ direction. The stylus is first sent to a position that clears the left side of the square (in line N9) and is brought to a position in *Z* that allows probing in line N10. Line N11 temporarily disables feedrate override. Line N12 actually drives the stylus into the left (*X*–) side of the square.

The overshoot and droop value for *X*+ probing is then calculated and stored in permanent common variable #501 in line N13. This is done by subtracting the current distance between the spindle center line and the center of the hole (the absolute value of #5061 renders this distance) from half the true square width plus the stylus radius.

Line N14 enables feedrate override and line N15 moves the probe stylus away in *X*. Line N16 moves the stylus above the calibration fixture in *Z*. The stylus is then moved back to its starting position in line N17. Line N18 begins the same series of movements for determining the overshoot and droop value for *X*- probing. This is followed by the commands needed to determine overshoot and droop values for *Y*+ (beginning at line N27) and *Y*– (beginning at line N36) probing.

When finished, the original values of fixture offset number six are reset in lines N45 and N46. The probe is turned off in line N47 and the program is ended in line N48.

A trick with calibration feedrate

Our calibration program (as well as all of our example probing programs) uses a hard-and-fixed 30.0-in. (762-mm)-per-minute feedrate. However, if you have *any* concern that this feedrate may have to change (to improve accuracy, for efficiency, based on changing applications, etc.), programming this kind of hard-and-fixed feedrate for calibration and probing is unwise. If you develop many probing programs using a hard-and-fixed feedrate, consider how many program changes you will have to make if you ever wish to change the probing feedrate.

It may be wiser to use a permanent common variable as a system constant for feedrate. For example, suppose you use variable #506 as the probing (and calibration) feedrate. Whenever the G31 command is used, reference this permanent common variable as the feedrate (with F#506). If, of course, you are using the two-touch probing cycle, use two permanent common variables, one for the initial touch and another for the final touch. This technique allows you to easily manipulate the probing feedrate without having to change many feedrate commands in the calibration and probing programs.

When do you calibrate spindle probes?

Calibration is necessary on a more regular basis with spindle probes than with any other form of touch probe. It should go without saying that calibration is necessary when the probing system is first installed. Other obvious times when calibration must be done is whenever a substantial mishap (crash) occurs that may affect machine positioning, or when you change the stylus. There are also less obvious times when calibration must be done. You may, for example, have different probing applications that require differences in accuracy and efficiency. You will need to calibrate whenever you wish to change probing feedrate (increasing feedrate to achieve better efficiency, decreasing feedrate to achieve better accuracy). You may even need to calibrate as the weight of your workholding setup substantially changes, especially if your control experiences a great amount of lag time before it can attain position. Again, we bow to your probe manufacturer's recommendations regarding other instances in which probe calibration is required.

LOCATING A PROBED SURFACE

Before we can show any complete application programs, you must understand how to attain a probed surface's position after spindle probe stylus triggering. Though somewhat tedious, this is actually quite easy to do—it takes little more than simple arithmetic. First, you must know the #500 series permanent common variables used to store stylus radius, overshoot and droop values, and the probe's length. As you know, these permanent common variables will vary from one probe manufacturer to another. All applications shown in this chapter use the same variables shown during our discussion of spindle probe calibration.

#500 — Stylus radius
#501 — *X+* direction overshoot and droop
#502 — *X–* direction overshoot and droop

#503 — *Y*+ direction overshoot and droop
#504 — *Y*– direction overshoot and droop
#505 — Probe length (after deflection)

Incidentally, if your probe manufacturer has not already done so, it would be wise to actually label each permanent common variable used with probe calibration to avoid the possibility of someone unwittingly overwriting the values. As you know from Chapter 5, the SETVN command (standing for *set variable name*) can be used for this purpose. Unfortunately, with most control models, the SETVN command allows only eight characters, meaning you cannot be very wordy. If your calibration values are stored as we have shown, load and run this program. It need only be run once. After that you can delete it from memory.

```
O9030 (Permanent common variable-setting program)
SETVN 500 [RADIUS]
SETVN 501 [X+ O&D]
SETVN 502 [X– O&D]
SETVN 503 [Y+ O&D]
SETVN 504 [Y– O&D]
SETVN 505 [PRB LEN]
M30
```

All probing you do must use the same probing cycle type (one-touch or two-touch) as you used when calibrating and the probing feedrate must be the same as that used during calibration. Again, all examples shown in this chapter assume the one-touch probing cycle (as in the calibration programs just shown) and all probing is done at 30.0 in. (762 mm) per minute (F30.0), just as is in the spindle probe calibration programs. Remember, in your own probing systems you must be prepared for variations in these functions.

LOCATING A SURFACE IN *X* OR *Y*

Figure 9-2 will help you understand how to locate a surface position after probing in *X* or *Y*. It shows the condition of the probe stylus at the instant the machine's position is determined.

In this example, we are driving the stylus into an *X* surface, probing in the *X*+ direction. This means that the *X*+ overshoot and droop value, stored in permanent common variable #501 must be referenced in the calculation that attains the probed surface. Based on our method of calibration, we simply subtract the *X*+ overshoot and droop value (in #501) from the spindle center line location determined after triggering (in #5063). This determines the *X*-axis center line location of the *stylus ball*. We must then add the stylus ball radius (in #500) to come up with the sur-

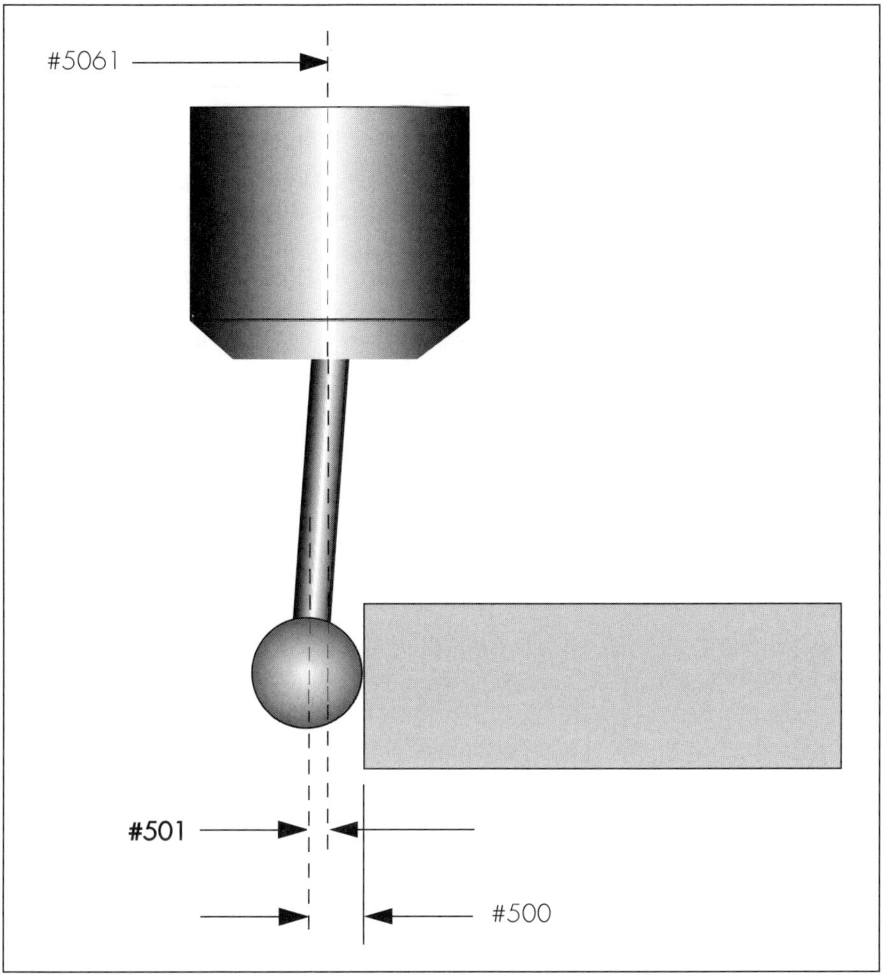

Figure 9-2. Stylus condition at the instant machine position is determined.

face location along the *X* axis. Assuming the stylus is positioned within about 0.50 in. (12.7 mm) of the surface, here are the custom macro B commands to drive the stylus into the surface and determine the surface location in *X*.

G91 G31 X0.75 F30.0 (Drive probe into *X* surface)
#100 = #5061 – #501 + #500 (Store *X* surface in common variable #100)

None of the four *XY* calibration values you will recall really have a *polarity* (based on our calibration method). They simply represent the distance between the stylus ball after deflection and the center line of the spindle. Unless there is an excessive amount of droop (possibly caused by a slightly bent stylus), all *XY* calibration values will be positive. Your surface location calculation must reflect this point. When probing in the *X*+ direction, as shown in Figure 9-2, the overshoot and droop value (in #501) must be *subtracted* from the spindle center line position to come up with the stylus ball center line. However, if probing in the *X*– direction, the related value (in #502) must be *added* to the spindle center line position to come up with the stylus ball center line location. These points apply to *Y*-axis probing as well. Again, you must confirm and thoroughly understand how your probe manufacturer uses overshoot and droop calibration values before you can locate surfaces in your own probing programs. Most methods are similar to those we show.

LOCATING A PROBED SURFACE IN *Z*

The same principles apply when probing in the *Z* direction. However, you must remember the impact tool-length compensation has on system variable #5063. As long as tool-length compensation is *not* instated (in a canceled state), system variable #5063 is taken to the *spindle nose position*, not to the tip of the stylus. In this case, the probe's length (in permanent common variable #505) must always be subtracted (since it was stored by the calibration program as a plus number) from #5063 to calculate the probed surface. For example, the commands

```
G91 G31 Z – 0.75 F30.0 (Drive stylus into Z surface)
#100 = #5063 – #505
```

will store the probed *Z* surface in common variable #100. Since the value of permanent common variable #505 reflects the probe length after deflection, there is no need to consider any *Z*– direction overshoot.

It is a good idea to confirm that tool-length compensation is canceled (commonly with G49) before probing in the *Z* axis if you expect to reference #505 during the surface calculation. You can even check the current condition of tool-length compensation from within your custom macro program. For most controls equipped with custom macro B, system variable #4008 gives you access to the current state of tool-length compensation. If its value is zero (0), tool-length compensation is currently canceled. (You will see this technique in an example program a little later.)

Depending upon the application, there may be times when you wish to reference the probe tip (end of stylus ball) when programming, as you would any cutting tool. This is especially helpful when developing totally automatic probing routines. To do this, you can instate tool-length compensation for the probe (just as you would any cutting tool) on the probe's first *Z*-axis approach movement. Prior to doing this, you can even easily (and automatically) transfer the probe length into the appropriate tool-length compensation offset. Here is an example: the probe is in tool station number five, and you wish to use offset number five as the probe's length. The commands

```
.
.
.
#2005 = #505 (Store length of probe in offset number five)
G00 X5.0 Y4.0 (Move to XY position)
G43 H05 Z0.25 (Rapid the probe stylus to Z0.25)
.
.
.
```

transfer the probe's length to offset five, move the probe to the correct *XY* location, and instate tool-length compensation for the probe. From this point on, *Z*-axis movements, as well as the machine's *Z* position after triggering (#5063) will reference the tip of the stylus ball instead of the spindle nose.

Again, the application will determine which of these two *Z*-axis-related methods best suits your needs. And as we show specific examples, we will discuss these points further.

TAILORING A UNIVERSAL PROBE-TRIGGERING CUSTOM MACRO PROGRAM

You can often make it easier to handle complex probing applications by dividing them into their most basic parts. Our first *divide and conquer* suggestion is related to making it easy to drive the probe into any surface and calculate the surface location. This is one of the most basic probing functions—one that will be performed (probably several times) in all probing programs.

You should agree that the surface-locating commands and location calculations will be repeated often. Though not terribly difficult to write, they are at the root of many tedious and error-prone commands in your probing programs, and of course mistakes will cause inaccurate probing. Every time you probe, you'll have at least four tedious commands to write. If you intend to use the two-touch probing cycle, you will have even more. If you will be probing in a similar fashion on a regular basis, you can minimize programming and totally eliminate the potential for making calculation mistakes in your probing programs by developing a *universal probe-triggering custom macro*.

Though this universal probe-triggering program will remain intact, you can tailor it to your specific needs. To illustrate, suppose you are commonly positioning your spindle probe stylus within about 0.50 in. (12.7 mm) from the surface to be probed (a common distance). At this point, you always disable feedrate override and give the probing command to drive the stylus into the surface to force triggering with a 0.75 in. (19.05 mm) incremental departure. Then you always calculate the surface location. After surface calculation, you retract the probe 0.25 in. (6.35 mm) from the surface just probed. You then always enable feedrate override. While these commands are relatively simple to write, they are quite tedious and open the door to making mistakes. Instead of writing all of these commands every time you drive the stylus into a surface, you can instead invoke the universal probe-triggering custom macro program.

This program, once created and verified, can be called over and over again. The program will eliminate the potential for errors related to:

- Remembering to include the feedrate override disable/enable commands;
- Remembering to include the G91, G31, or feedrate value in the probing command;
- Correctly stating the calculation command;
- Retracting in the correct direction; and
- Remembering to enable feedrate override.

Additionally, if you do a great deal of probing, this universal probe-triggering custom macro can actually save memory space, since the same commands will be executed over and over again.

For the sample universal probe-triggering custom macro program, we use program number O9501. To specify the axis and direction for probing, we use the letter address argument A (for axis). The number one (1.) will represent the *X* axis, two (2.) the *Y* axis, and three (3.) the *Z* axis. We can even specify the direction of probing (plus or minus) with this argument. For example, once the universal probe-triggering custom macro program is written, the command

```
G65 P9501 A-1.
```

will tell the control to probe in the *X*– direction. If A is set to one (A1.), probing will be done in the *X*+ direction. If A is set to minus two (A – 2.) probing will be done in the *Y*– direction. If set to two (A2.) probing will be done in the *Y*+ direction. And if A is set to three (+ or –3.), probing will be done in the *Z*– direction. Our program currently assumes that tool-length compensation is canceled, meaning the probe length (#505) is subtracted from #5063 to determine the surface location. Later, we will show another version of this program that includes a flag argument to specify that this is what we really want to do. Remember from Chapter 2 that *every argument* in the G65 call statement must have a decimal point (A1 with no decimal point would be assigned as A0.001). Also remember that the local variable which represents A in the custom macro is #1, meaning #1 must be used in the custom macro to reference the value of A.

Depending on how diversified you wish your universal probe-triggering custom macro to be, there may be other arguments you wish to pass (or preset) to specify how probing must be done. You may, for example, wish to vary the movement amount in the actual probing command (from 0.75 in. [19.05 mm]). Or you may wish to vary the retract amount. You may even wish to specify the probing cycle type you are using (one-touch or two-touch) which would, of course, also require two sets of #500 calibration values. Using your own arguments in the G65 call statement, you can pass any appropriate information.

Once probe-triggering is completed, this program must return the surface location to the calling program. Since common variables are not reset until the power is turned off, any common variable (in the #100 series) can be used for this purpose. We'll use common variable #100 for our universal probe-triggering custom macro program.

Here is the universal probe-triggering custom macro, limited by the criteria we have just described.

```
O9501 (Universal probe triggering macro)
#3004 = 2 (Disable feedrate override)
IF [#1 EQ 1] GOTO 1 (X+ direction probing detected)
IF [#1 EQ – 1.] GOTO 2 (X– direction probing detected)
IF [#1 EQ 2.] GOTO 3 (Y+ direction probing detected)
IF [#1 EQ – 2.] GOTO 4 (Y– direction probing detected)
IF [#1 EQ 3] GOTO 5 (Z– direction probing detected)
IF [#1 EQ – 3.] GOTO 5 (Z– direction probing detected)
#3004 = 0 (Problems exist with A word, so enable feedrate override)
#3000 = 100 (LETTER A IMPROPERLY SET)
N1 (Begin X+ direction probing)
G91 G31 X0.75 F30.0 (Drive stylus into surface)
#100 = #5061 – #501 + #500 (Calculate X surface position)
G91 G01 X – 0.25 F50.0 (Retract 0.25 in. [6.35 mm] from probed surface)
GOTO 99 (Exit custom macro)
N2 (Begin X– direction probing)
G91 G31 X – 0.75 F30.0 (Drive stylus into surface)
#100 = #5061 + #502 – #500 (Calculate X surface position)
G91 G01 X0.25 F50.0 (Retract 0.25 in. [6.35 mm] from probed surface)
GOTO 99 (Exit custom macro)
N3 (Begin Y+ direction probing)
G91 G31 Y0.75 F30.0 (Drive stylus into surface)
#100 = #5062 – #503 + #500 (Calculate Y surface position)
G91 G01 Y – 0.25 F50.0 (Retract 0.25 in. [6.35 mm] from probed surface)
GOTO 99 (Exit custom macro)
N4 (Begin Y– direction probing)
G91 G31 Y – 0.75 F30.0 (Drive stylus into surface)
#100 = #5062 + #504 – #500 (Calculate Y surface position)
G91 G01 Y0.25 F50.0 (Retract 0.25 in. [6.35 mm] from probed surface)
GOTO 99 (Exit custom macro)
```

```
N5 (Begin Z– direction probing)
G91 G31 Z – 0.75 F30.0 (Drive stylus into Z surface)
#100 = #5063 – #505 (Calculate Z surface position, assumes tool-length compen-
   sation is canceled)
G91 G01 Z0.25 F50.0 (Retract 0.25 in. [6.35 mm] from probed surface)
N99 #3004 = 0 (Enable feedrate override)
M99 (End of custom macro)
```

Notice that this custom macro begins by disabling feedrate override, then a series of IF statements determines which axis and direction is being probed. Note also that if all IF statements are determined to be false, feedrate override is enabled and an alarm is sounded (with the #3000 command), alerting the probe programmer that the letter A has been omitted or improperly set in the call statement (maybe the decimal point is missing). Then comes a set of commands for each of the alternatives. After each G31 command, common variable #100 is calculated according to the direction of probing specified. Then the stylus is retracted from the probe surface by 0.25 in. (6.35 mm) (in all cases), and the GOTO statement causes the control to exit the custom macro. Feedrate override is enabled and the custom macro is ended with an M99.

It is very important that you understand this universal probe-triggering custom macro because we use it throughout the balance of this chapter in our probing program examples.

To use this universal probe-triggering custom macro program, the probe programmer will simply invoke custom macro O9501 from the probing program, filling in the A word with the appropriate probing axis and direction. The very next command in the probing program can reference the current value of #100, which has been set to the surface location by program O9501.

Importantly, this *divide and conquer* approach can be used with any redundant probing-related function. Just as computer programmers use subroutines and functions to minimize the potential for mistakes and shorten the overall length of their programs, so can probe programmers do so when developing the various programs related to the probing system. We will show several other uses for this approach throughout this chapter.

PROGRAMMING FOR SETUP HELP APPLICATIONS

At long last, you should be ready to study specific application programs. As discussed in Chapter 7, there are times when spindle probes can help with machining center setups. In this section, we go through the applications once again, but this time we show the actual probing programs. This should bolster your understanding of spindle probe programming, and show you much of what it takes to write probing programs for your own applications.

HELP WITH PROGRAM ZERO ASSIGNMENT

You know from chapter seven that a spindle probe can be used to find the location of the program zero position, regardless of where the program zero point is placed (corner of rectangular workpiece, center of hole or boss, center of slot or rail, etc.). Once located, the corresponding fixture offset can be set automatically by the probing program. For setups that require the program zero point to be determined on line, this kind of probing program can dramatically minimize the length of time needed for the program zero measurement and eliminate the potential for calculation and entry mistakes.

An outside corner pickup probing program

The first version of this spindle probe program, while it is quite limited, will correctly locate a corner of a rectangular workpiece held in a table vise. The particular corner this probing program will locate is the left surface in *X* (*X*– side), the lower surface in *Y* (*Y*– side), and the top surface in *Z* (*Z*+ side). When program zero is found, it also will place the correct program zero assignment values into fixture offset number one. Figure 9-3 shows the application.

To keep this first program as simple as possible to understand, we will set some rather limiting criteria: the operator will first (manually) load the probe into the spindle and (manually) bring the probe stylus into a position about 0.50 in. (12.7 mm) in each axis away from the corner to be located. Figure 9-3 shows the stylus starting position for this program. We'll also say that this probe has already

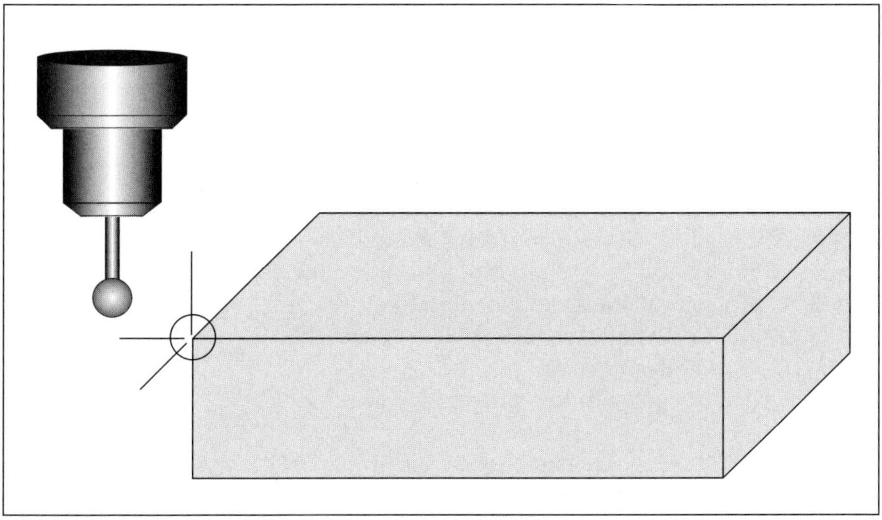

Figure 9-3. Initial position for the spindle probe prior to activating corner pickup program.

been calibrated and that the overshoot and droop values are stored in the #500 series permanent common variables as described earlier in this chapter. Additionally, this example uses the *universal probe-triggering custom macro.*

If you truly understand the material presented to this point, you should be able to understand this probing program with relative ease.

```
O9060 (Outside corner pickup program for lower left corner in XY and top surface in Z )
IF [#4008 EQ 0] GOTO 1 (Test if tool-length compensation is in canceled state)
#3000 = 100 (CANCEL TLC)
N1 #5221 = 0 (Set fixture offset number one's program zero point to the machine's reference point in X )
N2 #5222 = 0 (Set fixture offset number one's program zero point to the machine's reference point in Y )
N3 #5223 = 0 (Set fixture offset number one's program zero point to the machine's reference point in Z )
N4 G54 (Invoke fixture offset number one)
N5 S2000 M19 (Orient spindle to ensure consistency for droop)
N6 M13 (Activate probe)
N7 G91 G01 Y0.75 F50.0 (Incrementally, fast feed in Y+)
N8 Z – 0.75 (Still incrementally, fast feed in Z to allow contact with X– surface)
N9 G65 P9501 A1. (Invoke probe-triggering custom macro in X+ direction)
N10 #5221 = #100 (Store X surface location into the X value of fixture offset number one)
N11 G91 G01 Y – 0.75 F50.0 (Fast feed down in Y to clear corner)
N12 X0.75 (Fast feed to allow probing in Y )
N13 G65 P9501 A2. (Invoke probe-triggering custom macro in Y+ direction)
N14 #5222 = #100 (Store Y surface location into Y value of fixture offset number one)
N15 G91 G01 Z0.75 F50.0 (Fast feed up in Z to clear top surface)
N16 Y0.75 (Fast feed to allow probing in Z )
N17 G65 P9501 A – 3. (Invoke probe-triggering custom macro in Z– direction)
N18 #5223 = #100 (Store Z surface location into Z value of fixture offset number one)
N19 G54 (Invoke fixture offset number one)
N20 G90 G01 X0 Y0 F50.0 (Leave spindle centered on corner in XY )
N21 M14 (Turn probe off)
N22 M30 (End of corner pickup program)
```

Note again that sequence numbers are only included in our example programs for the sake of being able to reference line numbers during these clarifications. To conserve memory space, most probe programmers do not use them unless a statement label is required.

Since tool-length compensation will affect the value of #5063, the first two commands of this program test to ensure that tool-length compensation is canceled. If it is not, an alarm will sound. From the discussion in Chapter 5 on system variables, you know that almost every current state can be determined from within a custom macro B program. It happens that (for this control) system variable #4008 renders the current state relative to tool-length compensation. If this system variable is currently set to zero, tool-length compensation is in a canceled state, as it must be for this program to work. If tool-length compensation is not canceled, the IF statement at the beginning of this program will be evaluated as false, in which case the alarm (caused by the #3000 command) will sound.

Since we are working with fixture offset number one, and since this fixture offset is the one actually being set during this corner pickup program, we need not store the current values of this offset for the purpose of resetting them at the end of the custom macro (as we needed to during the calibration programs). However, since #5061, #5062, and #5063 render the position of the machine relative to program zero, and since the machine's reference position (zero return position) will be the point of reference for the fixture offset setting values, *the program zero point must be set to the machine's reference position.* This is being done in lines N1 through N3. Line N4 invokes this fixture offset.

Line N5 orients the spindle to ensure consistent droop; line N6 turns on the probe.

Since this program is initially locating the *X* surface, lines N7 and N8 position the stylus so that it is ready to touch the left side of the workpiece. As long as the setup person manually positioned the stylus within 0.50 in. (12.7 mm) of the corner to begin, the stylus should be about 0.50 in. away from the surface in *X* and 0.25 in. (6.35 mm) below the top surface in *Z*.

In line N9, notice how easy the universal probe-triggering macro now makes it to probe in the *X*+ direction. Program O9501 (shown earlier) will disable feedrate override, drive the probe into the surface in the *X*+ direction (selected by A1. in line N9), store the *X* surface position in common variable #100, retract 0.25 in. (6.35 mm) from the surface just probed, and enable feedrate override.

Line N10 takes the current value of #100, which has been set by the universal probe-triggering program as the distance from the machine's reference position to the program zero surface in *X*, and places it into the *X* register of fixture offset number one (specified by system variable #5221).

Line N11 moves the stylus clear of the corner in *Y* and line N12 positions the stylus in *X* so that it is ready to probe the *Y* surface of the corner.

Line N13 invokes the universal probe-triggering custom macro a second time, this time telling it to probe in the *Y*+ direction (done with the A2. word). Again, feedrate override is disabled, the surface is probed, the position of this surface is stored in common variable #100, and feedrate override is enabled.

Line N14 stores the *Y* surface location relative to the machine's reference position (currently in #100) in the *Y* register of fixture offset number one (just as line N10 does for *X*).

Line N15 moves the stylus above the top surface. Line N16 moves the stylus in *Y* to a position at which the *Z* surface can be probed.

N17 invokes the universal probe-triggering custom macro once more, and the *Z* surface position is stored in common variable #100.

Line N18 sets the *Z* value of fixture offset number one.

To leave the spindle in a logical position, line N19 instates fixture offset number one, line N20 moves the spindle center line right on the program zero location in *X* and *Y*.

Line N21 turns the probe off and line N22 ends the probing program.

Limitations yet to be overcome. This corner pickup program has at least three notable limitations. First, there is no way to manipulate the fixture offset number being used. It simply uses fixture offset number one. Second, the program picks up only one corner. (By now you can probably think of a way to develop a more powerful probing program that would locate any of the corners, based on one more piece of input data, possibly a *corner number.*) And third, this probing program requires a great deal of manual intervention. There may be times when you wish to automate your program zero measurements and program the spindle probe to go to its desired initial position, possibly as part of a program used to help complete the machine setup.

Allowing the fixture offset number to be specified. Most program zero-setting probing programs result in a fixture offset being set. It often will be necessary to use one of the available fixture offsets. In fact, when machining multiple workpieces in the setup, it will be necessary to assign several fixture offsets before production can be run. For this reason, you will likely need your program zero-assigning probing programs to set any one of the fixture offsets.

This can be handled in at least two ways. The easiest way assumes your control permits offsets to be entered by programmed command. Most controls use the G10 command for this purpose (discussed in Chapter 8). As you know, the P word of the G10 command specifies which fixture offset is being set. Using G10, it is very easy to simply set a variable corresponding to the fixture offset number you wish to assign. For example, if you wish to set fixture offset number one, the variable specifying the fixture offset can be set to a value of one.

If the program is to be activated manually, early on in the program (perhaps even the very first command), a common variable can be set to the fixture offset number being assigned. With this technique, the operator must remember to set this value (manually editing the program) before the program is activated. Additionally, since the fixture offset must be invoked with the corresponding G word (G54 for fixture offset number one, G55 for fixture offset number two, etc.), you must reference the common variable value in the fixture instating G word by adding the value 53 to the common variable value. Here is the corner pickup program shown earlier, modified to use G10 to set the fixture offset and allow the value of the fixture offset to be specified.

O9060 (Outside corner pickup program for lower left corner in *XY* and top surface in *Z*)
N1 IF [#4008 EQ 0] GOTO 3 (Test if tool-length compensation is in canceled state)
N2 #3000 = 100 (CANCEL TLC)
N3 #101 = 1 (Fixture offset number being set. Operator sets this value before activating)
N4 G90 G10 L2 P#101 X0 Y0 Z0 (Set desired fixture offset number's program zero point to the machine's reference point in *X*, *Y* and *Z*)
N5 G[53 + #101] (Invoke desired fixture offset)
N6 S2000 M19 (Orient spindle to ensure consistency for droop)
N7 M13 (Activate probe)
N8 G91 G01 Y0.75 F50.0 (Incrementally, fast feed in *Y*+)
N9 Z – 0.75 (Still incrementally, fast feed in *Z* to allow contact with *X*– surface)
N10 G65 P9501 A1. (Invoke probe-triggering custom macro in *X*+ direction)
N11 G90 G10 L2 P#101 X#100 (Store *X* surface location into the *X* value of chosen fixture offset)
N12 G91 G01 Y – 0.75 F50.0 (Fast feed down in *Y* to clear corner)
N13 X0.75 (Fast feed to allow probing in *Y*)
N14 G65 P9501 A2. (Invoke probe-triggering custom macro in *Y*+ direction)
N15 G90 G10 L2 P#101 Y#100 (Store *Y* surface location into *Y* value of chosen fixture offset)
N16 G91 G01 Z0.75 F50.0 (Fast feed up in *Z* to clear top surface)
N17 Y0.75 (Fast feed to allow probing in *Z*)
N18 G65 P9501 A – 3. (Invoke probe-triggering custom macro in *Z*– direction)
N19 G90 G10 L2 P#101 Z#100 (Store *Z* surface location into *Z* value of chosen fixture offset)
N20 IF [#101 EQ 0] GOTO 23 (If the common fixture offset is being set, do not move)
N21 G[53 + #101] (Invoke desired fixture offset)
N22 G90 G01 X0 Y0 F50.0 (Leave spindle centered on corner in *XY*)
N23 M14 (Turn probe off)
N24 M30 (End of corner pickup program)

Notice in lines N4, N11, N15, and N19, the P-word of the G10 command is being set to the fixture offset number stored in common variable #101. This ensures that the fixture offset specified by the setup person at the beginning of the program (line N1) in the #101 setting command will be the same as the fixture offset number actually set by the G10 command. Additionally, lines N5 and N21 invoke this fixture offset with the required G-code (G54 through G59). Notice that the actual value of the G-code is specified as a variable in the G[53 + #101] word. If you add 53 to the value of #101, you come up with the G-code number related to the fixture offset being used.

What if you don't have G10? Unfortunately, if your control does not allow the G10 command, specifying which fixture offset must be set gets harder. If this is the case, you must use the same series of system variables you use to assign fixture offsets (the #5200 series on most controls). Since there is no intuitive correlation between the fixture offset number and the #5200 series system variable numbers, you must set up a kind of cross-reference table to make the fixture offset number correspond to the required system variables. Note that many probe manufacturers do not assume that the G10 option is installed in your control. They handle the assignment of fixture offsets with system variables.

Since several commands will be required, and since these commands will be repeated in every program zero-finding custom macro, once again it is wise to use *divide and conquer* techniques and create a separate program, one we'll call a *universal fixture offset number-finding custom macro*.

As with the universal probe-triggering custom macro, one value must be retrieved from the universal fixture offset number-finding custom macro (the system variable number related to the *X* value of the fixture offset). Again, this value will be stored in a common variable. We'll use common variable #101. For *Y* and *Z* setting, we'll simply add (one or two) to the value in #101 to come up with the needed system variable number corresponding to the *Y* and *Z* register.

Additionally, we need to be able to specify to the universal fixture offset number-finding custom macro which fixture offset we are using. We'll do this with the letter address F (representing *fixture* offset) in the calling command to the universal fixture offset number-finding custom macro. The command

G65 P9502 F2.

for example, specifies that it is fixture offset number two that we are using in this program zero-assigning probing program. If activating this program manually, as is the case with our outside corner pickup macro, we can make it easy to for the setup person to enter the fixture offset number by incorporating another common variable setting command at the beginning of the program zero-finding custom macro (our example uses common variable #102 for this purpose). The setup person must remember to assign this value (manually) prior to running the custom macro.

Following is another version of the outside corner pickup macro. It is modified to use the universal fixture offset number-finding custom macro, and allowing the fixture offset number to be specified even though G10 is not available on the control using this program.

O9060 (Outside corner pickup program for lower left corner in *XY* and top surface in *Z*)
N1 IF [#4008 EQ 0] GOTO 3 (Test if tool-length compensation is in canceled state)

N2 #3000 = 100 (CANCEL TLC)
N3 #102 = 4. (Fixture offset number to be set. Setup person must set this value)
N4 G65 P9502 F#102 (Call universal fixture offset number-finding custom macro to determine correct system variables)
N5 #[#101] = 0 (Set desired fixture offset number's program zero point to the machine's reference point in *X*)
N6 #[#101 + 1] = 0 (Set desired fixture offset number's program zero point to the machine's reference point in *Y*)
N7 #[#101 + 2] = 0 (Set desired fixture offset number's program zero point to the machine's reference point in *Z*)
N8 G[53 + #102] (Invoke fixture offset number one)
N9 S2000 M19 (Orient spindle to ensure consistency for droop)
N10 M13 (Activate probe)
N11 G91 G01 Y0.75 F50.0 (Incrementally, fast feed in *Y*+)
N12 Z – 0.75 (Still incrementally, fast feed in *Z* to allow contact with *X*– surface)
N13 G65 P9501 A1. (Invoke probe-triggering custom macro in *X*+ direction)
N14 #[#101] = #100 (Store the *X* surface location into the *X* value of desired fixture offset)
N15 G91 G01 Y – 0.75 F50.0 (Fast feed down in *Y* to clear corner)
N16 X0.75 (Fast feed to allow probing in *Y*)
N17 G65 P9501 A2. (Invoke probe-triggering custom macro in *Y*+ direction)
N18 #[#101 + 1] = #100 (Store the *Y* surface location into *Y* value of desired fixture offset)
N19 G91 G01 Z0.75 F50.0 (Fast feed up in *Z* to clear top surface)
N20 Y0.75 (Fast feed to allow probing in *Z*)
N21 G65 P9501 A – 3. (Invoke probe-triggering custom macro in *Z*– direction)
N22 #[#101 + 2] = #100 (Store the *Z* surface location into *Z* value of desired fixture offset)
N23 IF [#101 EQ 0] GOTO 26 (If common offset is being set, do not move)
N24 G[53 + #102] (Invoke desired fixture offset)
N25 G90 G01 X0 Y0 F50.0 (Leave spindle centered on corner in *XY*)
N26 M14 (Turn probe off)
N27 M30 (End of corner pickup program)

In any command that references the fixture offset setting (N5, N6, N7, N14, N18, and N22), notice how we are using common variable #101 to determine which system variable is needed. In line N3, for example, we are setting the fixture offset number being used to four (fixture offset number four). When program O9502 is invoked in line N4, The value of common variable #101 will be set to 5281 (you'll see how in the next program). So in N5, the #[#101] word will be taken as #5281, which is the correct system variable needed for assigning the *X* value of fixture offset number four. In line N6, the #[#101 + 1] word will be taken as #5282 (the *Y* register of fixture offset number four). And in line N7, the #[#101 + 2] word will be taken as #5283 (the *Z* register of offset number four).

The following program is the universal fixture offset number-finding custom macro that will actually return (in common variable #101) the proper *X*-axis system variable for the fixture offset selected. Local variable #9 represents argument F coming from the call statement in the outside corner pickup custom macro.

```
O9502 (Universal fixture offset number-finding custom macro)
IF [#9 EQ 0] GOTO 1 (Test if common fixture offset)
IF [#9 EQ 1.0] GOTO 2 (Test if fixture offset number one)
IF [#9 EQ 2.0] GOTO 3 (Test if fixture offset number two)
IF [#9 EQ 3.0] GOTO 4 (Test if fixture offset number three)
IF [#9 EQ 4.0] GOTO 5 (Test if fixture offset number four)
IF [#9 EQ 5.0] GOTO 6 (Test if fixture offset number five)
IF [#9 EQ 6.0] GOTO 7 (Test if fixture offset number six)
#3000 = 101 (PROBLEM WITH F) (If this command is read, generate alarm)
N1 #101 = 5201 (Set #101 to system variable number [X value] of common fixture offset)
GOTO 99 (Exit custom macro)
N2 #101 = 5221 (Set #101 to system variable number [X value] of fixture offset number one)
GOTO 99 (Exit custom macro)
N3 #101 = 5241 (Set #101 to system variable number [X value] of fixture offset number two)
GOTO 99 (Exit custom macro)
N4 #101 = 5261 (Set #101 to system variable number [X value] of fixture offset number three)
GOTO 99 (Exit custom macro)
N5 #101 = 5281 (Set #101 to system variable number [X value] of fixture offset number four)
GOTO 99 (Exit custom macro)
N6 #101 = 5301 (Set #101 to system variable number [X value] of fixture offset number five)
GOTO 99 (Exit custom macro)
N7 #101 = 5321 (Set #101 to system variable number [X value] of fixture offset number six)
N99 M99 (End of custom macro)
```

Admittedly, all of this may be getting a bit complicated, especially considering that you will have to remember how the universal fixture offset number-finding custom macro works each time you use it. You also will have to remember not to use common variable #101 for any other purpose during your program zero-finding custom macros. For these reasons, it is a good idea to thoroughly document any custom macro you write, especially your *universal* custom macros that are used often by other programs.

Picking up other corners. Though we have been improving it, our corner pickup probing program still picks up only one corner of a rectangular workpiece. Of course, your setup people may have to pick up any of the other corners, possibly even in the same setup. You can make your outside corner pickup probing program much more powerful by allowing any of the corners to be located. Though there are probably many ways to handle this, our method accomplishes this by assigning a *corner number* to each potential outside corner. Figure 9-4 shows the corner numbering.

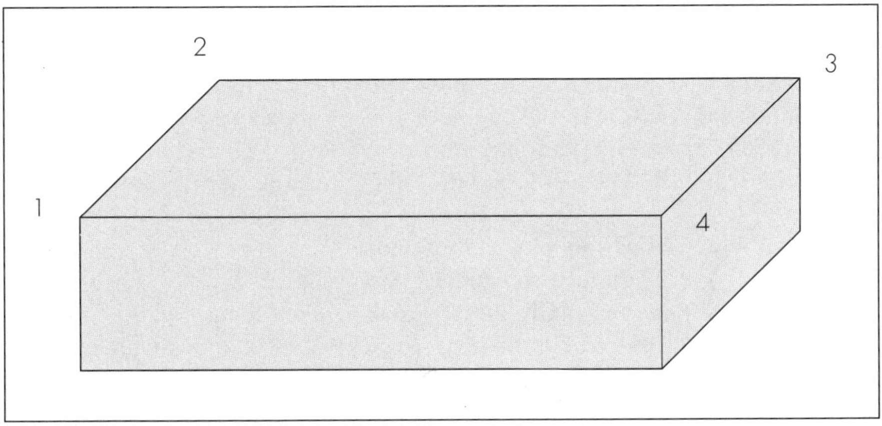

Figure 9-4. Corner number assignments for corner pickup probing program.

The setup person can easily specify which corner is being touched with a simple common variable, assigned early in the corner pickup program. Our example uses common variable #102 for this purpose. Based on the four possibilities, four IF statements can be given to determine which corner is being picked up. And based on the findings of the IF statements, the correct corner can be probed. As you will see, our example program is quite lengthy. It can be dramatically shortened by applying some rather advanced techniques involving *XY* exchange and mirror image. With these features, the same set of probing commands can be used for all four corners. Probe manufacturers tend to take advantage of these features to keep their standard programs as short as possible, which saves memory space. However, we do not use these features to make it as easy as possible to understand the program. Our example retains the ability to specify which fixture offset is being set and uses the easier G10 method of actually setting fixture offsets.

```
O9060 (Outside corner pickup program for lower left corner in XY and top sur-
   face in Z)
IF [#4008 EQ 0] GOTO 1
```

#3000 = 100 (CANCEL TLC)
N1 #101 = 1 (Fixture offset number being set. Setup person sets this value prior to activating)
N2 #102 = 1 (Corner number to pickup. Setup person sets this value prior to activating)
N3 G90 G10 L2 P#101 X0 Y0 Z0 (Set desired fixture offset number's program zero point to the machine's reference point in *X*, *Y*, and *Z*)
N4 G[53 + #101] (Invoke desired fixture offset)
N5 S2000 M19 (Orient spindle to ensure consistency for droop)
N6 M13 (Activate probe)
N7 IF [#102 EQ 1.] GOTO 12 (Begin for lower left corner)
N8 IF [#102 EQ 2.] GOTO 26 (Begin for upper left corner)
N9 IF [#102 EQ 3.] GOTO 40 (Begin for upper right corner)
N10 IF [#102 EQ 4.] GOTO 54 (Begin for lower right corner)
N11 #3000 = 102 (BAD CORNER NUMBER) (Alarm activates if #102 is improperly set)
N12 (Begin lower left corner pickup movements)
N13 G91 G01 Y0.75 F50.0 (Incrementally, fast feed in *Y*+)
N14 Z – 0.75 (Still incrementally, fast feed in *Z* to allow contact with *X*– surface)
N15 G65 P9501 A1. (Invoke probe-triggering custom macro in *X*+ direction)
N16 G90 G10 L2 P#101 X#100 (Store *X* surface location into the *X* value of chosen fixture offset)
N17 G91 G01 Y – 0.75 F50.0 (Fast feed down in *Y* to clear corner)
N18 X0.75 (Fast feed to allow probing in *Y*)
N19 G65 P9501 A2. (Invoke probe-triggering custom macro in **Y**+ direction)
N20 G90 G10 L2 P#101 Y#100 (Store *Y* surface location into *Y* value of chosen fixture offset)
N21 G91 G01 Z0.75 F50.0 (Fast feed up in *Z* to clear top surface)
N22 Y0.75 (Fast feed to allow probing in *Z*)
N23 G65 P9501 A – 3. (Invoke probe-triggering custom macro in *Z*– direction)
N24 G90 G10 L2 P#101 Z#100 (Store *Z* surface location into *Z* value of chosen fixture offset)
N25 GOTO 99 (Exit custom macro)
N26 (Begin upper left corner pickup movements)
N27 G91 G01 Y – 0.75 F50.0 (Incrementally, fast feed in *Y*–)
N28 Z – 0.75 (Still incrementally, fast feed in *Z* to allow contact with *X*– surface)
N29 G65 P9501 A1. (Invoke probe-triggering custom macro in *X*+ direction)
N30 G90 G10 L2 P#101 X#100 (Store *X* surface location into the *X* value of chosen fixture offset)
N31 G91 G01 Y0.75 F50.0 (Fast feed up in *Y* to clear corner)
N32 X0.75 (Fast feed to allow probing in *Y*)
N33 G65 P9501 A – 2. (Invoke probe-triggering custom macro in *Y*– direction)

N34 G90 G10 L2 P#101 Y#100 (Store *Y* surface location into *Y* value of chosen fixture offset)
N35 G91 G01 Z0.75 F50.0 (Fast feed up in *Z* to clear top surface)
N36 Y – 0.75 (Fast feed to allow probing in *Z*)
N37 G65 P9501 A – 3. (Invoke probe-triggering custom macro in *Z*– direction)
N38 G90 G10 L2 P#101 Z#100 (Store *Z* surface location into *Z* value of chosen fixture offset)
N39 GOTO 99 (Exit custom macro)
N40 (Begin upper right corner pickup movements)
N41 G91 G01 Y – 0.75 F50.0 (Incrementally, fast feed in Y–)
N42 Z – 0.75 (Still incrementally, fast feed in *Z* to allow contact with *X*– surface)
N43 G65 P9501 A – 1. (Invoke probe-triggering custom macro in *X*– direction)
N44 G90 G10 L2 P#101 X#100 (Store *X* surface location into the *X* value of chosen fixture offset)
N45 G91 G01 Y0.75 F50.0 (Fast feed up in *Y* to clear corner)
N46 X – 0.75 (Fast feed to allow probing in *Y*)
N47 G65 P9501 A – 2. (Invoke probe-triggering custom macro in *Y*– direction)
N48 G90 G10 L2 P#101 Y#100 (Store *Y* surface location into *Y* value of chosen fixture offset)
N49 G91 G01 Z0.75 F50.0 (Fast feed up in *Z* to clear top surface)
N50 Y – 0.75 (Fast feed to allow probing in *Z*)
N51 G65 P9501 A – 3. (Invoke probe-triggering custom macro in *Z*– direction)
N52 G90 G10 L2 P#101 Z#100 (Store *Z* surface location into *Z* value of chosen fixture offset)
N53 GOTO 99 (Exit custom macro)
N54 (Begin lower right corner pickup movements)
N55 G91 G01 Y0.75 F50.0 (Incrementally, fast feed in *Y*+)
N56 Z – 0.75 (Still incrementally, fast feed in *Z* to allow contact with *X*– surface)
N57 G65 P9501 A – 1. (Invoke probe-triggering custom macro in *X*– direction)
N58 G90 G10 L2 P#101 X#100 (Store *X* surface location into the *X* value of chosen fixture offset)
N59 G91 G01 Y – 0.75 F50.0 (Fast feed down in *Y* to clear corner)
N60 X – 0.75 (Fast feed to allow probing in *Y*)
N61 G65 P9501 A2. (Invoke probe triggering custom macro in *Y*+ direction)
N62 G90 G10 L2 P#101 Y#100 (Store Y surface location into *Y* value of chosen fixture offset)
N63 G91 G01 Z0.75 F50.0 (Fast feed up in *Z* to clear top surface)
N64 Y0.75 (Fast feed to allow probing in *Z*)
N65 G65 P9501 A – 3. (Invoke probe-triggering custom macro in *Z*– direction)
N66 G90 G10 L2 P#101 Z#100 (Store *Z* surface location into *Z* value of chosen fixture offset)
N99 IF [#101 EQ 0] GOTO 102 (If a common fixture offset is being set, do not move)

N100 G[53 + #101] (Invoke desired fixture offset)
N101 G90 G01 X0 Y0 F50.0 (Leave spindle centered on corner in *XY*)
N102 M14 (Turn probe off)
N103 M30 (End of corner pickup program)

Again, this program is quite long. However, it involves nothing more than the same techniques already shown, and it should still be relatively easy to understand.

Automating your corner pickup probing program. Even the most recent version of the outside corner pickup program still requires that the operator manually move the stylus 0.50 in. (12.7 mm) from the corner being located. As long as only one corner is being located, and as long as a setup person is available to perform this function (not an unattended system), this corner pickup program will work quite nicely. However, many probing applications require more automatic operation. Here we show a method of calling the corner pickup custom macro program from another program.

As you know from Part One of this book, when a custom macro B program is called from another program, an M99 must be used with which to end the custom macro, meaning that the program ending word (currently M30) for our corner pickup program must be changed to M99 so that a completely automatic operation can be executed.

To actually activate the corner pickup program, a G65 calling command can be given. If there are any arguments to pass (such as fixture offset number or which corner number is being picked up), the use of letter address arguments and local variables can be used.

For example, suppose you have a setup that requires the measuring of four program zero locations. The programmer knows the *approximate* location of each relative to the machine's reference position, but wishes to use the spindle probe to measure the precise location of each program zero point. Figure 9-5 shows the setup requiring the program zero measurements.

Notice that corner number one is the program zero location for the two leftmost workpieces and corner number four is the program zero location for the two

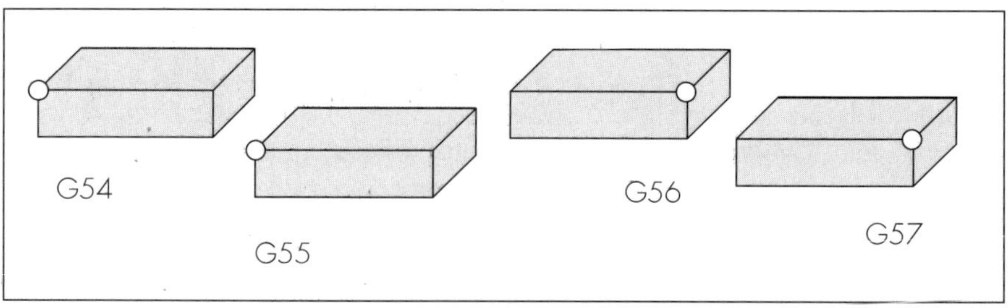

Figure 9-5. Program setup that automatically picks up four program zero points using the corner-finding probing program.

rightmost workpieces. Since the setup person, fixture designer, or programmer knows the approximate distances from the machine's reference position to each program zero point, we use the G53 command (motion relative to reference position) to cause motion to the initial position for probing each corner.

The setup program that will successfully measure all four workpieces follows. Note that the letter address F specifies the fixture offset number and C specifies the corner number. (The program still assumes that the probe is manually loaded into the spindle prior to activation. Of course, a simple tool-changing command could be easily added to allow automatic probe loading. Program zero points refer to those shown in Figure 9-4.)

O0001 (Setup program to measure four program zero points)
N005 G53 X – 25.0 Y – 15.0 (Rapid in *XY* to beginning position of leftmost workpiece)
N010 G53 Z – 12.0 (Move to probing position in *Z*)
N015 G65 P9060 C1. F1. (Measure corner one and store results into fixture offset number one)
N020 G53 Z – 8.0 (Move up to clear obstructions)
N025 G53 X– 20.0 Y – 12.0 (Move to second workpiece)
N030 G53 Z – 12.0 (Move to probing position in *Z*)
N035 G65 P9060 C1. F2. (Measure second workpiece position, store results in fixture offset number two)
N040 G53 Z – 8.0 (Move up to clear obstructions)
N045 G53 X – 10.0 Y – 11.0 (Move to third workpiece)
N050 G53 Z – 12.0 (Move to probing position in *Z*)
N055 G65 P9060 C4. F3. (Measure third workpiece position, store results in fixture offset number three)
N060 G53 Z – 8.0 (Move up to clear obstructions)
N065 G53 X – 3.0 Y – 12.0 (Move to fourth workpiece)
N070 G53 Z – 12.0 (Move to probing position in *Z*)
N075 G65 P9060 C4. F4. (Measure fourth workpiece position, store results in fixture offset number four)
N080 G91 G28 Z0 (Return to machine's *Z*-axis reference position)
N085 M30 (End of program)

Our final version of the corner pickup program allows automatic activation. The argument C is being referenced as local variable #3 (preset to corner one) and F is being referenced as #9 (preset to fixture offset number one).

O9060 (Outside corner pickup program for lower left corner in *XY* and top surface in *Z*)
IF [#3 NE #0] GOTO 200
#3 = 1.

N200 IF [#9 NE #0] GOTO 1
#9 = 1.
N1 #101 = #9 (Fixture offset number being set)
N2 #102 = #3 (Corner number to pickup)
N3 G90 G10 L2 P#101 X0 Y0 Z0 (Set desired fixture offset number's program zero point to the machine's reference point in *X*, *Y*, and *Z*)
N4 G[53 + #101] (Invoke desired fixture offset)
N5 S2000 M19 (Orient spindle to ensure consistency for droop)
N6 M13 (Activate probe)
N7 IF [#102 EQ 1.] GOTO 12 (Begin for lower left corner)
N8 IF [#102 EQ 2.] GOTO 26 (Begin for upper left corner)
N9 IF [#102 EQ 3.] GOTO 40 (Begin for upper right corner)
N10 IF [#102 EQ 4.] GOTO 54 (Begin for lower right corner)
N11 #3000 = 102 (BAD CORNER NUMBER) (Alarm if #102 is improperly set)
N12 (Begin lower left corner pickup movements)
N13 G91 G01 Y0.75 F50.0 (Incrementally, fast feed in *Y*+)
N14 Z – 0.75 (Still incrementally, fast feed in *Z* to allow contact with *X*– surface)
N15 G65 P9501 A1. (Invoke probe-triggering custom macro in *X*+ direction)
N16 G90 G10 L2 P#101 X#100 (Store *X* surface location into the *X* value of chosen fixture offset)
N17 G91 G01 Y – 0.75 F50.0 (Fast feed down in *Y* to clear corner)
N18 X0.75 (Fast feed to allow probing in *Y*)
N19 G65 P9501 A2. (Invoke probe-triggering custom macro in *Y*+ direction)
N20 G90 G10 L2 P#101 Y#100 (Store *Y* surface location into *Y* value of chosen fixture offset)
N21 G91 G01 Z0.75 F50.0 (Fast feed up in *Z* to clear top surface)
N22 Y0.75 (Fast feed to allow probing in *Z*)
N23 G65 P9501 A – 3. (Invoke probe-triggering custom macro in *Z*– direction)
N24 G90 G10 L2 P#101 Z#100 (Store *Z* surface location into *Z* value of chosen fixture offset)
N25 GOTO 99 (Exit custom macro)
N26 (Begin upper left corner pickup movements)
N27 G91 G01 Y – 0.75 F50.0 (Incrementally, fast feed in *Y*–)
N28 Z – 0.75 (Still incrementally, fast feed in *Z* to allow contact with *X*– surface)
N29 G65 P9501 A1. (Invoke probe-triggering custom macro in *X*+ direction)
N30 G90 G10 L2 P#101 X#100 (Store *X* surface location into the *X* value of chosen fixture offset)
N31 G91 G01 Y0.75 F50.0 (Fast feed up in *Y* to clear corner)
N32 X0.75 (Fast feed to allow probing in *Y*)
N33 G65 P9501 A – 2. (Invoke probe-triggering custom macro in *Y*– direction)
N34 G90 G10 L2 P#101 Y#100 (Store *Y* surface location into *Y* value of chosen fixture offset)
N35 G91 G01 Z0.75 F50.0 (Fast feed up in *Z* to clear top surface)

N36 Y – 0.75 (Fast feed to allow probing in *Z*)
N37 G65 P9501 A – 3. (Invoke probe-triggering custom macro in *Z*– direction)
N38 G90 G10 L2 P#101 Z#100 (Store *Z* surface location into *Z* value of chosen fixture offset)
N39 GOTO 99 (Exit custom macro)
N40 (Begin upper right corner pickup movements)
N41 G91 G01 Y – 0.75 F50.0 (Incrementally, fast feed in *Y*–)
N42 Z – 0.75 (Still incrementally, fast feed in *Z* to allow contact with *X*– surface)
N43 G65 P9501 A – 1. (Invoke probe-triggering custom macro in *X*– direction)
N44 G90 G10 L2 P#101 X#100 (Store *X* surface location into the *X* value of chosen fixture offset)
N45 G91 G01 Y0.75 F50.0 (Fast feed up in *Y* to clear corner)
N46 X – 0.75 (Fast feed to allow probing in *Y*)
N47 G65 P9501 A – 2. (Invoke probe-triggering custom macro in *Y*– direction)
N48 G90 G10 L2 P#101 Y#100 (Store *Y* surface location into *Y* value of chosen fixture offset)
N49 G91 G01 Z0.75 F50.0 (Fast feed up in *Z* to clear top surface)
N50 Y – 0.75 (Fast feed to allow probing in *Z*)
N51 G65 P9501 A – 3. (Invoke probe-triggering custom macro in *Z*– direction)
N52 G90 G10 L2 P#101 Z#100 (Store *Z* surface location into *Z* value of chosen fixture offset)
N53 GOTO 99 (Exit custom macro)
N54 (Begin lower right corner pickup movements)
N55 G91 G01 Y0.75 F50.0 (Incrementally, fast feed in *Y*+)
N56 Z – 0.75 (Still incrementally, fast feed in *Z* to allow contact with *X*– surface)
N57 G65 P9501 A – 1. (Invoke probe-triggering custom macro in *X*– direction)
N58 G90 G10 L2 P#101 X#100 (Store *X* surface location into the *X* value of chosen fixture offset)
N59 G91 G01 Y – 0.75 F50.0 (Fast feed down in *Y* to clear corner)
N60 X – 0.75 (Fast feed to allow probing in *Y*)
N61 G65 P9501 A2. (Invoke probe-triggering custom macro in *Y*+ direction)
N62 G90 G10 L2 P#101 Y#100 (Store *Y* surface location into *Y* value of chosen fixture offset)
N63 G91 G01 Z0.75 F50.0 (Fast feed up in *Z* to clear top surface)
N64 Y0.75 (Fast feed to allow probing in *Z*)
N65 G65 P9501 A – 3. (Invoke probe-triggering custom macro in *Z*– direction)
N66 G90 G10 L2 P#101 Z#100 (Store *Z* surface location into *Z* value of chosen fixture offset)
N99 IF [#101 EQ 0] GOTO 102 (If a common fixture offset is being set, do not move)
N100 G[53 + #101] (Invoke fixture offset number one)
N101 G90 G01 X0 Y0 F50.0 (Leave spindle centered on corner in *XY*)
N102 M14 (Turn probe off)
N103 M99 (End of corner pickup program)

Admittedly, we have spent a lot of time and effort discussing just one setup help-related application for spindle probes. Nonetheless, many of the techniques shown can be applied to a variety of probing applications.

A hole center-finding probing program

This probing program uses many of the same techniques shown to this point. An operator (or programmed command) can bring the stylus close to the hole center in *X* and *Y*, and slightly below the work surface in *Z* (note that the *Z*-axis program zero point is *not* being set by this example program). At this point, the hole center-finding program can be invoked (manually or by a programmed command). The spindle probe stylus is then driven into each surface of the hole (*X*+, *X*–, *Y*+, and *Y*–) to find each surface location, then the hole's center line position can be easily calculated based on these surface locations. The program zero assigning values can then be placed into the related fixture offset. Figure 9-6 shows the application for our example.

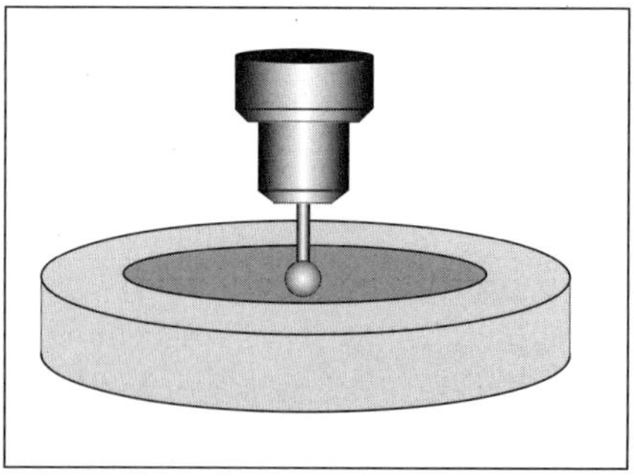

Figure 9-6. Spindle probe application for hole center-finding probing program.

The version of the hole center-finding program we show here will be manually activated and will store the results of the program zero measurement in the selected fixture offset (selected by common variable #101). Note that this program also requires that the setup person enter the *approximate* hole diameter (in #102) so that the stylus can be quickly brought to within a small distance (0.50 in. [12.7 mm]) of each surface to be probed. Note that only if the hole is more than 1.25 in. (31.75 mm) in diameter will the time-saving positioning movements be made. Also note that if the hole is less than 0.375 in. (9.525 mm) in diameter, the retract distance (0.25 in. [6.35 mm]) in the universal probe-triggering macro will cause the stylus to hit during the retract motion. This can be avoided in one of two ways. Either the retract distance in program O9501 can be made smaller, say 0.05 in. (1.27 mm). Or (as this program does), an alarm can be sounded if the hole is less than 0.375 in. (9.525 mm) in diameter.

O9061 (Hole center-finding probing program)
N1 #101 = 1 (Fixture offset number to be used)
N2 #102 = 3.0 (Approximate hole diameter)
N3 IF [#102 GT 0.375] GOTO 5 (Test if hole is too small for retract movement)
N4 #3000 = 103 (HOLE TOO SMALL) (Generate alarm if hole is less than 0.375 in. [9.525 mm])

N5 S200 M19 (Orient spindle to ensure consistent droop)
N6 M13 (Turn probe on)
N7 G90 G10 L2 P#101 X0 Y0 Z0 (Set machine's reference position as program zero point)
N8 G[53 + #101] (Invoke desired fixture offset)
N9 IF [#102 LE 1.25] GOTO 11 (Test if fast approach movement is required)
N10 G91 G01 X – [#102/2 – 0.5] F50.0 (If required, feed up close to surface)
N11 G65 P9501 A – 1. (Call universal probe-triggering custom macro, probe *X*–)
N12 #111 = #100 (Store left side *X* position)
N13 G91 G01 X[#102/2 – #500] F50.0 (Fast feed back to approximate hole center in *X*)
N14 IF [#102 LE 1.25] GOTO 16 (Test if fast approach movement is required)
N15 G91 G01 X[#102/2 – 0.5] F50.0 (If needed, fast feed up close to surface)
N16 G65 P9501 A1. (Call universal probe-triggering custom macro, probe *X*+)
N17 #112 = #100 (Store right side *X* position)
N18 G90 G10 L2 P#101 **X[[#111 + #112]/2]** (Set fixture offset *X* value)
N19 G91 G01 **X – [#112 – [[#111 + #112]/2]]** F50.0 (Fast feed back to precise center in *X*)
N20 IF [#102 LE 1.25] GOTO 22 (Test if fast approach movement is required)
N21 G91 G01 Y – [#102/2 – 0.5] F50.0 (If needed, fast feed up close to surface)
N22 G65 P9501 A – 2. (Call universal probe-triggering custom macro, probe *Y*–)
N23 #111 = #100 (Store front side *Y* position)
N24 G91 G01 Y[#102/2 – #500] F50.0 (Fast feed back to approximate hole center in *Y*)
N25 IF [#102 LE 1.25] GOTO 27 (Test if fast approach movement is required)
N26 G91 G01 Y[#102/2 – 0.5] F50.0 (If needed, fast feed up close to surface)
N27 G65 P9501 A2. (Call universal probe-triggering custom macro, probe *Y*+)
N28 #112 = #100 (Store back side *Y* position)
N29 G90 G10 L2 P#101 **Y[[#111 + #112]/2]** (Set fixture offset *Y* value)
N30 G91 G01 **Y – [#112 [[#111 + #112]/2]]** F50.0 (Fast feed back to precise center in *Y*. Spindle center line [not necessarily stylus ball center line] will be on hole center)
N31 M14 (Turn probe off)
N32 M30 (End of custom macro)

If you understand our previous discussions, you should be able to understand what is going on in this hole center-finding program. The only commands you may be questioning are those related to actually calculating the hole center position and moving to its location. In lines N18, N19, N29, and N30, we simply add the first surface position (#111) to the second surface measurement (#112) and divide the result by two. In all cases, this will render the center location of the hole being located. Additionally, tests are added to determine whether the hole is large enough to probe (lines N3 and N4) and whether a fast feed approach is required (lines N9, N14, N20, etc.).

Other program zero-finding applications

You should now be ready to tackle just about any program zero-locating probing program. Though we show no more programs, we do wish to at least list some other possibilities and briefly discuss how you would approach them.

Center of round boss. Using much the same approach as just shown for finding the center of a hole, you can easily develop a probing program to locate the center of a round (or rectangular) boss. The only major difference is in the approach to each probing surface. You might elect to have the setup person specify (with a common variable argument at the beginning of the program), the approximate diameter of the boss so the probing program can determine how far to move the probe in each approach movement.

Center of slot or key. Possibly you have qualified workholding fixtures that, once placed on the table, require little in the way of measurements. However, these fixtures locate from the machine tool's key slots, meaning you will have to know the center line location of each slot. From what you know so far, you should be able to write a probing program that locates the center of any key slot (*X* or *Y*).

For original positioning, you could have the setup person manually bring the stylus to the approximate center of the slot. To determine which kind of slot is being measured (*X* or *Y*) a common variable set by the setup person early in the custom macro could be used. For example, if set to a value of 1., a slot parallel to the *X* axis is being measured. If set to 2., a slot parallel to *Y* is being measured. To actually find the center of the slot, techniques similar to those used to find the center of a hole or boss can be used.

Special locators on fixtures. When faced with odd-shaped workpieces, the tool designer will often find the best or easiest surface of the workpiece from which to locate. Locators (also called *bump stops*) are placed in strategic positions to allow the workpiece to be sufficiently positioned, supported, and clamped. Most CNC programmers will choose the workpiece surfaces supported by the three most important locators (one for each axis) as the program's zero point.

Knowing each locator's approximate position (from the fixture drawing), the probe programmer can easily develop a special probing program for the fixture being probed. The setup person can be told to manually position the stylus to within 0.50 in. (12.7 mm) of the first locator to be probed. When activated, the probing program could be totally automatic, probing and moving from one important locator to the next, finding and storing the proper program zero assigning values.

COMPENSATING FOR DIFFICULT WORKHOLDING PROBLEMS

Spindle probes can provide more help during setup than simply helping to measure the program zero location. There may be times when, even from one cycle to the next, problems arise that make it nearly impossible to perfectly locate workpieces in the setup. In Chapter 7, we discussed many of the pros and cons of applying spindle probes to applications requiring their use in every cycle. Here we simply show the applications, with the presumption that it has been deemed the best choice to use them in this manner.

Compensating for shifting location surfaces

Castings and forgings are among the workpieces most prone to variations from one piece to the next. Generally speaking, the larger the casting or forging, the more the potential variation. With small workpieces, it may be possible to design hard tooling to compensate for the variation from one workpiece to the next with adjustable locators and targeting devices. However, as workpiece size increases, it becomes increasingly difficult (and expensive) to design hard tooling to hold and locate the workpiece properly.

However, these types of workpieces can be a very good application for spindle probes. Using essentially the same techniques to locate the program zero point shown in our last discussion, you can write a probing program to locate the surface(s) of the workpiece that are at variance from one to another. However, there is a subtle distinction. In our previous discussion, the spindle probe was used only one time (during setup). Here we use the spindle probe for every workpiece, meaning the spindle probe must be used in *every cycle*, and must be activated efficiently and automatically.

Figure 9-7 shows the round boss of a casting that varies in location from one workpiece to the next. While other machining operations may be performed on this workpiece, it is most important (for this particular workpiece) that the operations being machined on the boss be *central to the cored hole*. This means that it is not necessary to maintain close relationships to other machining operations being performed in other areas of the workpiece (in assembly, for example, maybe a cap is mounted to this round boss).

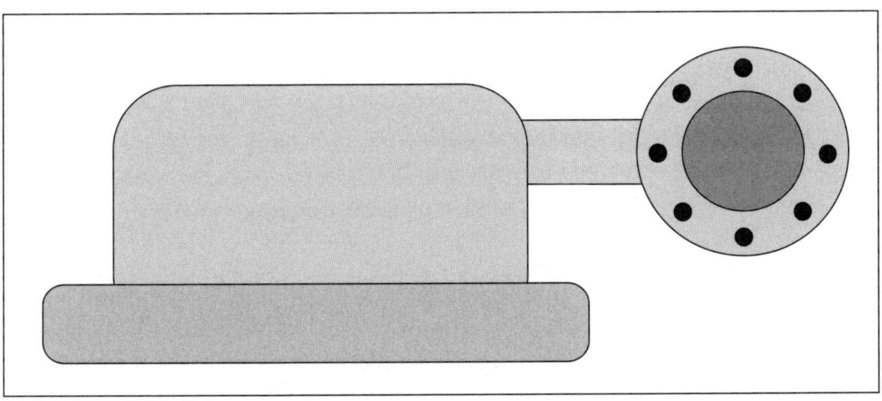

Figure 9-7. The round boss of this casting varies from one workpiece to the next, making it necessary that machining on the boss be central to the cored hole.

Here is a portion of the machining (CNC) program that calls the probe into the spindle and activates the probing program.

O0002 (Machining program for casting including boss)
N005 T01 M06 (Place first tool in spindle)
N010 G54 G90 S500 M03 (Select coordinate system, absolute mode, start spindle)
N015 G00 X5.0 Y2.0 (Rapid into position)
N020 G43 H01 Z.1 (Instate tool-length compensation)
N025 M08 (Start coolant flow)
N030 G81 R.1 Z – 0.4 F4.5 (Center drill first hole)
.
. (Other machining operations)
.
N350 G91 G28 Z0 M19 (Return to tool-change position, cancel tool-length compensation, orient spindle)
N355 M01 (Optional stop)
N360 T07 M06 (Place **probe** in spindle)
N363 #2007 = #505 (Store probe length in offset number seven)
N365 G54 G90 (Select coordinate system, absolute mode)
N370 G00 X8.0 Y5.0 (Rapid to approximate center of cored hole)
N375 **G43 H07 Z0.3** (Rapid stylus close to cored cavity)
N380 G01 **Z – 0.25** F50. (Fast feed 0.25 in. [6.35 mm] below cored cavity surface)
N385 G65 P9062 D5.0 F2.0 (Probe hole center, D = diameter, F = fixture offset number)
N390 G91 G28 Z0 M19 (Rapid to tool-change position)
N395 M01 (Optional stop)
N400 T08 M06 (Change tools to 3 in. [76.2 mm] face mill)
N405 **G55** G90 S800 M03 (Select coordinate system, absolute mode, start spindle)
N410 G00 X6.0 Y0 (Rapid to first position)
N415 G43 H08 Z0 (Rapid to work surface, instate tool-length compensation)
N420 M08 (Start coolant flow)
N425 G01 X3.25 F8.0 (Feed into workpiece)
N430 G02 I – 3.25 (Machine full circle to finish boss)
N435 G01 X6.0 (Feed off workpiece)
N440 G91 G28 Z0 M19 (Rapid to tool-change position, orient spindle)
N445 M01 (Optional stop)
N450 T09 M06 (Change tools to center drill)
N455 G55 G90 S700 M03 (Select coordinate system, absolute mode, start spindle)
N460 G00 X3.25 Y0 (Rapid to first hole location)
N465 G43 H09 Z.25 (Instate tool-length compensation)
N470 M08 (Start coolant flow)
N475 G81 R.1 Z – 0.4 F4.5 (Center drill first hole on bolt hole pattern)

```
N480 X2.2981 Y2.2981 (Machine second hole)
N485 Y3.25 (Machine third hole)
.
.
.
```

Note that in block N363, the probe's length is being stored in offset number seven. Remember that the machine's *Z*-axis position after probe triggering (which is stored in system variable #5063) is influenced by tool-length compensation. To make the probing program as easy as possible to use, we program the probe stylus tip, using tool-length compensation. Of course, this means that the probe's *Z* position found in the universal probe-triggering custom macro must reflect this difference in application. The surface location will be rendered as opposed to the spindle nose position. We show an example of the modified universal probe-triggering custom macro program a little later.

In line N385, we specify the custom macro calling command. The letter D (represented by #7 in the custom macro) is the approximate diameter of the cored hole and F (represented by #9 in the custom macro) is the fixture offset number into which the program zero position will be stored.

Note that after probing, fixture offset number two (in our case) is correctly set in line N405 and the subsequent machining operations will be in relationship to the round boss.

Next, we show the hole center-finding probing program (which was shown earlier) that can be used to locate the program zero point for the cored cavity. It has been modified to allow automatic activation and to allow the arguments (D and F) to be used by the custom macro. It has further been modified to include the location of the *Z* surface at the top of the boss.

O9062 (Hole center-finding probing program)
N1 #101 = #9 (Fixture offset number to be used)
N2 #102 = #7 (Approximate hole diameter)
N3 IF [#102 GT 0.375] GOTO 5 (Test if hole is too small for retract movement)
N4 #3000 = 103 (HOLE TOO SMALL) (Generate alarm if hole is under 0.375 in. [9.525 mm])
N5 S200 M19 (Orient spindle to ensure consistent droop)
N6 M13 (Turn probe on)
N7 G90 G10 L2 P#101 X0 Y0 Z0 (Set machine's reference position as program zero point)
N8 G[53 + #101] (Invoke desired fixture offset)
N9 IF [#102 LE 1.25] GOTO 11 (Test if fast approach movement is required)
N10 G91 G01 X – [#102/2 – 0.5] F50.0 (If required, feed up close to surface)
N11 G65 P9501 A – 1. (Call universal probe-triggering custom macro, probe *X*–)
N12 #111 = #100 (Store left side *X* position)

N13 G91 G01 X[#102/2 – #500] F50.0 (Fast feed back to the approximate hole center in *X*)
N14 IF [#102 LE 1.25] GOTO 16 (Test if fast approach movement is required)
N15 G91 G01 X[#102/2 – 0.5] F50.0 (If needed, fast feed up close to surface)
N16 G65 P9501 A1. (Call universal probe-triggering custom macro, probe *X*+)
N17 #112 = #100 (Store right side *X* position)
N18 G90 G10 L2 P#101 X[[#111 + #112]/2] (Set fixture offset *X* value)
N19 G91 G01 X – [#112 – [[#111 + #112]/2]] F50.0 (Fast feed back to precise center in *X*)
N20 IF [#102 LE 1.25] GOTO 22 (Test if fast approach movement is required)
N21 G91 G01 Y – [#102/2 – 0.5] F50.0 (If needed, fast feed up close to surface)
N22 G65 P9501 A – 2. (Call universal probe-triggering custom macro, probe *Y*–)
N23 #111 = #100 (Store front side *Y* position)
N24 G91 G01 Y[#102/2 – #500] F50.0 (Fast feed back to the approximate hole center in *Y*)
N25 IF [#102 LE 1.25] GOTO 27 (Test if fast approach movement is required)
N26 G91 G01 Y[#102/2 – 0.5] F50.0 (If needed, fast feed up close to surface)
N27 G65 P9501 A2. (Call universal probe-triggering custom macro, probe *Y*+)
N28 #112 = #100 (Store back side *Y* position)
N29 G90 G10 L2 P#101 Y[[#111 + #112]/2] (Set fixture offset *Y* value)
N30 G91 G01 Z0.75 F50.0 (Fast feed out of hole)
N31 Y0.75 (Fast feed up in *Y* to allow *Z* probing)
N32 G65 P9501 A – 3. C1.0 (Call universal probe-triggering custom macro, probe *Z*–)
N33 G90 G10 L2 P#101 Z[#100 – 0.1] (Store the *Z* surface minus stock in fixture offset)
N34 M14 (Turn probe off)
N35 M99 (End of custom macro)

Compensating for difficult setups

If you see your setup people struggling with a workholding setup or with workpiece loading, it may be a good time to evaluate the potential for using a spindle probe. If they are having trouble precisely locating fixtures or workpieces, whether the problem relates to precisely locating in a position or square with the table, it is likely that a spindle probe can help. Based on what you now know, you could, for example, develop a probing program to handle a large workpiece that is not mounted perfectly square with the table.

Perhaps a workpiece is so large that the operator is spending a great deal of time locating it precisely square with the table. You could develop a probing program that would quickly and easily establish the position of program zero, as well as the angular amount of coordinate rotation needed for the CNC program for each workpiece. The probing program would require about 30 seconds to run.

To handle this application, you would first need to determine the angular imperfection (how *out-of-square* the workpiece is with the table). This can be done simply by probing two surfaces at a known interval, say at an *X* position about 1 in. (25.4 mm) from the program zero point and the other precisely at 10 in. (254 mm) from the first probing point in *X*. By determining the *Y*-axis position at each point, you can easily calculate the *Y*-axis distance from one point to the other. Using simple right-angle trigonometry, you can easily calculate the workpiece's angle of imperfection. By probing just once more for the position at the left side of the workpiece, and by knowing the *Y*-axis position when this probing is done, you can calculate the true position in *X* and *Y* for the lower left-hand corner of the workpiece.

Stacking workpieces. Another time a spindle probe may be helpful in compensating for problems encountered during workpiece loading is in the *stacking* of workpieces. Some companies will sandwich two or more workpieces in a vise to allow the machining of several workpieces. However, if there is any deviation in thickness among the workpieces being machined, such deviation may cause sizing problems with the workpieces being machined. While the thickness of each individual workpiece may fall within the tolerance band, the *aggregate variation* will compound the problem when workpieces are stacked. If a hard-and-fixed CNC program is run without consideration for thickness variations, it is likely that one or more of the workpieces being machined will have to be scrapped.

To solve this problem, you could have the spindle probe find the overall distance from one end of the stacked workpieces to the other. The probing custom macro could then divide this overall distance by the number of workpieces being stacked to determine the *average* workpiece thickness in this setup. This average thickness could be stored in a common variable (in the #100 series or #500 series). During the machining program, instead of moving through hard-and-fixed movements, the movement commands can reference the value stored in the previously calculated variable. The machining program's movements then change with every cycle, based on the average thickness of workpieces in the current cycle.

While this method is still not perfect (the average thickness is not perfectly correct for every workpiece), at least the machining operations being performed should be within the tolerance band for each workpiece machined. That is, imperfections will no longer be compounded from one workpiece to the next.

PROGRAMMING IN-PROCESS GAGING APPLICATIONS

We discussed the pros and cons of in-process gaging during our discussion of probing applications in Chapter 7. A short review may be beneficial if there is any doubt as to whether your company will benefit from in-process gaging in your CNC process.

In truth, a spindle probe can be used to measure the value of just about any dimension given on a print. Moreover, certain surface relationship (also called geometric) tolerances, like parallelism and perpendicularity, can be measured with spindle probes. Though there are some attributes of a workpiece that a spindle probe cannot measure (surface finish, hardness, and others that do not result in a distance related-value), a spindle probe can provide a great deal of help with your measuring tasks.

For our in-process gaging applications, we will use the same calibration values as shown during the discussion of probe calibration in Chapter 3, and use the one-touch probing cycle. In fact, we will often use the same universal probe-triggering custom macro (O9501) shown earlier in this chapter to actually trigger the probe. We are assuming, of course, that your CNC control has a very short lag time (about 2 microseconds), which almost eliminates the problem of overshoot during probing and allows accurate probing even at relatively fast feedrates (our calibration program is triggered at 30 in. [762 mm] per minute). Of course, your probe manufacturer may recommend the two-touch probing cycle if your CNC control's lag time is excessive. We also will be using the inch mode with which to measure to keep our discussion quick to grasp for people in the U.S. However, you know from Chapter 8 that the metric mode will provide better resolution, a point that makes the metric mode more desirable for the higher accuracies required for in-process gaging.

In the first segment of this presentation we will simply show how you can figure out the values for dimensions being measured. Many of the same techniques shown thus far will be used. However, as you should remember from Chapter 7, coming up with a given dimension is often the least difficult part of in-process gaging. What you have the machine do based on the findings of in-process gaging can be dramatically more difficult. We'll address that issue after you know how to take measurements.

IN-PROCESS GAGING CONSIDERATIONS

While many of the same techniques carry over from the *help-with-setup* applications, in-process gaging does require some special considerations.

Point of reference

In almost all in-process gaging applications, the point of reference must be the program zero point for the workpiece being machined (*not* the machine's reference position). When measuring a hole location, for example, it is important that the dimension being measured reflect the dimension specified on the workpiece drawing. If an *X* position to the center of the hole is specified as 3.50 in. (88.90 mm), for example, the resulting in-process measurement should be 3.50 in. as well. For this to occur, the point of reference for in-process gaging *must* be the workpiece program zero. In the applications we show in this section, we assume that the correct coordinate system is instated (commonly with G54 through G59) prior to activating the in-process gaging program.

The need for automated measurements

Many of the applications having to do with help during setup require manual intervention. While this may be acceptable since the setup is made only once per production run, in-process gaging applications require a more automatic cycle. In many cases, in-process gaging must be done during *every* cycle. For this reason, all in-process gaging programs we show will be called with a G65 command (which is commonly part of the actual machining program). Additionally, to keep the commands related to calling in-process gaging programs as simple as possible, each one we show will include the position from which in-process gaging begins. The actual movements to get the probe to its starting position will be included in the in-process gaging custom macro program.

Programming Z-axis movements

As was evidenced by the automatic program zero measuring program shown earlier, it can be quite difficult to program the *Z*-axis movements of the probe if tool-length compensation is not used. For this reason, many probe programmers store the probe's length (as has been referenced to this point with permanent common variable #505) in a normal tool-length compensation offset. In fact, this could be automatically done at the time of calibration if the probe calibration program is written accordingly.

For the examples shown in this discussion of in-process gaging, we will assume that the probe is in tool station number seven, and that its length (already considering overshoot) is stored in tool offset number seven, as are all tool-length compensation values. Our universal probe-triggering custom macro also must be changed to reflect this usage difference.

While there are several ways to handle this, we simply modified the universal probe-triggering custom macro in such a way that it will still work in exactly the same fashion for all applications shown to this point (when the probe length is referenced during this program in #505). Yet, with the addition of one argument in the calling program (letter C for *compensation*), we can specify that tool-length compensation is being used, so this custom macro will not reference #505 in the position calculation during *Z*– probing. If C is left out of the call statement, (as it has been in all examples shown to this point), the custom macro will reference the probe's length from #505. If C is set to 1. (C1.), the custom macro will not reference #505.

In reality, we are setting the default condition such that #505 will be referenced as the probe's length, giving you the additional option of setting C to 1. if you wish to tell the custom macro not to reference #505. Depending on your own application, this may be reversed. If you find that you are using your spindle probes almost exclusively for in-process gaging, you may elect to set the default (in the custom macro) such that it will not reference #505 during the machine position calculation.

Here is the modified universal probe-triggering custom macro program. Note that only a few commands have changed.

```
O9501 (Universal probe-triggering macro)
IF [#3 NE #0] GOTO 21
#3 = 0.
N21 #3004 = 2 (Disable feedrate override)
IF [#1 EQ 1] GOTO 1 (X+ direction probing detected)
IF [#1 EQ – 1.] GOTO 2 (X– direction probing detected)
IF [#1 EQ 2.] GOTO 3 (Y+ direction probing detected)
IF [#1 EQ – 2.] GOTO 4 (Y– direction probing detected)
IF [#1 EQ 3] GOTO 5 (Z– direction probing detected)
IF [#1 EQ – 3.] GOTO 5 (Z– direction probing detected)
#3004 = 0 (Problems exist with A word, so enable feedrate override)
#3000 = 100 (LETTER A IMPROPERLY SET) (Generate A word alarm)
N1 (Begin X+ direction probing)
G91 G31 X0.75 F30.0 (Drive stylus into surface)
#100 = #5061 – #501 + #500 (Calculate X surface position)
G91 G01 X – 0.25 F50.0 (Retract 0.25 in. [6.35 mm] from probed surface)
GOTO 99 (Exit custom macro)
N2 (Begin X– direction probing)
G91 G31 X – 0.75 F30.0 (Drive stylus into surface)
#100 = #5061 + #502 – #500 (Calculate X surface position)
G91 G01 X0.25 F50.0 (Retract 0.25 in. [6.35 mm] from probed surface)
GOTO 99 (Exit custom macro)
N3 (Begin Y+ direction probing)
G91 G31 Y0.75 F30.0 (Drive stylus into surface)
#100 = #5062 – #503 + #500 (Calculate Y surface position)
G91 G01 Y – 0.25 F50.0 (Retract 0.25 in. [6.35 mm] from probed surface)
GOTO 99 (Exit custom macro)
N4 (Begin Y– direction probing)
G91 G31 Y – 0.75 F30.0 (Drive stylus into surface)
#100 = #5062 + #504 – #500 (Calculate Y surface position)
G91 G01 Y0.25 F50.0 (Retract 0.25 in. [6.35 mm] from probed surface)
GOTO 99 (Exit custom macro)
N5 (Begin Z– direction probing)
G91 G31 Z – 0.75 F30.0 (Drive stylus into Z surface)
IF [#3 EQ 1] GOTO 11
#100 = #5063 – #505
GOTO 12
N11 #100 = #5063 (Calculate Z surface position)
N12 G91 G01 Z0.25 F50. (Retract 0.25 in. [6.35 mm] from probed surface)
```

```
N99 #3004 = 0 (Enable feedrate override)
M99 (End of custom macro)
```

Now, the calling command for *Z*-axis probing when in-process gaging is:

```
G65 P9501 A – 3. C1.
```

Again the C1. is telling the control that tool-length compensation is being used to handle the probe's length, (meaning #5063 is the distance from the program zero point to the probed surface in *Z*) and not to reference permanent common variable #505 in the machine position calculation command.

Presenting measured values

We first showed how to take measurements. Then we showed how to handle the results of taking measurements. During each discussion on how to take a given measurement, the measured value will simply be placed in a permanent common variable. Remember (from Part One) that you can view the current value of any common variable on the display screen. Additionally, you have the ability to actually place a short message next to the first 20 permanent common variables (#500 through #519) to document the current meaning of the variable. We will use this technique for all of our examples related to actually taking measurements. For example, the command

```
SETVN 510 [HOLE DIA]
```

places the message HOLE DIA next to permanent common variable #510, making it quite clear that this variable represents the diameter of the hole just measured.

MEASURING DIMENSIONS

Measuring dimensions commonly requires little more than some ingenuity on the part of the spindle probe programmer. While it can sometimes be quite challenging to develop the best and most efficient manner of measuring dimensions (one that requires the fewest stylus-triggering movements and the least amount of wasted time, while still rendering accurate results), most problems you will encounter during measurement are easily solved. Admittedly there may be other (better, faster, and more accurate) ways to take any measurement we show in this section, but bear in mind that it is our most basic intention to help you understand *how* to take measurements. In many cases, we sacrifice the highly sophisticated techniques to keep our methods easy to follow. With a little experience, you will likely be able to improve on the techniques we show.

Measuring a hole's diameter

The following machining program actually mills a 2-in. (50.8-mm)-diameter counterbored hole and then measures the hole diameter just machined. Notice that the custom macro calling command at the end of this program specifies the hole position and reference diameter of the hole to be measured. While it is still rather limiting, it demonstrates nicely how your in-process gaging custom macros can be easily called from machining programs.

O0005 (Calling program)
N005 T01 M06 (Change tools, place 1-in. [25.4-mm] endmill in position)
N010 G54 G90 S500 M03 T02 (Select coordinate system, absolute mode, start spindle, get next tool ready)
N015 G00 X3.0 Y3.0 (Rapid to center of hole to be milled)
N020 G43 H01 Z.1 (Instate tool-length compensation, move 0.1 in. [2.54 mm] above work surface)
N025 G01 Z – 0.5 F2.5 (Plunge endmill into hole)
N030 G42 D31 Y4.0 F5.0 (Mill to beginning of circular move)
N040 G02 J – 1.0 (Mill full circle)
N045 G01 G40 Y3.0 (Move back to center)
N050 G91 G28 Z0 M19 (Return to tool-change position)
N055 M01 (Optional stop)
N060 ...
.
. (Other machining operations)
.
N405 M00 (BLOW CHIPS FROM HOLE TO BE MEASURED)
N410 T07 M06 (Place probe in spindle)
N413 G54 G90 (Select coordinate system, absolute mode)
N415 G00 X3.0 Y3.0 (Rapid to hole to be measured)
N420 G43 H07 Z.1 (Instate tool-length compensation for probe, rapid up to 0.1 in. [2.54 mm] from surface)
N425 G65 P9081 X3.0 Y3.0 D2.0 Z – 0.2 (Measure hole)
N430 G91 G28 Z0 M19 (Return to tool-change position)
N435 M30 (End of program)

The hole diameter-measuring custom macro (O9081) will measure the hole diameter and store the result in permanent common variable #511. Note that our method of measuring the hole diameter is quite similar to the method shown earlier to locate the hole's *X* and *Y* position.

O9081 (Hole diameter measuring)
N1 G90 G01 X#24 Y#25 F50.0 (Fast feed to starting position)
N2 G01 Z#26 (Fast feed to probing surface)

N3 #102 = #7 (Hole diameter specified on print)
N4 IF [#103 GT 0.375] GOTO 6 (Test if hole is too small for retract movement)
N5 #3000 = 103 (HOLE TOO SMALL) (Generate alarm if hole is less than 0.375 in. [9.525 mm])
N6 S200 M19 (Orient spindle to ensure consistent droop)
N7 M13 (Turn probe on)
N8 IF [#102 LE 1.25] GOTO 10 (Test if fast approach movement is required)
N9 G91 G01 X – [#102/2 – 0.5] F50.0 (If required, feed up close to surface)
N10 G65 P9501 A – 1. (Call universal probe-triggering custom macro, probe *X*–)
N11 #111 = #100 (Store left side *X* position)
N12 G91 G01 X[#102/2 – #500] F50.0 (Fast feed back to approximate hole center in *X*)
N13 IF [#102 LE 1.25] GOTO 15 (Test if fast approach movement is required)
N14 G91 G01 X[#102/2 – 0.5] F50.0 (If needed, fast feed up close to surface)
N15 G65 P9501 A1. (Call universal probe-triggering custom macro, probe *X*+)
N16 #112 = #100 (Store right side *X* position)
N17 G91 G01 X – [#112 – [[#111 + #112]/2]] F50.0 (Fast feed back to precise center in *X*)
N18 IF [#102 LE 1.25] GOTO 20 (Test if fast approach movement is required)
N19 G91 G01 Y – [#102/2 – 0.5] F50.0 (If needed, fast feed up close to surface)
N20 G65 P9501 A – 2. (Call universal probe-triggering custom macro, probe *Y*–)
N21 #111 = #100 (Store front side *Y* position)
N22 G91 G01 Y[#102/2 – #500] F50.0 (Fast feed back to approximate hole center in *Y*)
N23 IF [#102 LE 1.25] GOTO 25 (Test if fast approach movement is required)
N24 G91 G01 Y[#102/2 – 0.5] F50.0 (If needed, fast feed up close to surface)
N25 G65 P9501 A2. (Call universal probe-triggering custom macro, probe *Y*+)
N26 #112 = #100 (Store back side *Y* position)
N27 G91 G01 Y – [#112 – [[#111 + #112]/2]] F50.0 (Fast feed back to precise center in *Y*. Spindle center line [not necessarily stylus ball center line] will be on hole center)
N28 M14 (Turn probe off)
N29 SETVN 511 [HOLE DIA] (Set variable name for #511)
N30 #511 = [#112 – #111] (Store hole diameter in #511)
N31 M99 (End of custom macro)

Measuring the depth of a pocket

Our calling program will pass the position in *X* and *Y* for the measuring to begin. This *X* and *Y* position will be slightly below (*Y*– of) the pocket being measured, allowing the top *Z* surface to be touched first. The print dimension for the pocket depth in *Z* also will be specified (as D for depth). A *Z* value in the calling command will specify the position in *Z* from which probing will begin (above the top Z surface of the pocket). Here is an example calling command. We will store the result of the measurement in permanent common variable #511 and label it accordingly.

N455 G65 P9082 X3.0 Y2. Z0.25 D.50 (Call pocket depth measuring probing program)

Here is the pocket depth measuring program.

O9082 (Pocket depth measuring program)
N1 G90 G01 X#24 Y#25 F50.0 (Fast feed to measuring start position)
N2 G01 Z#26 (Fast feed to *Z* starting point)
N3 S200 M19 (Orient spindle to ensure consistent droop)
N4 M13 (Activate probe)
N5 G65 P9501 A – 3. **C1.0** (Probe top surface, note that C specifies that tool-length compensation is being considered in the *Z*-axis machine position calculation)
N6 #111 = #100 (Store top *Z* position)
N7 G91 G01 Y0.75 F50.0 (Move above pocket in *Y*)
N8 G01 Z – [#7] (Fast feed to 0.25 in. [6.35 mm] above pocket bottom)
N9 G65 P9501 A – 3. **C1.0** (Probe pocket bottom surface in *Z*)
N10 #112 = #100 (Store bottom pocket position)
N11 G90 G01 Z#26 (Fast feed out of pocket in *Z*)
N12 M14 (Turn off probe)
N13 **SETVN 511 [POCK DEP]** (Set variable name for #511)
N14 **#511 = #112 – #111** (Store pocket depth in #511)
N15 M99 (End of custom macro)

Other dimensions to measure

By now, you should be starting to see that any dimension that can be measured with conventional gaging equipment can be measured with a spindle probe. At this point, it's good to try a few exercises on your own: Based on the calling command description, write the specific custom macro for each application and store the results in the permanent common variables as specified and label them accordingly.

The length and width of a pocket. Calling command:

G65 P9083 X1.0 Y1.0 U4.0 W2.0 Z – 0.25

Where:

X and *Y* are the lower left corner of the pocket (*X*– and *Y*– sides),
U is the pocket length along *X*,

W is the pocket width along *Y*, and
Z is a position below the work surface that allows probing to be done.

The center position and diameter of a round boss. Calling command:

G65 P9084 X3.5 Y2.5 D4.0 Z0

Where:

X and *Y* specify the center position of the boss (from the print)
D is the specified boss diameter (from the print); and
Z is the top surface of the boss. Assume that you can position the stylus to within 0.25 in. (6.35 mm) of this surface and that you can move the stylus below this surface by 0.25 in. (while outside the boss diameter) without fear of interference.

Store the *X* center position in permanent common variable #511, the *Y* center position in #512, and the diameter of the boss in #513 and label them accordingly.

WHAT TO DO WITH THE RESULTS OF MEASURING

As stated earlier, measuring the desired workpiece dimensions is only half the battle. Once measured, your in-process gaging program must *do* something based on the results of your measuring. In all examples shown so far, we simply stored the results of each measurement in permanent common variables. While this can be quite helpful, most companies performing in-process gaging want much more. At the very least, most want their in-process gaging operation to report (automatically) to the company's statistical process control (SPC) system. More likely, they want the behavior of the machine tool to be altered, based on the findings of the in-process gaging program.

Reporting probe findings to your SPC system

Just as all current automatic gaging fixtures have the ability to send data (the current gage readings) through a serial communications port to your SPC system, so can a custom macro program send data through the CNC control's RS-232C communications port. Truly, anything the custom macro program knows can be reported in this manner. This includes the value of any variable, along with a special message specifying what the variable represents.

As we discussed in Chapter 5, two custom macro words are used for this purpose. BPRNT does not allow decimal-point values to be reported in their entirety, making DPRNT the more powerful and more commonly used command.

Once a measured value is determined and stored in a custom macro variable, its value can be sent out to the SPC system. Of course, the SPC system software must be properly configured and ready to receive the data. Here we can only address how you can format your transmission commands with custom macro. You must determine how the data must be transmitted by studying your SPC system specifications.

Here is an example. In the previous hole diameter-measuring custom macro, we stored the diameter of the hole in permanent common variable #511 with the command:

N30 #511 = [#112 – #111]/2 (Store hole diameter in #511)

After this command, the following commands send the value of #511 through the communications port.

POPEN (Opens the serial port for transmission)
DPRNT HOLE DIAMETER #511 24
PCLOS (Close the serial port after transmission)

As we learned in Chapter 5, the POPEN command opens the serial port (RS-232C port) for transmission. The DPRNT command will send the words HOLE DIAMETER and then the value of permanent common variable #511 with a two/four decimal format (two places before the decimal point and four places after the decimal point). The PCLOS command closes the serial port. These commands assume, of course, that a cable is connecting the RS-232C port to the SPC system and that the SPC system is up and ready to receive this data.

Causing the machine to behave differently

As stated, most companies that use spindle probes for in-process gaging want more than simple SPC reporting. Most want the machine to behave differently, based on the findings of a given measurement. For example, many companies apply the *tolerance band* from the print to each dimension being measured. If the workpiece is within the tolerance band, possibly nothing would be done. Or possibly an offset would be changed that would cause the next workpiece to be machined right in the *middle* of its tolerance band. Or, if true SPC techniques are applied, possibly the user would want this dimension to follow a *trend* before an offset adjustment is made. Perhaps the offset would be changed and the tool machining the surface recalled for re-machining. If the surface is not within the tolerance band and cannot be saved, perhaps an alarm would be sounded.

The possibilities for making the machine behave in different ways are almost unlimited. As you may agree, generating custom macros with the intelligence to handle in-process gaging applications can be quite challenging, especially if several dimensions are involved.

Divide and conquer—handling the tolerance band issue. If you do in-process gaging often, it is likely that you will be applying the same custom macro commands over and over again. You will have to set up logic for the various possibilities you wish your custom macro to handle. It is likely that these commands will be required for *every* measurement taken. And as you know from discussions earlier in the book, whenever you must repeat commands on a regular basis, it is a good idea to create a universal custom macro that can be used each time the commands are required. We can call this custom macro the *universal tolerance band custom macro*.

Before you can actually write your own version of the universal tolerance band custom macro, you must first determine the various alternatives available to you to handle the results of in-process gaging findings. This alone is no easy task. Though we will set some rather limiting criteria for our own universal tolerance band custom macro, it should make absolutely clear what you can expect.

We will set up for only three possibilities.

1. The dimension is within the tolerance band. In this case, we will allow production to continue in its normal manner. However, we will adjust the tool's offset to bring the next workpiece in the middle of the tolerance band; this being the case, it is likely that the offset will be changed after each measurement.
2. The dimension is too small.
3. The dimension is too large.

Consider possibilities 2 and 3. With one of them, there is still stock left to be machined, meaning that it may still be possible to save the workpiece. For example, if measuring a pocket depth, and the pocket depth comes out too shallow (condition two), we could recall the milling cutter and re-machine the pocket. With the other case, too much stock has already been machined and the workpiece has to be scrapped.

Even with this *simple* set of alternatives, note that we are already running into complications. First of all, how you handle conditions 2 and 3 will vary, based on the kind of dimension involved. With the measuring of *inside* shapes, like holes and pockets, condition 2 is the one that allows the workpiece to be salvaged. However when measuring outside shapes, like the diameter of a boss, *condition 3* is the one that will allow the workpiece to be salvaged.

It may be possible to adjust the machine's behavior based on the measurement results with a simple offset change. For Z-axis problems, a *tool-length compensation* offset will be involved. For *XY* problems, at least those machined with a milling cutter, possibly a *cutter-radius compensation* offset will be involved. Remember that our universal tolerance band custom macro will calculate the overall required adjustment. With applications involving cutter-radius compensation, it will be necessary to divide this value by two.

To complicate matters further, there are those machining operations that cannot be altered, at least not automatically, by any kind of programmed command. When finish-boring on a machining center, for example, an operator will be required to make a manual adjustment when hole diameters go out of tolerance. For these operations, possibly the best our in-process gaging custom macro can do is generate an alarm if the workpiece is out of tolerance, while printing the out-of-tolerance amount to a permanent common variable.

In your own in-process gaging applications, you will probably want more than these three alternatives. But don't be too quick to add more. You must really think through each alternative you add to avoid letting the application grow into a terribly complicated program that no one can write.

For our example, we use the universal tolerance band custom macro to simply determine which of the three conditions exist and set a common variable (#102) to 1, 2, or 3, corresponding to the three conditions. It also will calculate the amount of adjustment necessary to bring the tolerance to the middle of its tolerance band. It will still be up to the *in-process gaging program* (commonly part of the machining program) to determine what to do based on these findings.

An example of the calling command for our universal tolerance band custom macro is

```
G65 P9503 D#111 S3.245 B3.250
```

With argument D, we pass the actual dimension as it has been measured by the spindle probe. This value is specified with a common variable, since the measured value will change with every use of the probe. With S, we pass the small limit (low end of the tolerance band). With B, we pass the big limit (high end of the tolerance band).

Here is the universal tolerance band custom macro. Note that D is represented by local variable #7, S by #19, and B by #2.

```
O9503 (Universal tolerance band custom macro)
IF [#7 LT #19] GOTO 1 (Test if dimension is too small)
IF [#7 GT #2] GOTO 2 (Test if dimension is too large)
#102 = 1 (Dimension is within the tolerance band)
#103 = [[#19 + #2]/2] – #7 (Calculate offset adjustment amount)
GOTO 99 (Exit custom macro)
N1 (Dimension is too small)
#102 = 2 (Dimension too small)
#103 = [[#19 + #2]/2] – #7 (Calculate offset adjustment amount)
GOTO 99 (Exit custom macro)
N2 (Dimension is too large)
#102 = 3 (Dimension too large)
#103 = [[#19 + #2]/2] – #7 (Calculate out of tolerance amount)
N99 M99
```

A full example. Here we present a full (yet still very simple) example. We are first machining a round pocket with an endmill to a 2-in. (50.8-mm) diameter. We then call for in-process gaging with the spindle probe, using the custom macro shown earlier for determining the diameter of a hole. We then use the universal tolerance band custom macro to determine correctness of the current hole size (stored by the hole-measuring custom macro in #511). If the hole is within the tolerance band, we allow the machine to continue in its normal manner (after adjusting the cutter-radius compensation offset [#31] to bring the next milling operation to the middle of the tolerance band). If the hole is too small (out-of-tolerance), we will adjust the offset and re-machine. If the hole is too large (scrapped), an alarm will be sounded. Even though we consider this to be a relatively *simple* application for in-process gaging, notice how involved the program is.

O0005 (Example of in-process gaging)
N005 T01 M06 (Change tools, place 1-in. [25.4-mm] endmill in position)
N010 G54 G90 S500 M03 T07 (Select coordinate system, absolute mode, start spindle, get next tool ready)
N015 G00 X3.0 Y3.0 (Rapid to center of hole to be milled)
N020 G43 H01 Z.1 (Instate tool-length compensation, move 0.10 in. [2.54 mm] above work surface)
N025 G01 Z – 0.5 F2.5 (Plunge endmill into hole)
N030 G42 D31 Y4.0 F5.0 (Mill to beginning of circular move)
N040 G02 J – 1.0 (Mill full circle)
N045 G01 G40 Y3.0 (Move back to center)
N050 G91 G28 Z0 M19 (Return to tool-change position)
N055 M01 (Optional stop)
N060 T07 M06 (Place probe in spindle)
N063 G54 G90 (Select coordinate system, absolute mode)
N065 G00 X3.0 Y3.0 (Rapid to hole to be measured)
N070 G43 H07 Z.1 (Instate tool-length compensation for probe, rapid up to 0.10 in. [2.54 mm] from surface)
N075 G65 P9081 X3.0 Y3.0 D2.0 Z – 0.2 (Measure hole [diameter is stored in #511])
N080 G65 P9503 D#511 S1.997 B2.003 (Check hole size, condition in #102, adjustment amount in #103)
N085 IF [#102 EQ 1] GOTO 100 (Test for condition 1)
N090 IF [#102 EQ 2] GOTO 115 (Test for condition 2)
N095 #3000 = 105 (HOLE TOO LARGE) (Generate alarm if condition three, hole too large)
N100 (Workpiece is within tolerance band)
N105 #2031 = #2031 + #103/2 (Adjust offset)
N110 GOTO 130 (Exit program)
N115 (Workpiece is out of tolerance on small side, repair is possible)
N120 #2031 = #2031 + #103/2 (Adjust offset)
N125 GOTO 005 (Re-machine with milling cutter)

```
N130 G91 G28 Z0 M19 (Return to tool-change position)
N135 M30 (End of program)
```

From lines N005 through N055, we machine the pocket in the normal manner. In lines N060 through N070, the spindle probe is loaded and positioned. In line N075 we measure the hole. In line N080, we determine the hole's condition and amount of offset needed to put the hole in the middle of its tolerance band.

Beginning with line N085, notice how much logic must remain in the main program for this supposedly *simple* example. We must still determine, based on the measurement results, exactly what to do. If you have many similar machining operations to perform, you may be able to write more *universal* custom macros to handle your specific tasks, minimizing this kind of logic in the machining program.

There is still one minor problem with this kind of application. It did not present itself in our example because there is only one cutting tool (the endmill). Our program milled the pocket, immediately measured its diameter, and adjusted accordingly. If condition 2 exists and the hole must be re-machined, line N125 sends the control back to line N005. The endmill is recalled to re-machine. When finished, the in-process gaging program will be called again for another measurement (which may or may not be desirable). However, if there are more cutting tools after the endmilling operation, we would not want them recalled as well.

One way to prevent this tool recall is to use a *flag* to signal the current condition of the pocketing operation. For our example, we'll use common variable #121 as the flag. If #121 is set to zero (as it will be at the beginning of the machining program), we are not re-machining. If #121 is set to one (1.), we are re-machining. Combined with IF statements, we can determine at the end of the pocketing operation whether re-machining is taking place. If it is, we can cause the control to skip the balance of the machining operations and proceed directly to repeating the in-process gaging operation (or to the end of the program, whichever is preferable). Note that this technique adds even more logic to the calling program, complicating the program considerably (especially when several in-process gaging measurements are being taken).

Here is another version of the example program which stresses this technique. Notice the new commands, which are highlighted.

```
O0005 (Example of in-process gaging)
N003 #121 = 0 (Set flag to normal machining)
N005 T01 M06 (Change tools, place 1-in. [25.4-mm] endmill in position)
N010 G54 G90 S500 M03 T02 (Select coordinate system, absolute mode, start
   spindle, get next tool ready)
N015 G00 X3.0 Y3.0 (Rapid to center of hole to be milled)
N020 G43 H01 Z.1 (Instate tool-length compensation, move 0.10 in. [2.54 mm]
   above work surface)
```

N025 G01 Z – 0.5 F2.5 (Plunge endmill into hole)
N030 G42 D31 Y4.0 F5.0 (Mill to beginning of circular move)
N040 G02 J – 1.0 (Mill full circle)
N045 G01 G40 Y3.0 (Move back to center)
N050 G91 G28 Z0 M19 (Return to tool-change position)
N055 M01 (Optional stop)
N060 IF [#121 EQ 1] GOTO 450 (If re-machining, skip balance of machining operations)
N065 T02 M02 (Change tools)
N070 G54 G90 S300 M03 T03 (Select coordinate system, absolute mode, start spindle, get next tool ready)
N075 G00 X4.0 Y3.0 (Rapid to first position)
.
.
. (Balance of machining operations)
.
.
N445 M01
N450 T07 M06 (Place probe in spindle)
N455 G00 X3.0 Y3.0 (Rapid to hole to be measured)
N460 G43 H07 Z.1 (Instate tool-length compensation for probe, rapid up to 0.1 in. [2.54 mm] from surface)
N465 G65 P9081 X3.0 Y3.0 D4.0 Z – 0.2 (Measure hole, diameter stored in #511)
N470 G65 P9503 D#511 S1.997 B2.003 (Check hole size, condition in #102, adjustment amount in #103)
N475 IF [#102 EQ 1] GOTO 490 (Test for condition 1)
N480 IF [#102 EQ 2] GOTO 505 (Test for condition 2)
N485 #3000 = 105 (HOLE TOO LARGE) (Generate alarm if condition 3)
N490 (Workpiece is within tolerance band)
N495 #2031 = #2031 + #103/2 (Adjust offset)
N500 GOTO 520 (Exit program)
N505 (Workpiece is out of tolerance on small side, repair is possible)
N510 #2031 = #2031 + #103/2 (Adjust offset)
N513 #121 = 1 (Set re-machining flag to re-machining)
N515 GOTO 005
N520 G91 G28 Z0 M19 (Return to tool-change position)
N525 M30 (End of program)

Note that #121 is being set to zero (not re-machining) at the very beginning of the program. The first time the IF statement in line N060 is executed, the condition will be evaluated as false and the control will proceed with the balance of the machining. After measurement, if condition 2 is determined and re-machining is necessary, the flag gets set to 1. (in N513), then the control is sent to line N005.

The next time the IF statement in line N060 is executed, it will be evaluated as true and the balance of the machining operations will be skipped. The re-machining flag will continue to be set to 1. until the next cycle is run.

When do you need to measure?

In-process gaging takes time. Moreover, many companies do not perform 100% inspections. For these reasons, you may elect to invoke the in-process gaging functions on a *sampling* basis, not inspecting every workpiece. If this is what you elect to do, you will need a way to invoke in-process gaging only when it is required. For this you will need a part counter.

Divide and conquer—a part counter. If you perform in-process gaging on a sampling basis, every in-process gaging program you write will need some way to track *when* in-process gaging is required. Since these commands will be so often needed, it is wise to develop a *universal part-counting* custom macro program that will work for all applications.

If, for example, you wish to perform in-process gaging on every fifth workpiece, your part counter's call statement would look like this:

```
G65 P9505 C5.
```

In this command, the only argument (C) represents the number of workpieces you machine between inspections. A permanent common variable should be used with which to actually count, since you will need the current part count to be retained even after power off. We will use permanent common variable #520 as our part counter. This custom macro will return (in common variable #104) the condition of the part counter. If #104 is zero, no measurement is required. If #104 is 1., the in-process gaging commands must be invoked. Here is the universal part-counting custom macro.

```
O9505 (Universal part-counting custom macro)
SETVN 520 [PART CNT] (Label #520 to avoid overwriting)
#104 = 0 (Set measurement not needed state)
#520 = #520 + 1 (Step counter)
IF [#520 LT #3] GOTO 99 (If no need to measure, exit macro)
#104 = 1 (Measurement needed)
#520 = 0 (Reset counter)
N99 M99 (End of custom macro)
```

The hole-measuring machining program shown earlier, modified to measure the hole of every fifth workpiece, looks like this:

```
O0005 (Example of in-process gaging)
N003 #121 = 0 (Set flag to normal machining)
```

N005 T01 M06 (Change tools, place 1-in. [25.4 mm] endmill in position)
N010 G54 G90 S500 M03 T02 (Select coordinate system, absolute mode, start spindle, get next tool ready)
N015 G00 X3.0 Y3.0 (Rapid to center of hole to be milled)
N020 G43 H01 Z.1 (Instate tool-length compensation, move 0.10 in. [2.54 mm] above work surface)
N025 G01 Z – 0.5 F2.5 (Plunge endmill into hole)
N030 G42 D31 Y4.0 F5.0 (Mill to beginning of circular move)
N040 G02 J – 1.0 (Mill full circle)
N045 G01 G40 Y3.0 (Move back to center)
N050 G91 G28 Z0 M19 (Return to tool-change position)
N055 M01 (Optional stop)
N060 IF [#121 EQ 1] GOTO 450 (If re-machining, skip balance of machining operations)
N065 T02 M02 (Change tools)
N070 G54 G90 S300 M03 T03 (Select coordinate system, absolute mode, start spindle, get next tool ready)
N075 G00 X4.0 Y3.0 (Rapid to first position)

.

.

. (Balance of machining operations)

.

.

N445 M01
N446 G65 P9505 C5.0 (Count five parts, return condition in #104, if 0, no measurement)
N447 IF [#104 EQ 0] GOTO 520 (If no measurement is necessary, skip gaging)
N450 T07 M06 (Place probe in spindle)
N455 G00 X3.0 Y3.0 (Rapid to hole to be measured)
N460 G43 H07 Z.1 (Instate tool-length compensation for probe, rapid up to 0.10 in. [2.54 mm] from surface)
N465 G65 P9081 X3.0 Y3.0 D4.0 Z – 0.2 (Measure hole, diameter stored in #511)
N470 G65 P9503 D#511 S1.997 B2.003 (Check hole size, condition in #102, adjustment amount in #103)
N475 IF [#102 EQ 1] GOTO 490 (Test for condition 1)
N480 IF [#102 EQ 2] GOTO 505 (Test for condition 2)
N485 #3000 = 105 (HOLE TOO LARGE) (Generate alarm if condition 3)
N490 (Workpiece is within tolerance band)
N495 #2031 = #2031 + #103/2 (Adjust offset)
N500 GOTO 520 (Exit program)
N505 (Workpiece is out of tolerance on small side, repair is possible)
N510 #2031 = #2031 + #103/2 (Adjust offset)
N513 #121 = 1 (Set re-machining flag to re-machining)

```
N515 GOTO 005
N520 G91 G28 Z0 M19 (Return to tool-change position)
N525 M30 (End of program)
```

CONCLUSION TO IN-PROCESS GAGING

While we have given you the raw tools you need to develop your own in-process gaging programs, we urge you to exercise caution—especially for the first few programs you write. Be sure to keep your first few programs as simple as possible, handling only the most important criteria, and ensuring that your custom macros stay short and easily understood. As you gain experience with in-process gaging, you will surely be able to improve on the techniques shown in this text, especially as you gain knowledge specifically related to your company's own in-process gaging needs.

PROGRAMMING UTILITY APPLICATIONS

There are times, especially in unattended applications, when you may need the spindle probe to perform functions that an operator is not available to do. These functions could include checking to see whether chips have been adequately cleared from a drilled hole prior to tapping, for example, or checking that adequate clearance has been made for subsequent machining operations, or even workpiece identification in flexible manufacturing systems (FMS).

With your knowledge of what spindle probes can do, you should be ready to approach almost any utility-related application. For this reason, and since applications in this category tend to be quite diverse, we show just one example of a utility-related application.

DEVELOPING A WORKPIECE IDENTIFIER

In identifying this application, let's suppose you have a fixture designed to hold 10 different component workpieces of an assembly. To make this as ideal an application as we can, we'll say the machine is *dedicated* to running only the workpieces held in this fixture, and all programs related to the 10 workpieces are permanently stored in the control's memory. Let's further stipulate that you have had problems with the operator calling up the wrong program from time to time, and you want to make the system more fail-safe by applying your spindle probe.

If production quantities are rather high (let's say over 100 parts), you might (rightly) point out that this application fits more into the setup help category. However, let's say that *only one workpiece* is run per production run, and we want the spindle probe to automatically determine which workpiece is in the fixture and invoke the corresponding CNC program. Cycle times are quite long, and you want the operator to simply load a workpiece, activate the cycle, and then start doing other things while the CNC cycle is running.

The first step in developing this kind of application is specifying one (or more) workpiece attributes you can use to determine which workpiece is in the fixture. To keep it simple, we'll say that the position of a given small hole on a flat face (currently in the raw material) changes with each workpiece in the family. The goal will be to get our spindle probe to find out where the hole is located and determine the appropriate CNC-related program to call.

Following is a table that shows the relationship of the hole's location to the program number to be called.

X0.25 Y0.25 — O0001
X0.50 Y0.25 — O0002
X0.75 Y0.25 — O0003
X1.0 Y0.25 — O0004
X1.0 Y0.50 — O0005
X0.75 Y0.50 — O0006
X0.50 Y0.50 — O0007
X0.25 Y0.50 — O0008
X0.25 Y0.75 — O0009
X0.50 Y0.75 — O0010

Admittedly, we have kept the determination of which program to call very simple. In reality, you may have to study your application for some time to determine the relationship of some workpiece attributes to the CNC program number that must be called.

We'll say the small identification hole in the raw material is 0.25 in. (6.35 mm) in diameter and 0.30 in. (7.62 mm) deep. It is in the top surface of the workpiece, which is the program zero position in *Z*. The probing program will move the stylus to a position above the first potential hole location and then drive the stylus a distance that allows the stylus to touch the hole bottom. If the stylus reaches the hole depth, we know which program to call. If it does not, we proceed to the next potential hole location and try again.

Here is a portion of a program that will determine which program to call. Notice that the *program number* is being saved as a variable. Fixture offset number one has already been assigned when this program is executed.

```
O9091 (Workpiece identifying custom macro)
T07 M06 (Place probe in spindle)
G54 G90 (Select coordinate system, absolute mode)
M13 (Activate probe)
S200 M19 (Orient spindle)
G00 X0.25 Y0.25 (Rapid to first potential hole location)
G43 H07 Z0.25 (Instate tool-length compensation, the probe length is stored in
   offset number one)
```

```
G91 G31 Z – 0.75 F30.0 (Drive probe into surface)
IF [#5063 GT – 0.05] GOTO 1 (If this is not the hole, go to next position)
#101 = 1 (Hole is here, store program number to call in #101)
GOTO 50 (Exit the identification routine)
N1 G90 G01 Z0.25 F75.0 (Fast feed away from surface)
G00 X0.5 Y0.25 (Move to next potential hole location)
G91 G31 Z – 0.75 F30.0 (Drive probe into surface)
IF [#5063 GT – 0.05] GOTO 2 (If this is not the hole, go to next position)
#101 = 2 (Hole is here, store program number to call in #101)
GOTO 50 (Exit the identification routine)
N2 G90 G01 Z0.25 F75.0 (Fast feed away from surface)
G00 X0.75 Y0.25
G91 G31 Z – 0.75 F30.0 (Drive probe into surface)
IF [#5063 GT – 0.05] GOTO 3 (If this is not the hole, go to next position)
#101 = 3 (Hole is here, store program number to call in #101)
GOTO 50 (Exit the identification routine)
N3 G90 G01 Z0.25 F75.0 (Fast feed away from surface)
G00 X1.0 Y0.25
G91 G31 Z – 0.75 F30.0 (Drive probe into surface)
IF [#5063 GT – 0.05] GOTO 4 (If this is not the hole, go to next position)
#101 = 4 (Hole is here, store program number to call in #101)
GOTO 50 (Exit the identification routine)
N4 G90 G01 Z0.25 F75.0 (Fast feed away from surface)
G00 X1.0 Y0.50
G91 G31 Z – 0.75 F30.0 (Drive probe into surface)
IF [#5063 GT – 0.05] GOTO 5 (If this is not the hole, go to next position)
#101 = 5 (Hole is here, store program number to call in #101)
GOTO 50 (Exit the identification routine)
N5 G90 G01 Z0.25 F75.0 (Fast feed away from surface)
G00 X0.75 Y0.50
G91 G31 Z – 0.75 F30.0 (Drive probe into surface)
IF [#5063 GT – 0.05] GOTO 6 (If this is not the hole, go to next position)
#101 = 6 (Hole is here, store program number to call in #101)
GOTO 50 (Exit the identification routine)
N6 G90 G01 Z0.25 F75.0 (Fast feed away from surface)
G00 X0.50 Y0.50
G91 G31 Z – 0.75 F30.0 (Drive probe into surface)
IF [#5063 GT – 0.05] GOTO 7 (If this is not the hole, go to next position)
#101 = 7 (Hole is here, store program number to call in #101)
GOTO 50 (Exit the identification routine)
N7 G90 G01 Z0.25 F75.0 (Fast feed away from surface)
G00 X0.25 Y0.50
G91 G31 Z – 0.75 F30.0 (Drive probe into surface)
```

```
IF [#5063 GT – 0.05] GOTO 8 (If this is not the hole, go to next position)
#101 = 8 (Hole is here, store program number to call in #101)
GOTO 50 (Exit the identification routine)
N8 G90 G01 Z0.25 F75.0 (Fast feed away from surface)
G00 X0.25 Y0.75
G91 G31 Z – 0.75 F30.0 (Drive probe into surface)
IF [#5063 GT – 0.05] GOTO 9 (If this is not the hole, go to next position)
#101 = 9 (Hole is here, store program number to call in #101)
GOTO 50 (Exit the identification routine)
N9 G90 G01 Z0.25 F75.0 (Fast feed away from surface)
G00 X0.50 Y0.75
G91 G31 Z – 0.75 F30.0 (Drive probe into surface)
IF [#5063 GT – 0.05] GOTO 10 (If this is not the hole, go to next position)
#101 = 10 (Hole is here, store program number to call in #101)
GOTO 50 (Exit the identification routine)
N10 G90 G01 Z0.25 F75.0 (Fast feed away from surface)
#3000 = 106 (HOLE NOT FOUND)
N50 M14 (Turn off probe)
G90 G01 Z0.25 F75.0 (Fast feed out of hole)
M98 P#101 (Call correct program to machine workpiece)
M30
```

Since after probing #5063 will contain the current position of the stylus relative to program zero (the top of the workpiece), we can easily test to see if a hole exists. If the surface is above a position of –0.05 in Z, we can assume the hole is not in this location and can proceed to the next potential hole location. If the IF statement is false, the value of #5063 is more minus than –0.05, and we can assume the hole is in this location.

In this case, the program number for this hole location is stored in common variable #101. In the M98 command at the end of this program, note that the P word specifies the current value of #101. If #101 is currently equal to five, it will be just as if the M98 command includes the word P5.

Index

A

B

C

D

M

N

O

P

Q

R

S

T

U

V

W

Z